# Further Praise For *A Dangerous Master*

"It would be hard to find a more thoughtful, better prepared guide through the difficult terrain of emerging technologies than Wendell Wallach, as he demonstrates yet again in this comprehensive, erudite, and highly readable book. This is a must-read volume."

—BRADEN ALLENBY,
professor of civil and environmental engineering,
Arizona State University

"Wendell Wallach has done all of us a service. He has alerted us in detail, and provocatively, that there are dangers as well as gains in our national romance with innovative technologies. His account of the troubled technology romance is well told, and it is one we need to hear."

—DANIEL CALLAHAN,
president emeritus, The Hastings Center

"*A Dangerous Master* does a masterful job of describing in an accessible but precise manner the emerging technologies, and their profound and fascinating ethical and social implications."

—GARY MARCHANT,
regents' professor of law,
Sandra Day O'Connor College of Law

# A DANGEROUS
# MASTER

How to Keep Technology from
Slipping Beyond Our Control

## WENDELL WALLACH

BASIC BOOKS
A Member of the Perseus Books Group
New York

CARTOON CREDITS

Page 16: P. Clay Bennett / © 2001 The Christian Science Monitor (www.CSMonitor.com). Reprinted with permission.

Page 92: Jim Day, "How Electric Cars Really Work." Reprinted with permission from Cagle, Cartoons, Inc.

Page 151: Sidney Harris "then a miracle occurs." Reprinted with permission from ScienceCartoonsPlus.com

Page 252: Scott Santis / © 2001 Tribune Content Agency, LLC. All Rights Reserved. Reprinted with permission. Scanned Image courtesy of the Rob Rogers Collection, The Ohio State University Billy Ireland Cartoon Library & Museum

Published by Basic Books
A Member of the Perseus Books Group

Books published by Basic Books are available at special discounts for bulk purchases in the United States by corporations, institutions, and other organizations. For more information, please contact the Special Markets Department at the Perseus Books Group, 2300 Chestnut Street, Suite 200, Philadelphia, PA 19103, or call (800) 810-4145, ext. 5000, or e-mail special. markets@perseusbooks.com.

Text design by *BackStory Design*

Cataloging-in-Publication data for this book is available from the Library of Congress.
ISBN: 978-0-465-05862-4 (hardcover)
ISBN: 978-0-465-04053-7 (eBook)

10 9 8 7 6 5 4 3 2 1

*For Gert, who wanted his children to follow him into medicine or become scientists. However, he spent our time together at the dinner table discussing history, politics, and ethics.*

*To the members and the presenters at meetings of the Technology and Ethics study group at Yale University's Interdisciplinary Center for Bioethics. You have been my mentors.*

# CONTENTS

"Technology is a good servant but a dangerous master."
—CHRISTIAN LOUS LANGE

"If we don't change direction soon, we'll end up where we're going."
—PROFESSOR IRWIN COREY

# { 1 }

# Navigating the Future

THE SOFT-SPOKEN AND CELEBRATED BIOCHEMIST OTTO RÖSSLER seemed an unlikely candidate to make a dramatic public announcement that starting up the Large Hadron Collider (LHC) would mark the last days of planet Earth. At a January 2008 gathering in Berlin, Germany, he proclaimed that the LHC, the largest particle collider in the world, could plausibly create a tiny black hole into which the planet would disappear.

Rössler was not alone; a few other scientists joined his campaign to stop the LHC from being turned on. They were a distinct minority.

It would be easy to dismiss Otto Rössler as a crazy scientist, attention seeker, or misguided alarmist unwilling to accept the consensus of the vast majority of his peers. However, years before he criticized CERN (the European Organization for Nuclear Research that manages the LHC) I met Otto at three academic conferences in Baden-Baden, Germany. He impressed me as an unusually gentle, thoughtful, and careful man, and not a person whose opinion I could ignore easily. Rössler was also expressing concerns that were worth heeding—or at least hearing out.

When physicists describe black holes, they commonly use the example of a collapsed star, which is compressed into a small, dense mass that has so much gravitational force, not even light can escape its pull. The creation of tiny black holes by the collision of high-energy particles in the LHC is theoretically possible, and Rössler implored CERN to suspend any operation of the LHC until physicists conducted more preliminary research to prove that a dangerous black hole could not be created. On August 26, 2008, ten days before the LHC was to be turned on, citizens of the EU filed a legal suit at the European Court of Human Rights in Strasbourg to halt its use.

Large science projects, particularly those like the LHC that are funded by public money, have always been a lightning rod for criticism. Everyone—from science detractors, to kooks, to those who argue the money could be spent better elsewhere—seems to have an opinion as to why such projects are not worthwhile. This is especially true for experiments in theoretical physics whose results may lack any immediate application. Yet even while their theories are difficult to understand, the physicists involved with the LHC have been successful at convincing government leaders of the research's importance. Twenty individual countries plus the European Union provided the primary funding for CERN. In addition, six observer states (including the U.S.) contributed to build the LHC. By the time of the 2008 lawsuit, more than six billion dollars had already been spent. With so much public money allocated to study their theories, one can easily imagine the pressure on the lead scientists at CERN to justify the value of their research. Any decision by CERN officials to halt further work would have required tremendous courage.

The proposed danger was serious and eminent scientists voiced their concerns. It would have been totally irresponsible for CERN to dismiss that danger out of hand. A committee of physicists was appointed to study the validity of the theories Rössler employed to challenge the safety of the LHC. This approach is accepted practice among scientists, but the determinations of such committees are based upon methods that are incomprehensible to the wider public—whose fate was, in this case, potentially at risk. One important reason the committee rejected the proposed danger was their conclusion that any tiny black hole created would be unstable and would decay in a matter of moments. A short-lived black hole could pose no real threat. The disintegration would be due to release of energy in the form of Hawking radiation (proposed by and named after Stephen Hawking). Most physicists embrace the existence of Hawking radiation, but it had not been proven because there was no equipment sensitive enough to perform the necessary experiments. In other words, experiments designed to test a collection of theories, including the existence of the elusive Higgs boson, were declared dangerous according to another theory (Rössler's), and that other theory was rejected by scientists because of one additional theory (Hawking radiation) that was widely accepted but also unproven.

CERN fought the lawsuit to halt the research. The European Court of Human Rights immediately dismissed the suit. The LHC was turned on. On September 13, 2008, the physicist and best-selling author Michio Kaku declared in the *Wall Street Journal*, "If you can read this sentence, congratulations! You have survived the official opening of the Large Hadron Collider

(LHC), which an army of critics claim might create mini black holes that will devour the earth."

This was not the first time that scientists had gone ahead with an experiment that could conceivably lead to humanity's extinction. The July 1945 Trinity test of the first nuclear weapon in the New Mexico desert is the most famous example. Prior to the test, key physicists working on the Manhattan Project speculated as to whether the explosion would set off a chain reaction in the earth's atmosphere. Calculations determined that an extinction level event would not happen, and yet the frightening prospect could not be absolutely ruled out. According to reports, the Nobel Prize-winning physicist Enrico Fermi, whose sense of humor bordered on the macabre, offered to take bets on whether a fire in the atmosphere would occur and, if so, would it consume just New Mexico or the entire world? It didn't destroy either, of course, but even after the success of Trinity, further study was required before the possibility of a more powerful bomb igniting the atmosphere could be definitively ruled out.

Otto Rössler continues to assert that the possibility remains, even if of low probability, that the LHC could create a stable tiny black hole. Furthermore, he criticizes CERN for failing to put in place precautionary measures. Is it possible that, despite its safe operation so far, the LHC could still produce a black hole into which the earth disappears? The seventeen-mile long LHC, configured in a ring, was designed to create collisions of high-energy particles. Two high-energy particle beams are directed at each other in an attempt to reproduce conditions similar to the micromoments following the Big Bang that created our universe roughly 13.8 billion years ago. Six quadrillion (that's six followed by fifteen zeros) collisions of particles within the LHC were performed by CERN over one three-year period. Only a few of those collisions actually provided the telltale statistically significant evidence for the existence of the Higgs boson, a breakthrough announced in July 2012. Nonetheless, considering the sheer number of collisions per year, even an extremely low probability that one goes awry is cause for genuine concern.

Rössler also reproached CERN for failing to update its official safety report after the LHC was in full operation. A report that none of the initial six quadrillion collisions created a stable tiny black hole, nor would one ever occur, might be reassuring. But CERN, as Rössler notes, has no means to know whether a stable black hole has been created until one clearly sucks up a considerable amount of matter, which could take some time. Furthermore, "never" is a difficult word for good scientists, who always wish to stay open to the possibility of evidence that might refute even those theories that have been elevated to the status of physical laws. Perhaps, for example, under some as yet unobserved situation, an apple might actually float up off a tree rather than fall to the ground, and thereby challenge the claim that gravity is a law that applies in absolutely all circumstances.

Physicists talk in terms of probabilities and uncertainties, not absolutes. In the world of quantum physics, the field that describes the behavior of subatomic particles, probabilities frame robust theories that lead to predictable results. However, a safety report from CERN couched in the language of probabilities would be less than reassuring to the public, and would provide fodder for its critics. Nevertheless, fear that a report might be misunderstood would not exempt CERN from either minimizing the importance of safety or concealing the results of its evaluations.

In February 2013, after three successful years of operation, CERN shut down the LHC. It is undergoing a major upgrade and will reopen in 2015. I presume the scientists at CERN were diligent and did not knowingly place humanity at risk. But I find it totally unsatisfactory when scientists and courts that simply defer to their judgments are the only parties involved in making decisions that affect us all. Nonetheless, in the case of the LHC, the determination of the scientific experts holds more weight than it should in other fields of study. Physics is an essentially lawful science in which it is possible to make a mathematical determination of an experiment's safety, presuming, of course, that we adequately understand the laws governing the physical system being studied. But under certain circumstances even lawful activity leads to unpredictable events.

There is, however, a much larger problem here that goes beyond physics into other fields such as computer science, genetics, neuroscience, and geo-engineering. It is becoming increasingly difficult to comprehend the risks entailed in emerging forms of scientific research and in the adoption of innovative technologies. These fields of scientific investigation are entangled with human activity and with the environment. This adds additional layers of complexity and uncertainty. A scientific determination of a technology's safety is insufficient to calculate the wide-ranging societal impact of tools and techniques that may have already been introduced into daily life.

Furthermore, even plausible outcomes that have an extremely low probability of occurring should not be ruled out when one analyzes billions and quadrillions of incidents. And that is exactly what we must do when we consider the number of calculations made by computers, the variations in the combinations of different genes, the multiplicity of patterns in neuronal firings, or the atmospheric conditions from which weather patterns emerge.

Uncertainty in the development, progress, and societal impact of an emerging technology is nothing new. However, our expanding reliance on complex systems whose risks we do not or cannot calculate is troubling. The rate of scientific discovery and adoption of innovative technologies outpaces the ability of governments to regulate their development. In addition, emerging technologies pose not only new variants of standard risks, such as public health and environmental threats, but also involve unique societal issues. For example, the availability of technological enhancements, such as growth hormones in sports, can expand dangerous experimentation and cheating, exacerbate inequalities, and disrupt industries. Collectively, all these concerns increase opportunities for tragedies in which a newly deployed technology is complicit.

Not only is it increasingly difficult to assess the risks of emerging technologies; even having realistic conversations can be a challenge. In emerging fields of research, endless speculation and hype foster the illusion of inevitable progress on both laudable and anxiety-provoking fronts. Yet even commonly repeated possibilities such as personalized medicine, designer babies, brain simulations, and smarter-than-human computers and robots will be extremely difficult to fully realize. Filtering out the hype from the reality is itself a complicated task. For non-experts, discerning the difference between a thought experiment, a speculative theory, a testable hypothesis, and a proven fact can be close to impossible. Thus, it is difficult to know which risks truly require our attention.

Rössler's theory regarding the dangers of LHC research has been rejected. But recently, two law professors raised similar concerns regarding new research at the Relativistic Heavy Ion Collider (RHIC) accelerator located in Brookhaven, New York. Without Rössler's scientific eminence, they trod an even more difficult road in getting the issues they raise taken seriously. Eric Johnson and Michael Baram published a 2014 opinion piece calling for a commission to study the safety of the fifteen-year-old RHIC (pronounced "Rick"), which is second in size to the LHC. RHIC is undergoing an upgrade in order to perform experiments in energy ranges far outside those for which it was originally designed. Johnson and Baram express concern that the new experiments will create a subatomic object called a "strangelet," which under certain circumstances could ignite a chain reaction. In the words of Sir Martin Rees, Astronomer Royal of the

U.K., that chain reaction would turn the earth into "an inert hyperdense sphere about one hundred metres [109 yards] across."

The problem of strangelets was also raised before the RHIC opened. After media attention in July 1999, the Brookhaven director appointed a four-member committee to study the matter. Their report two months later concluded the RHIC was safe. Later critics, including Judge Richard Posner, in his book *Catastrophe: Risk and Response*, have charged that the committee was made up of individuals who would either participate in or had a vested interest in the research. Martin Rees wrote that the committee "seemed to have aimed to reassure the public . . . rather than to make an objective analysis." In any case, the original study from 1999 was based upon assumptions that RHIC would only be in operation for ten years, and used for experiments much less powerful than those to be conducted after the latest upgrade.

Johnson and Baram's call for a commission to investigate the safety of new experiments at RHIC has been largely ignored. As lawyers, they were criticized for questioning the judgment of scientific experts. But therein lies the problem. In serious matters of safety, such as those of the LHC and RHIC, it could be argued that the scientists start with a bias in favor of the research moving forward, so much so that they might dismiss or overlook risks. Beyond the immediate determination of safety, many scientists have a form of tunnel vision that makes them disinterested in, or insensitive to, the societal impact of their research. Even when a scientist appointed to a safety panel is deeply aware of the broader implications, that alone does not give her the authority to make decisions that will affect us all. On the other hand, non-experts sensitive to the ramifications of a new technology may lack the specialized knowledge necessary to evaluate whether their concerns are well-founded. Committed citizen advocates can also be myopic—they often politicize the practice of science, which holds its own dangers. So the onus is not just on the scientists, it's also on non-experts and ordinary citizens. There is seldom any incentive for the experts and non-experts to communicate with each other or bridge the huge chasm in their understanding. Even should they want to work toward some form of mutual understanding, there is no blueprint for how they might do so.

Determining when to bow to the judgment of experts and whether to intervene in the deployment of a new technology is certainly not easy. How can government leaders or informed citizens effectively discern which fields of research are truly promising and which pose serious risks? Do we have the intelligence and means to mitigate the serious risks that can be anticipated? How should we prepare for unanticipated risks?

The fears raised by LHC or RHIC experiments fall within a short list of technologies that might theoretically end human existence. Others include nuclear warfare or the science fiction scenario of super-intelligent machines that, in *Terminator*-like fashion, become intent on annihilating humanity. However, most of the dangers and disruptions arising from the use of innovative tools and techniques do not threaten human existence. Those risks can be very large, for example, a worldwide pandemic that kills millions, or an economic disruption that topples governments. Most risks are relatively small, such as a malfunctioning device that harms individuals, one by one. However, a significant number of innovations are expected to have a broad societal impact. Many will be beneficial, the impact of some will be controversial, and others will be damaging. A single technology can even be beneficial, controversial, and damaging at the same time. For example, surveillance cameras and data mining ostensibly enhance security while simultaneously undermining freedom and privacy. Contrary to Benjamin Franklin's admonishment that those willing to sacrifice freedom for security deserve neither, citizens in the U.S. and the U.K. apparently seem quite willing to accept the trade-off.

Given these trade-offs for substantial benefits, the risks of even disruptive technologies are often judged acceptable. But which stakeholders should be consulted before the judgment is made as to whether the trade-offs are truly justified? And who should make the final determination?

*A Dangerous Master* examines the challenge of predicting and managing the potential harms that result from the adoption of emerging technologies, and weighs those against the anticipated benefits. Our discussion will make some mention of threats to humanity's existence, while focusing primarily upon significant but non-existential risks. Different kinds of risks require that differing stakeholders participate in the decision-making process. And certainly different levels of risk demand different forms of oversight.

Navigating the future of technological possibilities is a hazardous venture. It begins with learning to ask the right questions—questions that reveal the pitfalls of inaction, and more importantly, the passageways available for plotting a course to a safe harbor.

## TECHSTORM

Consciousness moves across the face of time, first highlighting one aspect of life, and then another. In our present chapter of history the spotlight rests upon technology and information. Time races by and gigabytes of information clog digital and neuronal highways. Individually and collectively we are under tremendous pressure to absorb, assimilate, experiment, engage, and make choices with little opportunity to reflect on the ramifications and human costs of selecting unwisely. A juggernaut of change in the form of genetic engineering, mood- and character-altering drugs, nanotechnology, and advanced forms of artificial intelligence threaten to redesign our minds and bodies and redefine what it means to be human. Pressures are building to embrace autonomous computers and robots that replace the flawed decision-making of humans, and to adopt bioengineered fuels as a new source of energy.

A never-ending stream of new gadgets and near-term possibilities are transforming daily life and perhaps human destiny. Each discovery can contribute to tangible benefits such as a cure for a disease or a new way to communicate. However, even the most beneficial discovery can be misused, have undesirable side effects, or undermine institutions and time-honored values. With the adoption of innovative tools and techniques, there have always been trade-offs. If driverless cars actually reduce highway fatalities that would certainly be beneficial. On the other hand, should we be under pressure to give up the privilege of driving? That trade-off is not on par with a technology that threatens annihilation, but it certainly could alter human activity in untold ways. Living a few extra, healthy years would be great, but extending life to an average of 150 years could be considered both a blessing and a curse for individuals and society. Robots that perform onerous tasks are easy to embrace. Smart robots that fill more and more human jobs, and might eventually surpass human intelligence, pose a host of economic and societal challenges.

The cumulative impact of new technologies expected to be available over the next few decades is difficult to imagine and likely to be unsettling. I refer to the incessant outpouring of groundbreaking discoveries and tools as a techstorm. While rain showers nurture plant life, a forceful and never-ending downpour can have a destructive impact.

Technological development is similar to an economy in that both can either stagnate or overheat. Vocal advocates for the rewards of technology call for more and more governmental policies to create funding for research and to lower product liability as a means to stimulate innovation. However, development in a few fields can accelerate toward a rate dangerously beyond our control. Arguably, that is the case with the expansion of information technologies and the manner in which they increasingly afford opportunities to undermine privacy and property rights. The genie is out of the bottle.

Ours is not the first era to experience an incessant outpouring of new technologies. The tools and emerging techniques that gave form to the Industrial Revolution had an even greater impact, and were more disruptive than what we are experiencing today. Crowded cities, unsafe and unsanitary working conditions, long working hours, and the use of children as laborers were among the prominent harmful features of the Industrial Revolution's techstorm. Nonetheless, while it was not initially apparent to those living in the squalor of overcrowded slums, the standard of living greatly improved. Often the destructive impact of a new technology precedes the reaping of its benefits.

The Information Age is still in its infancy and different enough from the Industrial Revolution that it is hard to make meaningful comparisons. In future decades, advances in genomics and nanotechnologies will garner even more attention than information technologies. These fields will not only change the world around us, but will also alter our bodies as well as facilitate intimate connections between our senses and external devices. The cumulative impact of the converging effects from many fields of innovation will contribute toward an incessant techstorm that will demand people constantly adapt in order to keep pace, reap benefits, and avoid harm.

Measuring the overall impact of emerging technologies will need to wait for at least another thirty to fifty years. In the meantime, it will be the foresight evident in the actions we take, which will determine whether the Information Age truly advances human well-being and understanding. Benefits take care of themselves. *A Dangerous Master* will address what can go wrong in order to introduce a conversation about various approaches to minimize risks. More importantly, this conversation should be about societal goals—not just what is possible, but what priorities we want our leaders and our society to focus upon.

Throughout the journey of technological and societal transformation, squalls, storms, and torrential cloudbursts will appear. Experiments in genetic engineering can go wrong. Promising theories will lead investors down blind alleys. Long before we see the anticipated benefits from genomics and nanotechnologies, the release of a powerful toxic substance outside of a laboratory might occur by accident or by malevolent intent.

Harmful outcomes challenge assumptions and raise questions as to whether the benefits are worth the costs. Critics will contend that short-term profits and speculated benefits do not justify immediate ills or future risks.

Nevertheless, the proposed benefits of new technologies address so many of today's problems that eschewing research holds little attraction. Even after filtering out the more speculative promises, the potential rewards of technological advances likely to be realized in the near future are huge. But if we fail to deal with anticipated ills, we both invite harm and threaten public support for the continuation of those areas of research whose rewards might otherwise be enjoyed. When no clear means exist for mitigating the risks of a technology, critics and citizen groups will oppose further research.

Bowing to political and economic imperatives is not sufficient. Nor is it acceptable to defer to the mechanistic unfolding of technological possibilities. In a democratic society, we—the public—should give approval to the futures being created. At this critical juncture in history, an informed conversation must take place before we can properly give our assent or dissent.

## INFLECTION POINTS

While it is still too soon to measure the broader effects of new technologies, we do have various opportunities to mitigate the damages and preserve the benefits. But in order to take advantage of these moments, or "inflection points," we first need to be able to recognize them. Inflection points are turning points in history followed by either positive or negative consequences. They provide windows of opportunity that allow us to assert a degree of control over the future we create. These windows can remain open for years, but in many situations they open and close quickly. Once a technology gets entrenched in the fabric of a society, making a course correction becomes extremely difficult. In failing to act on the opportunities that do emerge, we surrender the future to forces largely beyond our control.

Defusing the likelihood or intensity of a disaster occurring usually requires putting precautionary measures in place. There are many examples of inflection points where a disaster has been averted or a future disaster limited by an early response to a perceived danger. One recent example is the substantial upgrade in worldwide preparation for the outbreak of an influenza pandemic. A major pandemic can be horrific. The Spanish flu pandemic of 1918 infected five hundred million people and killed fifty to one hundred million (3-5 percent of the world's population).

Health officials worldwide took the opportunity to upgrade pandemic preparedness after the outbreak of two different varieties of flu. Between 2002 and 2013, more than 60 percent of the 630 confirmed cases of people who contracted bird flu (H5N1) died according to the World Health Organization (WHO). H5N1 is transmitted from poultry to humans, and, to date, there is little evidence of human-to-human transmission of the virus. In 2009, a mysterious new strain of the virus known as "swine flu" (H1N1) appeared in Mexico. By November, the WHO declared "482,300 laboratory confirmed cases" of H1N1, which included 6,071 deaths. There was fear among health officials that either the deadly avian flu would mutate into a form that transmitted from human to human, and thus spread rapidly, or that the H1N1 virus would mutate into an extremely deadly strain.

If fewer people contract avian flu, the prospect of a human-to-human mutation occurring is lowered. So the initial focus was upon preventing infection through the culling of flocks that have contracted bird flu; providing protective clothing and other safety measures for poultry workers; and education regarding hygiene and food preparation for communities likely to eat infected chickens or other bird species.

In fighting any new variation of a disease-causing virus, the time required to develop an effective vaccine and to produce that vaccine in sufficient quantities to inoculate a critical mass of the world's population, create serious bottlenecks. Reducing that time represents an opportunity to save millions of lives. In response to the avian flu and swine flu outbreaks, world health authorities took important measures to speed up research and the production of a vaccine. They include improvements in communications among public health officials around the world, additional research facilities, support to develop better diagnostic equipment, stockpiling of vaccines for known varieties of the flu, and the ongoing maintenance of vaccine production factories that will only be needed if and when a pandemic occurs.

Health authorities turned the concern caused by these two flu outbreaks into an opportunity to put in place precautionary measures to ward off potential future disasters. A pandemic equal to or exceeding the 1918 Spanish outbreak is still possible. Fortunately, the measures taken by world health officials in recent years have significantly decreased the prospect of a truly disastrous pandemic.

The power to alter the genome, and most specifically the human genome, signals a major inflection point in the history of humanity. Secondary inflection points can be thought of as adjustments in the rate of progress or shifts in the trajectory of the unfolding research. For example, the deciphering of the human genome could have proceeded slowly over decades, but was accelerated by large-scale public and private funding.

On June 13, 2013, the U.S. Supreme Court altered the course of research in human genomics by ruling that human genes cannot be patented. This ruling clarified and revised an earlier 1980 decision in Diamond v.

Chakrabarty, which effectively granted corporations the right to own genes as property and set off a "gold rush" to patent genes. By the time of the Court's 2013 decision, 41 percent of the genes in the human body had been patented. The right to patent human genes appeared outrageous to many critics, who argued that patent law was enacted to protect creative discoveries, not to own natural phenomena. In advance of their decision, it was not at all clear what the U.S. Supreme Court would do. There was precedent and tremendous political pressure to honor the costs incurred by private companies in conducting research to discover individual genes.

The case before the Court concerned the rights of Myriad Genetics, which held patents for two genes named BRCA1 (for BReast CAncer 1) and BRCA2. Through its patents, Myriad Genetics claimed far-reaching rights to not only market products that identify mutations in BRCA1 and BRCA2, but also for monopoly control over research on the genes and therapies that target the genes. The likelihood of breast cancer for women carrying mutations in BRCA1 and/or BRCA2 is high. Gene therapists commonly advise these women to have preventative mastectomies, the surgical removal of their breasts. The actress Angelina Jolie had a well-publicized prophylactic double mastectomy in 2013 after the discovery of a "faulty" BRCA1 gene indicated that she had an 87 percent chance of getting breast cancer at some time in her life, and a 50 percent risk that she might contract ovarian cancer. Ms. Jolie's mother died from ovarian cancer at fifty-six, and her aunt died from breast cancer at the age of sixty-one.

By a unanimous decision, the Justices of the Supreme Court ruled against Myriad and declared that isolated human genes could not be patented. Sanity prevailed. Future harms, from overcharging for gene therapies to limitations on who could perform research, were averted. The pace of genomic research was paradoxically slowed and accelerated by the decision. Private companies would find it more difficult to recoup their investment, while independent researchers were freed to study individual genes without going through companies that had patented the genes.

The Myriad case illustrates a well-recognized inflection point in the development of genomic research. The Court's action in clarifying, if not exactly overturning, a past precedent demonstrates that the course of new technologies is not at all beyond human control. Occasionally there are opportunities to make course adjustments. But once the court delivered its ruling, this particular inflection point disappeared. It would be extremely difficult to reopen the question of whether existing human genes can be patented.

## A COURSE CORRECTION

In the adoption of new tools and techniques, which risks are justified and which are foolhardy? The present pattern is to embrace every new tool that can be conceived of if someone wants to sell it and someone else has a use for it. The introduction of drones into domestic airspace provides a good example of this pattern. Companies developed drones for military applications, and now they are looking to expand their markets. In the U.S., Congress has been more concerned with instructing the Federal Aviation Administration to develop flight rules for introducing drones into domestic airspace, than in fostering a national conversation as to whether the benefits justify the losses. Does the use of drones for search and rescue, to facilitate police surveillance, to conduct research, and to enable one-day delivery of Amazon orders justify further loss of privacy, crowded skies, air accidents, damage to property, and swarms of camera-bearing mechanical insects buzzing overhead? A full public discussion would probably lead to the embrace of some uses for drones in domestic airspace, but with restrictions that would dramatically reduce their numbers. Unfortunately, that discussion isn't happening.

Throughout this book, I advocate for a more deliberative, responsible, and careful process in the development and deployment of innovative technologies. A cavalier attitude toward the adoption of technologies whose societal impact will be far-reaching and uncertain is the sign of a culture that has lost its way. Slowing down the accelerating adoption of technology should be done as a responsible means to ensure basic human safety and to support broadly shared values. For each project that is slowed, however, there will be costs: energy-saving devices whose implementation will be stalled, weapons systems that will not be perfected and deployed, and fortunes that will never be made. Dreamers whose hopes go unfulfilled will be disappointed. People who will die, but believed their illness should have been cured in their lifetime, will be angry. Hard choices must be made.

The course we are on is perilous on many fronts. There is serious need for a course adjustment. Subtle shifts can be introduced through the recognition of inflection points, through care in the selection of values that need to be maintained and reinforced, through the will to put in place good research and engineering practices, and through the development of institutions that comprehensively monitor, manage, and modulate technological development. *A Dangerous Master* will introduce recommendations as to how humanity can successfully navigate a perilous future.

## INEVITABILITY, DISQUIET, AND ACQUIESCENCE

My concerns regard the adoption of specific innovative technologies, not the process of scientific discovery. As history attests, government interference in scientific research undermines freedom on countless levels. Even so, in many fields governments should demonstrate more care in what they fund and what safeguards they require scientists to put in place.

Distinguishing activities that constitute scientific discovery from the deployment of new technologies can be somewhat arbitrary. A significant portion of research is directed at developing new tools, for example, more accurate diagnostic scanning systems, which will be used for both advancing scientific discovery and for therapeutic purposes. The Large Hadron Collider was not merely a research project, but also the deployment of a new technology. The testing of key theories could not be performed without the use of an extremely large particle accelerator. The creation of a new species in the laboratory is generally an accepted form of scientific discovery, but releasing that species into the environment outside the laboratory is a totally different matter.

Differentiating between scientific discovery and technologies that should not be developed or deployed requires a degree of maturity from engineers, government officials, leaders of industry, and military planners. To date, that maturity has often been sadly wanting, and is easier said than done. Free and open scientific discovery naturally leads to new technological possibilities. When a new form of weaponry appears possible, it will probably get the necessary funding for development long before military planners reflect upon its strategic value. Scientists and engineers get lost in the challenge of figuring out how to succeed in almost any conceivable venture. In striving to answer the question, "Can we do this?" too few ask, "Should we do this?"

For the past ten years, I have chaired the Technology and Ethics (T&E) study group at Yale University's Interdisciplinary Center for Bioethics. Founded in 2002, T&E is one of the longest running forums investigating the ethical, legal, and societal challenges posed by the emerging technologies. Through T&E and innumerable other venues, I have had the opportunity to meet and become friends with many of the scientists and engineers inventing the future, entrepreneurs developing and marketing innovative technologies, military leaders deploying new weaponry, and futurists hailing the benefits of transforming humanity. In nearly all cases, these are good people intent on bettering the human condition and helping the ill and the underprivileged among us. In addition, I have gotten to know much of the international community of policy planners, bioethicists, social scientists, legal scholars, theorists, science fiction writers, critics, and gadflies focusing upon the societal impact of adopting new technologies.

Everyone approaches the societal impact of the emerging technologies from a piecemeal perspective. Academics tend to be trapped in the silo of their own discipline. Policy-makers are heavily influenced by political and economic considerations, especially the desire to avoid blame if something goes wrong. Proponents for and critics of emerging technologies commonly approach controversies through the lens of one big defining idea. There is serious need for a more comprehensive and interdisciplinary conversation regarding the trajectory, impact, and management of the emerging technologies.

The discussions during our T&E meetings, while fascinating, are often deeply disturbing. Many of the risks are either unrecognized by policy planners or go unaddressed. Claims that all human problems will soon be solved technologically sound dangerously naïve. Techno-enthusiasts run the risk of indulging in a facile disregard for the lessons of history. This disregard is supported by a rather circular argument in which technology makes tools available to transcend all the limits placed upon past generations. Thus, the lessons of history do not apply to present-day challenges because a technological solution can be forged to solve each and every problem. This line of thinking is disturbing because it presumes any technological solution that can be conceived of will be realized.

Techno-optimists argue that scientific discovery and technological development cannot be separated. What becomes feasible will be realized. In their eyes, the inevitable course of technological innovations will lead to the near-future creation of designer babies and smarter-than-human robots. Perhaps, but even the inevitable can be slowed dramatically. Slower development affords opportunities for care in the design of safety mechanisms. It also alters the importance of various goals. With more time, inflection points come into focus and expand occasions to make course corrections. When I was eight years old, Walt Disney and Werner von Braun, then head of the U.S. space program, promised me a vacation on a proposed International Space Station, with side excursions to the moon. I'm still waiting. Von Braun's large wheel-like Space Station was never built because we, the American public, would not approve the funding to place such an expensive project ahead of other needs. Sixty years later, U.S. citizens would still love the chance to holiday in space, but few are ready to make this a national priority.

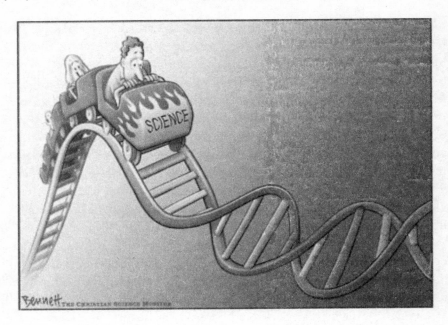

For thousands of years innovative technologies have served human aspirations. While most of us perceive technology as an engine of promise and productivity, there remains in many people a pervasive disquiet regarding specific fields of research, and bewilderment about the overall trajectory of technological development. This disquiet, and attempts to dispel it, are evident in the worldwide prohibition on human cloning, the banning of human growth hormones in sports, restrictions within the European Union on growing and importing genetically modified foods, and the contentious debate in the U.S. over embryonic stem cell research. International proposals to ban lethal autonomous weapons (killer robots) are currently under consideration, as is a ban on atmospheric experiments directed at mitigating the effects of global climate change.

Disquiet that derives from science fiction and hype can be dispelled, presuming one does the work of sorting out real prospects from speculation. Some disquieting concerns can be dismissed as a normal fear of the unknown. But often this sensibility reveals a form of intuitive intelligence. When crossing a busy street, disquiet functions as one of the tools the body has to alert the mind to a situation demanding its full attention. Likewise, a pervasive societal disquiet can indicate a challenge requiring resolution. The intelligence to discern appropriate measures, however, is hard to come by. Massive budgetary expenditures on biosecurity and cybersecurity have not come close to neutralizing the anxiety created by these new threats. The responses to threats such as cybercrime, cyber espionage, and cyber warfare appear to be little more than next steps in a spiraling escalation of technological vulnerabilities. In other words, the responses to these new threats are perceived as aspects of the problem, and unlikely to be effective as long-term solutions.

Self-driving cars provide an apt metaphor for apprehension and the sense of inevitability regarding the trajectory of technological development. Is technology both literally and figuratively moving into the driver's seat? While technology served our aspirations for thousands of years, it now appears that we must adapt our lifestyles and aspirations to what new technologies make possible. Have scientific possibilities and technological imperatives become the primary determinant of humanity's future? Are we each now expected to become technology's "good servant"?

I share in the disquiet over the trajectory of scientific discovery and technological development. However, I am also fascinated by what is actually being discovered and anxious to witness what scientists and engineers can create. Embracing the tension inherent in these two attitudes functions as a guiding force. From my perch in the field of technology and ethics, I am privileged to have time to explore innovative strategies to keep technology from slipping beyond our control. Indeed, other tensions, for example that between experts that are often too gung-ho, and non-experts who are prone to indulge in fears based upon speculation, can also be enlisted to provide fodder to explore alternative pathways forward.

Much of the disquiet comes to the surface through the long-standing debate as to whether enhancing human capabilities by technological means represents an opportunity or a defilement of human nature. For those immersed in the debate, there is evidence of some progress in the unmasking of false or superficial arguments. However, in a larger sense, our society has failed to have a satisfactory conversation about what is actually being created and which possibilities truly represent goals toward which we should be striving. The confrontation between those eager to embrace the means to enhance and alter human capabilities, and those disgusted by this prospect is, at this stage, a draw. A stalemate serves the interest of those invested in projects and ends that already have momentum. But some event, perhaps the cloning of a human or unanticipated neurological damage caused by a cognitive-enhancing drug, will likely alter that course.

Evidence of disquiet over human cloning or genetically modified foods makes it difficult to understand the comparatively smooth embrace of the Internet, smartphones, and other information technologies. No doubt disquiet emerges unevenly across different technologies, yet even the embrace of digital systems or gadgets carries a disturbing quality. The appalling story of a parent lost in her tablet computer, while greeting her young son with a perfunctory question about the day at school, may be apocryphal. But her periodic nods and screen tapping while he relates the events of the day bear the ring of truth. Devices with screens are mesmerizing. Playing with smart devices can override engagement with the people around us. Social media arguably offer a different way of socializing. Nevertheless, the repeated dopamine hits from beeps announcing the arrival of a text message or from winning a game of FreeCell, Candy Crush, or Angry Birds reinforce borderline addictive behavior. The computer trains us to be its servant.

Technological wizardry inspires awe and acquiescence. That acquiescence takes on a broader meaning in light of claims that technology marches along an inevitable course. In a narrative that has acquired considerable power, that course leads toward a future in which the role for mere humans is greatly diminished. Indeed, various prophets of technology proclaim that we have entered the first stage of an inevitable process during which the human species, as we have known it, will be invented out of existence. The masters of the future will be cyborgs, *techno sapiens,* and super-intelligent computers. Such claims would have less impact if the economic, political, and scientific forces driving technological development forward were not perceived as being all-powerful.

This book was born out of a 2011 conversation in which a friend conveyed his disturbance over what he perceived as the inevitable direction of research in genomics and artificial intelligence. His despondency and resignation troubled me. He had bought into the narrative where the future course of technological development was determined. It is understandable why many people believe the course of technological development has already been written. Perhaps the bread and circuses of technological wizardry function as a distraction that masks acquiescence and the sense of helplessness in confronting a process that appears to have already slipped beyond our control.

Fortunately, the course of technological development is not predetermined. Throughout this book, I will explore various challenges to the notion of technological inevitability. Claims of inevitability generally rest upon downplaying the difficulties in crossing major technological thresholds, and presumptions that theoretical solutions to all problems will bear out. Futuristic claims, such as the inevitability of vastly superior forms of artificial intelligence, underestimate the human capacity to exploit the vulnerabilities of any system that begins to demonstrate dangerous characteristics. New possibilities certainly become available as scientific knowledge deepens, but whether we embrace those possibilities, listen to disquiet, or take advantage of inflection points remains within our hands.

Even though we are not on an inevitable course, there is no broad plan in place to manage our deployment of new technologies. No one fully understands this momentous period in human history. Experts with specialized knowledge are in ready supply. Advocates for particular goals amplify their voices through the Internet, print, and TV or radio interviews. A few thoughtful souls share their visions for genomics, neuroscience, or nanotechnology. Talking heads joust with each other, while flying the colors of individual values such as freedom, security, equality, and responsibility. In the hope that only their accurate forecasts will be remembered, prophets dish out countless predictions. Busy responding to the latest crisis and struggling to advance one or two priorities, elected officials postpone attending to secondary concerns or ignore them altogether.

The unfolding technological landscape is confusing—a modern version of the Wild West, in which prospector engineers discover gold in the form of a new software application or approach to social networking. Once the prospector becomes a baron, he joins the ranks of business leaders protective of their economic niche and scornful of regulations that interfere with their profits, freedoms, and privileges.

On the one hand, there is need for a deeper and more comprehensive understanding of the forces and possibilities in play. On the other hand, any vision will be contingent. It is as if we are assembling a jigsaw puzzle of a technological landscape in which some pieces continue to change shape and others are missing altogether.

The most important missing pieces are the absence of a critical mass of informed citizens active in the conversation, a cadre of scholars dedicated to forging a comprehensive understanding, and forums through which the voices of informed citizens and scholars get factored into the decision-making of policy planners. Without these, the mechanistic unfolding of technological possibilities will prevail.

Young scholars are certainly interested in approaching the issues more comprehensively. But they are under pressure to justify their existence by establishing expertise in a specialized niche within the existing academic ecology. Universities pay lip service to the need for interdisciplinary understanding, but few scholars are ever rewarded for this pursuit.

Given the uncertainties and the recognition that decisions with momentous import must be made, both scholars and policy planners are calling for more public participation in the conversation. A role exists for scientifically literate citizens capable of asking good questions, listening closely, and then sharing their perceptions, insights, and intuitions. However, citizens perceive few venues through which to effectively join the conversation and express their concerns. Without a means for constructive participation, the value of being informed is diminished, and power is relinquished to those who already have it.

The last two chapters of this book will introduce approaches for nurturing a critical mass of informed citizens and interdisciplinary scholars by creating venues for their voices to be heard. But first, it will be necessary to lay foundations for a broader understanding of the fascinating and emerging technological landscape.

## A PLAN OF ATTACK

My perspective is that of an American. However, the challenges we confront involve every country and all of humanity. We need both national and international conversations about goals and concerns in the adoption of innovative technologies. There are fields of research and realms within public policy where other countries lead the way. Japan and South Korea, for example, play a prominent role in the development of artificial intelligence and robotics. The European Union has shown more initiative than other regions in forging interdisciplinary approaches to managing the societal impact of emerging technologies. Historically, the Chinese have been adept at taking the long view. This will be particularly important as China's political and economic influence grows on the world stage. What all cultures share is a meditation on the pace and trajectory of technological development. Individually and collectively, we must make decisions as to when these developments should be embraced, regulated, or rejected. In a more closely linked world, local restrictions on the development of new technologies will only be effective if backed up by international consensus. This will be discouraging to groups wishing to promote more parochial values. But it is empowering for those focusing on more universal concerns.

The need to harmonize technology policy internationally poses major challenges. Differences in forms of governance and value structures between nations cause many concerned scholars to believe that the emerging technologies cannot be regulated effectively. What one state restricts another state may allow. There is the risk that developers will simply move to those countries that place the least restrictions on their research. Government leaders desirous of reaping economic benefits as torchbearers in new fields will embrace areas of research that are controversial in other regions.

Certain instrumental values, such as safety and responsibility, resonate broadly across many cultures. Other more humanistic values including compassion, care for the needy, and the dignity of the individual human often get entangled with religious traditions. Unfortunately, many religions are intolerant of scientific views that challenge traditional beliefs. Will only the more instrumental values survive in a multicultural world? Fortunately, there exist other means to reinforce values, such as the United Nations Universal Declaration of Human Rights. But will this be sufficient? Or must a new values equation be forged to ensure that the pressures created by emerging technologies do not undermine respect for the individual?

Predictions as to what to expect over the next few decades of technological innovation fill periodicals and the Internet. Chapter 2 opens with one such dramatic prediction for you to assess. It describes the dangerous course we are on, and the need to redirect the trajectory of technological development.

Chapters 3–7 outline the three core reasons why emerging technologies will lead to tragedies and disasters. Increasing dependence upon unpredictable complex systems, the accelerating pace of development, and harms associated with specific fields of innovative research all give rise to new dangers. The convergence of global climate change and the spiraling demand for new sources of energy provide a backdrop for exploring how differing approaches entail trade-offs in risks and benefits. The application of genomic technologies to the creation of biological products, plants, and nonhuman animal species poses very different problems than in its application to the human genome.

The societal impact of radical life extension and human enhancements are respectively outlined in Chapters 8–10. Research ethics will slow and even stop certain types of experimentation that endanger human subjects. The contention that humans can and should be improved through technologies that enhance human capabilities, contributes to a broader narrative in which the mind and body are viewed as biological machines.

The biomedical industry and the military-industrial complex persist as leading drivers of an aggressive program of scientific research and technological innovation. Massive budgetary expenditures on health care and advanced military systems are supported by key assumptions, which will be questioned in Chapter 11.

Chapters 12–15 offer policies that can and should be put into place, values that must be reinforced, inflection points that should be recognized as calls to action, and the importance of supporting institutions and individuals vigilant in monitoring emerging technologies. Chapter 12 elucidates the far-reaching implications of an inflection point through the example of a proposed ban on killer robots. Chapter 13 turns to methods for embedding values in the design and engineering of new technologies. An institutional reform for coordinating the activities of the various regulatory bodies, professional associations, and nongovernmental organizations (NGOs) involved in the oversight of technological innovation will be described in Chapter 14. The concluding chapter outlines a role for scientifically literate citizens sensitive to the social impact of emerging technologies.

Our investigation can only offer suggestions and outline pathways forward. Ideas and understanding derive meaning through actions. The final chapter of this story must be written upon the world at large by our individual and collective initiatives. Do we surrender to the blind forces that have been set in motion or do we learn how to evolve consciously and responsibly? Navigating the future requires attention, care, and the willingness to make some hard choices.

# { 2 }

# A Prediction

SOCIAL DISRUPTIONS, PUBLIC HEALTH AND ECONOMIC CRISES, environmental damage, and personal tragedies all made possible by the adoption of new technologies will increase dramatically over the next twenty years. Some of these events will result in the death of many people. This prediction is not meant to be melodramatic or to generate fear. Nor am I interested in thwarting the progress of the many scientific paths of research that will improve our lives. I offer this warning in the hope that, through a little foresight and planning, and the willingness to make some hard choices, many of the dangers will be addressed. Unfortunately, there is little evidence that we or our governments have the will, intelligence, or intention to make those hard choices. Indeed, there are reasons to believe that such crises are inevitable, that the pace of calamities involving new technologies will accelerate, and that the opportunity to give direction to the future of humanity is slipping away.

Thalidomide babies, Chernobyl, the explosion at a Union Carbide chemical factory in Bhopal, India, the *Challenger* space shuttle, and the BP oil spill evoke images of tragedies in which technology was complicit. To this list add the use of harmful technologies consciously designed for destructive purposes: the gas chambers at Auschwitz, the firebombing of Dresden, Hiroshima and Nagasaki, use of Agent Orange in Vietnam, sarin gas attacks in the Tokyo subways, and the dangers posed by the proliferation of cruise missiles and biological weapons.

Many new risks are posed by technologies under development. A failure of, or a cyber attack upon, critical information systems will cause a major banking crisis or a sustained loss of electricity. A nanomaterial used in many consumer products will be discovered to cause cancer. Students will suffer brain damage as they mix drugs, each of which is intended to give them a competitive edge in their studies. An enraged teenager will kill her father with a plastic gun produced on a $275 3D printer. In his home laboratory, a psychopath or terrorist will brew a designer pathogen capable of starting a worldwide flu pandemic. Autonomous weapon systems will kill civilians, and may even start new wars. An island nation, threatened by rising tides, will engineer local climate and cause a drought in neighboring regions.

In our 2009 book, *Moral Machines: Teaching Robots Right From Wrong*, my co-author Colin Allen and I made a similar prediction about a catastrophic event caused by a computer system taking actions independent of direct human oversight. We sketched a fictionalized example of how a disaster caused by computers might unfold. The scenario entailed a series of plausible incidents based upon present-day or near-term computer technology. Collectively, these occurrences triggered a spike in oil prices, a failure in the electrical grid, a Homeland Security alert, and the unnecessary loss of lives.

A real-life incident created by computers occurred at 2:45 PM on March 6, 2010. The Dow Jones Industrial Average took a steep dive and then recovered in a matter of minutes. What has been named the *flash crash* is the biggest intraday point decline (998.5 points, 9 percent) in the history of the Dow Jones Industrial Average. High-speed automated trading by computer systems that buy and sell shares was a key contributing factor. At the time of the crash, high-frequency traders tendered at least 60 percent of all transactions. By some estimates roughly half of all trades today are made automatically by computers that tender buy and sell orders algorithmically once mathematically determined thresholds are crossed. The percentage exaggerates the overall importance of computerized market activity in that high-frequency trades take the form of two transactions—a buy and a sell order occurring within a few minutes or even a fraction of a second of each other.

The *flash crash* unduly robbed some investors while rewarding others, and undermined confidence in the operations of the stock exchanges. To avert a future *flash crash,* additional circuit breakers that automatically kick in when errant trades are detected were built into the stock markets. Nevertheless, on August 1, 2012 a "rogue algorithm" from Knight, a company that specializes in computer-driven trading, tendered buy and sell orders for millions of shares in 148 different companies before the circuit breakers halted trading. The major harm was to the trading company Knight and its clients, who lost $440 million in less than an hour. But this was just one more in a series of events that have reinforced an image in the minds of small investors that the robots are already in control of financial markets, and that the investment game is fixed.

The reliance on computers by financial markets goes well beyond systems responsible for high-frequency trading. They also played a role in the earlier real estate collapse of 2008. Computers enabled the assembly of complex derivatives that bundled bad loans together with good loans. Once the real estate market collapsed, it became impossible for anyone to evaluate the worth of the derivative shares held by banks. Even heads of large banks could not determine the viability of their own institutions. What is interesting about the role of computers in the derivative crisis is that the machines did not fail. They functioned as critical infrastructure supporting a faulty banking system. Yet their very existence enabled the creation of the complex derivative market that helped cause the crisis, and certainly exacerbated its impact. In the following chapter, we will explore further the inherent dangers of relying on complex systems whose activity and impact cannot be fully predicted.

The role computers played in the real estate collapse was noted by a few analysts, but the emphasis has been more upon assigning blame to greedy bankers making bad bets and uncovering fraudulent activity by crooks such as Bernie Madoff. The *flash crash* and the "rogue algorithm" were initially attributed to human error. Computers have escaped blame.

From our present vantage point, it is impossible to know all the potential dangers. Bad agents intent on using a technology to pursue destructive and illegal goals will be responsible for much of the risk. Those risks range from individual tragedies to public health crises; from the societal and ethical impact of technologies that enhance human capabilities to the loss of privacy, property, and liberty; and from the collapse of critical infrastructure to the onset of a dystopian society.

The promoters of a cutting-edge technology submerge its dangers beneath enthusiasm for the potential benefits of the research. They boast that the blind will see and the lame will walk with the help of cameras and mechanical limbs wired into the nervous system. Wars will be won with the latest in high-tech weaponry. Deciphering an individual's genome will lead to personalized medical treatments for inherited diseases. Autonomous cars will have fewer accidents and free drivers to text message while traveling to and from work. Each of us will be happier, smarter, and even more moral if we elect to take a morning cocktail of cognitive enhancers. Of course, we know that investment bankers, venture capitalists, entrepreneurs, and even a few inventors will get rich along the way. Politicians will receive the patronage of the powerful, contracts for industries within their districts, and support from voters who receive jobs. Scientists and engineers will be rewarded with tenured professorships, well-financed laboratories, bright research assistants, and the occasional Nobel Prize or other prestigious award.

Short of one catastrophe that threatens human existence, the greatest challenge would be the convergence of many disasters from different quarters occurring within a short period of time. Technological development is intimately entangled with health care, environmental, and economic challenges. Usually technology provides solutions to these problems, but it can also create public health crises, damage to the environment, or, as discussed, economic disruption. If a confluence of disasters should occur, technology may not be implicated in them all. While many other scholars and reporters ably cover the consequences of failing to address challenges in these other spheres, this book addresses the technology side of the equation.

Social systems can manage only so much stress. Reacting to multiple concurrent disasters taxes and quickly overwhelms even robust institutions. The government may or may not bear direct responsibility for a crisis, but when it fails to effectively respond, it loses the confidence of citizens. In democratic countries, small failures in governance often lead to a turnover in the ruling party. But large failures undermine the public's faith in its governing system. One very large crisis or multiple crises could potentially bring on the collapse of major social institutions and a government.

Finding a resolution to each challenge as it arises provides the best method for staving off a future situation in which multiple crises arise simultaneously. This usually entails putting in place precautionary measures that can be costly. However, short of a self-evident instance in which a disaster is prevented, there is no good way to determine the efficacy of precautionary measures precisely because the disasters averted never actually occur. Furthermore, precautionary measures in the form of regulations and governmental oversight can slow the development of research whose overall societal impact will be beneficial. Thus, legislators are reluctant to ask businesses and citizens to make sacrifices. In recent years, politicians have gone one step further by pretending that problems from global climate change to high-frequency trading either do not exist or cannot be tamed.

Without precautionary measures we are left with the often unsatisfactory downstream attempt to solve a problem after a tragedy has occurred. Disasters do focus attention. Bovine Spongiform Encephalopathy, commonly known as mad cow disease, was a wake-up call for the European

Union. The meltdown of reactors at the Fukushima Nuclear Power Plant after the giant tsunami on March 11, 2011, alerted the Japanese people to failures that pervade their management of a potentially dangerous technology. Unfortunately, responses to a disaster tend to be more reactionary than well thought out. Four million four hundred thousand cattle were slaughtered in the UK in the attempt to stem Creutzfeldt–Jakob disease, the human variant of mad cow disease. Japan, which is heavily dependent on nuclear power generation, was forced to shut down all reactors by May 2012. After extensive testing, a few are slated to restart in 2015.

Governing in reaction to disasters is costly. Whether the costs incurred from waiting until a tragedy happens are greater than the losses incurred by zealous upstream regulation is a matter on which policy planners disagree. Time, however, will answer that question. If a convergence of multiple crises takes place, many of which result from unaddressed, foreseen problems, the answer could be the rapid onset of a dystopian future.

## VISION OR THE LACK THEREOF

Dystopian visions of the future have long sparked the imagination of science fiction novelists and moviemakers. Aldous Huxley's *Brave New World* (1931) and George Orwell's *Nineteen Eighty-Four* (1949) are cautionary tales that alert readers about futures to avoid. The cyberpunk subgenre was built upon William Gibson's classic *Neuromancer* (1984), and later, Neal Stephenson's *Snow Crash* (1992), which, in the context of near-future societies where only the technologically enhanced prevail, introduced the concept of *cyberspace* and the possibilities of *virtual reality* to a broader public. *The Terminator* (1984) and *The Matrix* (1999) dramatize a future in which artificial intelligence intentionally seeks to exterminate humans. Between computer-generated images in movies that look real, and hype over speculative possibilities, the public and the media can have great difficulty in discerning which proposed technologies are probable and which are highly unlikely. Even experts disagree about whose predictions are credible and which potential harms require attention.

Cautionary science fiction feeds the Pandora's box intuition—recognition of alleys best left unexplored and roads the intelligent traveler avoids. This intuition is often expressed through judgments that the human genome should not be tinkered with or that killer robots should not be built. The Pandora's box intuition is frequently used in support of irrational, anti-scientific courses of action, such as an unwillingness to vaccinate one's children against deadly diseases. Redirecting that intuition away from false problems toward challenges that can and should be addressed is a tall order.

The fear of technological mishaps is commonly countered by an ideology, particularly prevalent in the U.S., that for every danger posed by a new technology there will be a technological fix. A naïve faith in technological fixes can rise to the level of religious zeal. Call it technological solutionism.

Technological dystopias can have a stultifying effect on science and science policy. When science fiction scenarios appear likely in the public's mind there will be demands to prohibit even those areas of research that have a very low probability of going awry. In retrospect, few among us would have wanted to give up the past sixty years of advances in computer science and genomics based upon 1950s' fears of robot takeovers and giant mutant locusts.

Cognizant of depressing scenarios, the techno-optimists are working overtime to paint visions of a utopian future. The more attractive visions describe ways to conquer disease, end poverty, and restore the environment. However, when the visions rely upon the beneficence of superior enhanced humans or superhuman robots, they sound no less hollow than the destructive utopian ideologies that precipitated calamities throughout the twentieth century.

The story of the dangers inherent in the unchecked march of technological progress is one that many of my friends and colleagues—scientists and engineers, those requiring treatments for incurable diseases, transhumanists anxious to enhance their minds and bodies, and others hopeful of reaping rewards from cutting-edge research—do not want told. They are particularly fearful that any discussion of dangers will empower the growing anti-science coalition, which already rejects the theory of evolution and empirical evidence of global climate change for ideological, political, and religious purposes. Scholars within colleges and universities perceive free and open inquiry as a delicate enterprise that can be compromised through political and economic pressure. This is particularly true of older professors and administrators who witnessed the pall cast over university campuses in the early 1950s by Senator Joseph McCarthy's anti-Communist crusade. Indeed, in fear of the growth of an irrational anti-science movement, researchers and those who yearn for the benefits of scientific research have doubled their efforts to dampen the effect of criticism. In other words, even criticism contributes to driving forward the technological juggernaut.

I view the challenge differently. The promoters of new technologies need to speak directly to the disquiet over the trajectory of emerging fields of research. They should not ignore, avoid, or superficially dampen criticism to protect scientific enquiry. Furthermore, they should stop coddling secret research that results in, for example, new weapons systems whose dangers are only made known to an unsuspecting public after they are in use. The public, policy planners, and scholars must be given an opportunity to reflect on the societal impact of transformative technologies before they are deployed or marketed.

In recent years, there have been many calls from within the scientific community for researchers to play a role in conveying the societal impact of their innovations to the public. Regrettably, many scientists and engineers do not believe that the ethical and policy challenges arising from their work is their problem. They feel ill-equipped for, or disinterested in, a public dialogue. They leave the responsibility to address issues to others. But who are these others? Politicians? Leaders of industry? Bioethicists? Fundamentalist preachers? Tech-savvy young men and women who revel in the latest gadgets? Concerned citizens? Something more than a collection of semi-informed elected officials making decisions that will affect all of humanity is required.

Solomon, in all his wisdom, would be hard-pressed to parse out the potentially destructive possibilities from the positive fruits of technological innovation. Nevertheless, each of us is asked to give at least tacit approval to the glorious and unsettling march of technological progress. Unfortunately, most people have only a rudimentary understanding of the ideas and discoveries that will shape their own future. There is so much new science. Even those who craft public policy regarding the financing of research are highly dependent on the recommendations of experts. The difficulty lies in knowing what questions to ask and where to turn for a balanced perspective. All too often scientists extrapolate beyond their expertise and their understanding. The desire to get research funded can cloud their judgment. Hype from scientists, in the form of overly optimistic time frames for achieving the next technological breakthrough, encourages both unrealistic hopes and fears regarding scientific progress. The most persuasive voices commonly are those that stridently support or stridently oppose the development of a particular technology.

Decisions dependent upon the expertise of individuals with specialized forms of knowledge or special interests are not satisfactory for dealing with problems that can alter thousands and millions of lives. The complications posed by emerging technologies are beyond the intelligence of any one person to solve. That does not mean they are unsolvable. Individuals with different kinds of understanding can be brought together to share their insights, highlight issues, note vulnerabilities, shine a spotlight upon gaps in existing governance mechanisms, and work toward resolving the issues at hand.

Solomon did not have ready answers to the problems he solved. When confronted with two women claiming to be the mother of a child, he did not know who was telling the truth. But he did understand how to create a situation in which the truth could be revealed. When he proposed cutting the child in half, the responses from the two women made it clear who was the real mother. Good reporters and good lawyers know how to ask questions in a manner that brings a telling fact to light. Sometimes a subtle gesture divulges when a man is being dishonest, evasive, or laying claim to an understanding he does not have. The forms of intelligence brought to bear on elucidating difficult challenges are varied. Wise elders, good-hearted parents, and even street-smart hustlers are often sensitive to nuances that get missed by experts.

A forum that brings together experts capable of making difficult concepts accessible to jurists, journalists, and interested citizens would be one method for evaluating various proposals. The give-and-take can continue until a consensus about a way forward emerges. A first step toward such forums, or any vehicle for navigating the challenges of emerging technologies, requires the cultivation of scholars and concerned citizens steeped in an appreciation for the lay of the technological landscape. A critical mass of informed citizens would be helpful for balancing the influence of stakeholders with a vested interest in specific outcomes. Toward this end, *A Dangerous Master* maps a framework for the broad interdisciplinary understanding of the societal impact of emerging technologies.

Any prediction I make will fail to warrant lawmakers from taking action to put in place more rigorous mechanisms for managing technological innovation. After all, the history of forecasting is littered with predictions that were far off the mark. Nor should legislators react to public disquiet generated by overly optimistic hype countered by fearful melodramatic scenarios. Only through a full examination of emerging technologies will the reasons for concern come into focus. That examination will begin with a discussion of core issues and specific dangers that animated my sense that this book was needed. Once the areas of concern are illuminated, it becomes possible to perceive inflection points and appreciate both general and targeted means to defuse potential dangers. We can successfully navigate technological development, but only through concerted effort.

# { 3 }

# The C Words

ON DECEMBER 6, 1999, AFTER A SUCCESSFUL LANDING, A GLOBAL Hawk unmanned air vehicle (UAV) unexpectedly accelerated its taxiing speed to 155 knots (178 mph), and at a curve in the runway, veered off the paved surface. Its nose gear collapsed in the adjacent desert, causing $5.3 million in damage. At the time of the accident, the operators piloting the UAV from the safety of their control station had no idea why it had occurred. An Air Force investigation attributed the acceleration to software problems compounded by "a breakdown in supervision." A spokesperson for Northrop Grumman, the UAV's manufacturer, placed blame for the excessive speed totally upon the operators. There was certainly some delay in the operators' recognition of the UAV's unanticipated behavior. Even after they realized something was wrong, they did not understand which of its programmed subroutines the aircraft was following. The compensating instructions they did provide were either not received or not accommodated by the aircraft. That the operators failed to understand what the semi-autonomous vehicle was trying to do and provided ineffective instructions makes sense. But any suggestion that the aircraft could understand what it was doing or what the pilots were trying to do, implies that the UAV had some form of consciousness. That was, of course, not the case. The vehicle was merely operating according to subroutines in its software programming.

The Global Hawk UAV is only one of countless new technologies that rely upon complex computer systems to function. Most of the smart systems currently deployed and under development are not fully autonomous, but nor are they operated solely by humans. The UAV, like a piloted aircraft or a robotic arm that a surgeon can use to perform a delicate operation, is a sophisticated, partially intelligent device that functions as part of a team. This type of team exhibits a complex choreography between its human and nonhuman actors, and draws aspects of its intelligence from both. When operating correctly, the UAV performs some tasks independently and other tasks as a seamless extension of commands initiated by human operators.

In complex systems whose successful performance depends upon coordinating computerized and human decisions, the task of adapting to the unexpected is presumed to lie with the human members of the team. Failures such as the Global Hawk veering off the runway are initially judged to be the result of human error, and a commonly proposed solution is to build more autonomy into the smart system. However, this does not solve the problem. Anticipating the actions of a smart system becomes more and more challenging for a human operator as the system and the environments in which it operates become more complex. Expecting operators to understand how a sophisticated computer thinks, and anticipate its actions so as to coordinate the activities of the team, actually increases their responsibility. Designing computers and mechanical components that will be sufficiently adaptive, independent of human input, is a longer-term goal of engineers. However, many questions exist as to whether this goal is fully achievable, and therefore it is also unclear whether human actors can be eliminated from the operation of a large share of complex systems. In the meantime, the need to adapt when unforeseen situations arise will continue to reside with the human members of the team.

Complex systems are by their very nature unpredictable, and prone to all sorts of mishaps when met with unanticipated events. Even very well designed complex systems periodically act in ways that cannot be anticipated. Events that have a low probability of occurring are commonly overlooked and not planned for, but they do happen.

Designing complex systems that coordinate the activities of humans and machines is a difficult engineering problem. Just as tricky is the task of ensuring that the complex system is resilient enough to recover if anything goes wrong. The behavior of a computerized system is brittle—consider a Windows computer that locks up when a small amount of information is out of place—and seldom capable of adapting to new, unanticipated circumstances. Unanticipated events, from a purely mechanical failure, to a computer glitch, to human error, to a sudden storm, can disrupt the relatively smooth functioning of a truly complex system such as an airport.

I've concluded that no one understands complex systems well enough. As MIT professor and best-selling author Sherry Turkle once said to me in reference to managing complex systems, "Perhaps we humans just aren't good at this." Can we acquire the necessary understanding? To some degree, yes, but there appear to be inherent limitations on how well complex adaptive systems can be controlled. This is not a new problem. Debate has been going on for more than a half century as to whether governments and energy companies can adequately manage nuclear power plants. When the danger is known, generally extra care and attention are directed at managing the risks. But even so, many, but not all, of the nuclear accidents that have happened were caused by a lack of adequate care, stupidity, or both. A few nuclear accidents were the result of unexpected or unaddressed events. The meltdown of nuclear reactors at Fukushima, Japan was caused by a huge once-in-a-thousand-year tsunami.

The basic meaning of the word "complex" is easy to grasp. But why complexity functions as a game changer is more difficult to understand. This chapter will serve as an introduction to complexity, the kinds of complex systems already evident, others under development, and the challenge in recognizing which forms of complexity pose dangers. In the process of clarifying difficulties in monitoring and managing complex systems, the related C word "chaos" will also be elucidated. For our purposes, the inspection of complexity and chaos is central for appreciating challenges, difficulties, and dangers that might otherwise be overlooked. Complexity need not be treated as a roadblock, but it does warrant our attention if we are to safely navigate the adoption and regulation of new technologies.

There are already fields of scientific research in place for studying how complexity and chaos work. In the years following WWII, the interdisciplinary study of systems, and how the components within those systems behave, became increasingly important, and eventually evolved into a field called systems theory. Self-regulating systems can be natural, such as the human body; social systems such as a government or a culture; or technological inventions, such as cars or synthetic organisms. In everyday speech, words such as complexity and chaos are used loosely. In system theory, the words are used in a more precise way in order to lay a foundation for two related fields, complexity science and chaos science. But are these growing areas of scientific research helpful in wrestling with unpredictability and limiting the likelihood of unanticipated catastrophes?

Weather patterns are complex, as are the behavior of economic markets. And yet modest success has been achieved in predicting the behavior of both. The science of complex systems has made some progress in modeling the patterns that might emerge out of an interdependent set of diverse inputs.

Much of this progress relies upon computer simulations. Unanticipated factors, however, make predicting the actions of markets and weather a combination of probability and luck. The same holds true for predicting the behavior of many technologies we create. This is deeply troubling, because contemporary societies are increasingly reliant on truly complex technological systems such as computer networks and energy grids.

To appreciate the breadth of the problem, it is important to note that technological systems are not merely composed of their technological components. In our UAV example, the complex system includes the people who built, maintain, improve, operate, and interact with the technology. Collectively, the technological components, people, institutions, environments, values, and social practices that support the operation of a system constitute what has been called a sociotechnical system. Water purification facilities, chemical plants, hospitals, and governments are all sociotechnical systems. In other words, a specific technology—whether a computer, a drug, or a synthetic organism—is a component within a larger sociotechnical system. What is at stake is whether the overall system functions relatively well, not just the technological component. The focus in *A Dangerous Master* is on the technological components or processes, but the problems often arise out of the interaction of those components with other elements of the sociotechnical system.

Can complex systems ever be adequately understood or sufficiently tamed? Engineers strive to make the tools they develop predictable so they can be used safely. But if the tools we develop are unpredictable and therefore at times unsafe, it is foolhardy for us to depend upon them to manage critical infrastructure. There is an imperative to invest the resources needed (time, brainpower, and money) to develop new strategies to control complex technologies, and to understand the limits of those strategies. If planners and engineers are unable to limit harms than we, as a society, should turn away from a reliance on increasingly complex technological systems. The examples in this chapter have been selected to help clarify which aspects of complex adaptive systems can be better understood, and therefore better managed, and which cannot.

## COMPLEX AND CHAOTIC SYSTEMS

The individual components of a complex system can be the nodes in a computer network, neurons in the brain, or buyers and sellers in a market. When many components are responding to each other's actions, there is no surefire way to predict in advance the overall behavior of the system. To understand a complex system's behavior, it is necessary to observe how it unfolds one step at a time. Consider a game of chess. The possible actions within each move are limited, but it is nevertheless difficult to predict what will happen five, ten, or twenty moves ahead. Deep Blue, the IBM computer that beat the World Chess Champion Garry Kasparov in 1997, required tremendous computing power to calculate possible sequences of moves. Even then, it could only predict with decreasing probability the fifth or tenth moves in a series that created new branches at each step. Each move made by Kasparov would eliminate some branches while opening up additional possibilities.

A mathematical formula yields a result by working through the problem, but the algorithms that lie at the heart of computer technology define a process that must unfold step-by-step. A pattern can emerge over time out of a sequence of events, but that pattern may tell us very little about what to anticipate for a later sequence. Similar to the failure of the Global Hawk UAV, sophisticated future robots will act according to algorithmic processes, and their behavior when confronted with totally new inputs will be hard, if not impossible, to predict in advance. The kinds of damage semi-autonomous robots working in a warehouse shipping out Amazon orders could potentially cause might be judged acceptable, given the rewards. But the unfolding actions of a robot that serves as a weapons platform could unexpectedly, yet conceivably, lead to the loss of many lives.

To quote an old adage, "Baking a cake is easy. Building a car is complicated. Raising a child is complex." In complicated machinery, the way components interact is more or less understood. The behavior of a complicated system is predictable. Complicated systems can fail when a component wears out, but engineers have become good at predicting in advance how long before they will fail. When I was a youngster, the breakdown of individual components of a car, such as tires, could not be predicted, and so periodically there would be an unanticipated blowout. As automotive technology has advanced, cars have become more complicated and also more reliable. Incredibly complicated technologies can be extremely good at preventing accidents. Arguably, automobiles have evolved over the past century from complex, unpredictable systems to much more predictable yet extremely, complicated systems. Nonetheless, the continuing stream of recalls demonstrates that components still fail. Increasingly, computerized vehicles are also susceptible to software errors. In other words, accidents continue to occur due to factors that had not been anticipated before the vehicles were marketed.

Not all technological innovations are complicated or complex. But even simple technologies can be dangerous when they intervene in the behavior of preexisting complex systems, such as climate, natural environments, or the human body. The aerosol can and its chlorofluorocarbon (CFC) propellants, a rather simple technology, altered the ozone layer of the atmosphere in ways that no one could have foreseen. Looked at from another perspective, CFCs became one more element adding to the complexity of already complex and unpredictable weather patterns.

A system does not need a large number of components to be unpredictable. In 1887, the French mathematician and physicist Henri Poincaré (1854–1912) was studying a classical enigma known as the three-body problem. The problem was to discover how the gravitational pull of three large bodies in space would influence each other's orbits. Poincaré demonstrated that there was no discernible pattern. In other words, the orbits of the three bodies did not repeat the same pattern over and over again. Poincaré demonstrated that, even in a mechanical universe where all actions are determined by physical laws, activity could be chaotic and unpredictable.

Complexity and chaos are related and sometimes overlapping concepts. Both refer to systems whose activities are unpredictable and can be altered by small events. What distinguishes the two concepts is that chaos science or chaos theory looks at how the activity of components of a system can lead to dynamically rich but seemingly random behavior while following simple physical laws. Complexity science studies the dynamic behavior of interactions between large numbers of components or units within a system. In a chaotic system, the individual elements need not affect the behavior of other elements in any discernible way. In contrast, the nodes, components, or individuals in a complex system are coupled and adapt to each other's behavior.

Complex systems are robust and can self-organize. If an important firm in a stock market fails (think of the bankruptcy of Lehman Brothers during the financial crisis of 2008), the other players in the market will adjust and the market will continue on without that firm. In designing critical infrastructure, such as power plants, engineers build in backup systems and redundancies in order to improve the system's robustness, and its ability to withstand or quickly recover from a failure.

The actions of one or a small number of elements in either chaotic or complex systems can have repercussions that reverberate throughout the whole system, creating large events that can, at times, be catastrophic. In chaos theory, this is commonly referred to as the butterfly effect, in reference to the theoretical possibility that a butterfly flapping its wings in Mexico could, under the right circumstances, be amplified, leading to tornadoes in Texas. The phrase "right circumstances" is crucial here, for any slight variation in the initial conditions will alter the effect of the flapping wings.

An event whereby one simple alteration forces a complex system to reorganize is referred to as a tipping point. Tipping points are similar to phase transitions in physics when, for example, at a temperature below 32° F (0° C), liquid water freezes into crystals. The best example for understanding how small incidents can alter complex systems in ways that turn catastrophic comes from political history, not science. The assassination of Archduke Ferdinand of Austria and his wife in Sarajevo on June 28, 1914, triggered a series of events that started World War I. In 1914, the European powers were the constituent elements, and their political and military alliances the interconnections that collectively made up a complex political and military system. The Serbian assassin's motives were local, directed at splitting Slavic provinces from the Austro-Hungarian Empire. Once war broke out between Austria-Hungary and Serbia, however, countries throughout Europe were dragged into the conflict because of their network of alliances. Political leaders learned a great deal about the tight coupling of countries' alliances from the way World War I started, but little of that understanding has passed on to business leaders who tightly couple technological systems within a multinational corporation.

In both the spread of a disease or the emergence of a fad there can be a tipping point where adding just a few more individuals quickly gives rise to a widespread epidemic. The spread of a new technology, such as cell phones, passes through a tipping point where its use is initially restricted to a few early adopters and then rapidly embraced by large segments of a society. A worldwide pandemic occurs when the spread of a new virus passes through a tipping point well before the World Health Organization has developed an effective vaccine.

Any change in a complex system can require readjustments throughout the system and cause an array of problems. Assimilating a new manager, whose style and psychology differ from that of her predecessor, into a corporation or university administration will force hundreds of other workers in that system to adjust their activities. Aimee van Wynsberghe, a philosopher of technology at the University of Twente, points out how introducing a robot that performs a few tasks into the care practice of a hospital can alter the roles and responsibilities of a wide variety of staff members.

A few years back I was on a flight from Connecticut to California with a stop en route at the Dallas/Fort Worth Airport, a hub for American Airlines.

We were unexpectedly ushered off the plane because of a dramatic but short-lived storm. After reloading passengers, hundreds of planes lined up for takeoff. Three hours passed before our plane left the ground, and these delays, in turn, contributed to scheduling problems at airports all across the U.S. and Canada. Fortunately, civil aviation authorities have learned a great deal about handling disruptions from past experience and from the study of complex systems. The air transport industry is designed to be robust and capable of self-organizing as it adapts in response to almost any challenge. A storm, a crash, or terrorist attack will not by itself bring down the civil aeronautic system. However, even that process of reorganization can entail delays and serious disruptions, individual tragedies, and disasters. Airport delays go beyond being an inconvenience for those who miss flights. Disruptions, however, are the stuff of life. Of greater concern are those occasions when unpredicted events turn disastrous and even catastrophic for an entire system.

## DISASTERS

Complex systems turn disastrous for four reasons (or a combination thereof), not all of which can be easily prevented—if they can be averted at all. Incompetence or wrongdoing by managers or workers is the first reason. There was a long list of bad decisions leading up to, and during, the Chernobyl nuclear meltdown. Managers and workers had not been trained properly, and took incorrect remedial actions to arrest the breakdown. Chernobyl was an accident just waiting to happen. Failures of complex systems can also cause harm due to the unwillingness of profit-seeking executives to implement costly safety systems. The explosion of the Deepwater Horizon offshore oil rig killed eleven workers and spewed an estimated 4.9 million gallons of oil into the Gulf of Mexico between April and July of 2010. This accident could have been avoided or been much less damaging if officials at British Petroleum and Transocean (the responsible corporations) had not rushed to complete the job and in the process compromised safety procedures.

Design flaws or vulnerabilities account for a second reason complex systems fail. A flawed reactor design was a contributing factor in the Chernobyl disaster. Programming errors or bugs are particularly common. Often software vulnerabilities do not even come to users' attention until just the right convergence of events. For example, new versions of operating systems such as Windows contain millions of lines of code. They can initially be released with thousands of known bugs and many that remain unknown until reports come in from end users. Debugging complex programs is a continual process, as each bug fix may leave behind new vulnerabilities. As mentioned in Chapter 2, the Knight trading company lost $440 million on Wall Street in less than one hour. This loss was the result of introducing a software update that was not ready for prime time. Software for critical activities must undergo rigorous testing procedures before it is deployed. All too often, software fails to be tested sufficiently. Even thorough testing can miss serious vulnerabilities.

Furthermore, for each vulnerability there can be an exploiter, such as a hacker who elects to create a computer virus or break into a system for illegal purposes. In the flourishing field of cybercrime and cyber espionage, hacking for nefarious purposes has been transformed into a high art. We are all familiar with the antivirus programs, firewalls, and encryption codes meant to defeat destructive viruses and criminals bent on exploiting weaknesses in computers. Unfortunately, such security measures add additional layers of complexity that can undermine ease of use and, more importantly, increase the unpredictability of complex systems.

Lack of attention to critical features of any sociotechnical system can lead to catastrophes. Many accidents had already occurred at chemical plants throughout the world before a methyl isocyanate gas leak on the night of December 2, 1984, at a Union Carbide plant in Bhopal, India killed an estimated 3,700 people. The absence of a warning, the numbers of residences near the plant, the fact that the leak occurred at night, and the direction of a slow-moving wind all contributed to the seriousness of the disaster. A change in any one of these factors might have resulted in less than a hundred deaths; certainly a tragedy, but not the catastrophic loss of thousands of lives. In the thirty years since Bhopal, there has not been another similar chemical plant disaster. According to Charles Perrow, a retired Yale sociologist, "It is not because we have tried harder to play safe; if anything, the rate of serious chemical accidents has gone up since Bhopal, but we have not had the proper configuration of plant and environmental conditions to produce a catastrophe."

Sometimes accidents occur even when no one does anything wrong; these constitute the third reason for system failures. Charles Perrow named these events *normal accidents*. Perrow considers the 1979 partial nuclear meltdown and release of radioactive material at Three Mile Island in Pennsylvania a classic example of a normal accident. The disaster was caused by three different components failing at the same time. While the reactor's designers had planned for the failure of each of these components, they did not, and probably could not, have anticipated the simultaneous failure of all three. To plan for this contingency would have required the designers to analyze the impact of every conceivable combination of component failures. Given the many components in a sophisticated nuclear reactor, the number of possible combinations would exceed the sands on the beaches of the world. Perhaps the designers could have looked only at the simultaneous failure of critical components, but even this would have entailed untold time and expenditure. The best efforts of engineers will not catch every possibility.

In a 1996 article titled "Blowup," the best-selling author Malcolm Gladwell reviewed the literature analyzing the accidents at Three Mile Island and the 1986 explosion of the space shuttle *Challenger*. All of the crew died during the *Challenger* launch including Christa McAuliffe, a New Hampshire schoolteacher who was the first civilian invited to join a spaceflight. The *Challenger* explosion was traced to O-rings, circular flexible sealing gaskets that became brittle during a spate of cold weather in normally sunny Florida. Gladwell concluded that the common wisdom, that these accidents were anomalies that happened because various people failed to do their job, was wrong. Any procedures that might have caught the combination of parts that failed at Three Mile Island might well have missed very different possibilities that could also cause accidents. Writes Gladwell, "We have constructed a world in which the potential for high-tech catastrophe is embedded in the fabric of day-to-day life."

Given that disasters do not happen often, bad management, poor design, and normal accidents can all be considered low-probability events. But the fact that they are considered low probability is precisely the problem, and accounts for the fourth way complex systems fail. Often the probability of some untoward event has been underestimated and therefore no advanced planning was in place for its occurrence. Such events are called black swans because if you see a black swan you are likely to be surprised. Nassim Taleb, a Lebanese-American statistician and best-selling author, has championed the black swan theory as a way of highlighting why people are blind to the inevitability of low-probability events, and why these events, when they occur, are likely to have a broad impact. Taleb points out that the bell curve is not the best representative of the probability distribution for many contexts. In some situations the outliers actually have a higher probability of occurring than a standard bell curve distribution would suggest. This higher likelihood of outliers can be visualized as a fat tail or long tail on the ends of a distribution curve. Worse yet, in many real world situations we have absolutely no idea what the actual probabilities are, but irrationally dismiss the outliers until one occurs. The behavior of both individuals and institutions are riskier than they understand. An investment firm can engage a strategy that has worked year in and year out for decades, and then, in one fell swoop, the strategy bankrupts the firm. From the perspective of the consistent yearly earnings the strategy appeared to be a success, but from the perspective of the eventual and inevitable bankruptcy, it was a failure.

Complex technological systems, particularly computer systems, are quite likely to act in ways that are presumed to be low probability behaviors. As with a slot machine at a casino, the alignment of all the correct symbols to win the jackpot seldom happens, but it does take place occasionally. An event that has a low probability of happening is not beyond predictability, but when and where it will happen can usually not be predicted. Furthermore, we often have very little understanding of the possible courses of action available to a computer system or the actual probabilities of these actions being taken. This is because so many untold bits of information feed into the algorithmic processes through which the computer determines the action to initiate. In any given moment, the informational input can be unique. A closer look at the *flash crash*, first introduced in Chapter 2, will make this point clearer.

## COMPLEX TECHNOLOGICAL SYSTEMS

Between 2:42 and 2:47 PM on May 6, 2010, the Dow Jones Industrial Average dropped over six hundred points, adding to the more than three hundred points it was already down that day. The market had shed 9 percent, roughly a trillion dollars in value. At its low point, the CNBC commentator Erin Burnett reported that the stock of Proctor & Gamble (P&G, ticker symbol PG) was down 24 percent to $47 a share. Immediately, the market maven Jim Cramer, who was sitting beside her, said on-air, "that is not a real price, just go buy Proctor." As he explained why buying P&G was an intelligent investment, the market reversed course and went up three hundred points. Apparently others, including the computers that automatically buy and sell securities, also concluded that something was wrong and began buying. Another minute passed before Cramer stated, "the machines obviously broke, the system's broke big." The market continued to regain most of its losses over the next few minutes.

Jim Cramer went on to display a bit of prescience while discussing the events of the day, when he said that we would probably never know what happened. There are theories about what caused the *flash crash,* but no universally accepted explanation. The best guesses revolve around the effect of one large trade being magnified by the anomalies of computerized trading.

Computers that automatically tender buy and sell orders according to algorithms are complex systems. In financial markets, the trading of one computer influences that of others. All of these computers are entangled with human actors whose behavior the technology influences, and who influence the behavior of the technology. From a broader perspective, the modern financial market is a complex adaptive system made up of computers and corporate and individual human actors who are influenced by news, world events, and their own analysis. And in an even broader sense, the world's economy is a complex adaptive system whose behavior is influenced by an array of factors from weather, to political events, to activity within individual markets and individual corporations, and on to the decision-making of individual actors, including computers. In other words, there are systems within systems within systems. And there are feedback loops that influence the behavior of each system and its component actors.

The *flash crash* is only one in a growing list of events during which computerized high-frequency trading programs sent shock waves through markets around the world. With the luxury of twenty-twenty hindsight, analysts have attempted to track down the reasons for these dramatic and abnormal effects on market activity, and have suggested reforms to forestall the reoccurrence of similar situations. Market breakers, for example, can shut down trading once an atypical pattern has been detected. But whether breakers or any other market reform can truly stop freakish events remains unclear. It is quite possible that they merely serve to create the illusion that all is well. If investors do not feel that markets are reliable and fair, they will not return to trading. But reforms in any complex system where the actions of computers and humans are tightly coupled can introduce new challenges and new inequalities. A reform, for example, that fixes the unfair advantage of some traders whose technology enables them to get their orders processed a fraction of a second earlier, could, in turn, provide an unfair advantage to other traders.

System theorists surmise that occasional and unpredictably erratic activity is normal for complex systems. In other words, the machines were not broken during the *flash crash*; they did what they were supposed to do, just as the software in the Global Hawk UAV did exactly what it was programmed to do. Yet, from the perspective that financial markets should be rational and that the price of stocks should be somewhat tethered to the tangible value of the underlying company, there was no justification for the P&G stock to go down 24 percent during the *flash crash*. Nevertheless, stocks are commonly tethered to the overall market, which can go up or down on any particular day for reasons that have more to do with the psychology and short-term goals of investors than the health of the company. Human psychology is complex, and that alone can contribute to the unpredictability of any activity in which people are engaged. The same unpredictability holds true for the behavior of computers even though they lack minds of their own. What occurred on May 6, 2010 was not necessarily irrational, but was rather a confluence of factors that led to a low-probability incident representing one possibility within the distribution of possible incidents.

In the world of high-frequency trading, however, the sheer complexity of the system, combined with how quickly it acts, can speed up the likelihood of a low-probability event occurring. In a coin toss, the chance that the coin will come up heads is 50 percent, and for each individual flip the probability remains the same. Bad luck could cause one to lose money betting on one hundred tosses, but if you flip a coin enough times the average should converge on 50 percent heads and 50 percent tails. The likelihood of getting ten tails in a row is relatively low. However, if you toss the coin enough times, sooner or later you will get ten in a row. In fact, non-mathematicians seriously underestimate the likelihood of getting a long series of heads or tails when tossing a coin. On average, a series of ten tails will occur once in 1,024 coin tosses.

Now consider a computer tossing a simulated coin with one toss occurring every millisecond (one one thousandth of a second). The probability that a series of flips will yield ten tails will be the same for the computer as for a person who takes five seconds for each individual toss. Yet the simple fact that the computer is tossing coins more quickly than the human means that the computer comes up with ten consecutive tails once nearly every second. The person might toss a coin for an hour and a half to get ten tails once. Speeding up transactions speeds up the occurrence of outliers.

For short-term traders success in financial markets is increasingly dependent on speeding up transactions through the use of faster and faster computers. Both purchasing and selling the same financial instrument can all occur in less than a second, with the firm pocketing or losing the change in value during that minuscule period of time. The firm that gets data the quickest and tenders its order first will reap rewards. In 1815, Nathan Rothschild made a fortune buying British Government bonds because carrier pigeons informed him, well ahead of other investors, about the Duke of Wellington's victory over Napoleon at Waterloo. Computer-trading firms today are willing to invest vast sums in services that reduce the time to acquire information by fractions of a second. A few additional milliseconds can be the difference between a successful and a failed trade.

Eliminating humans from the decision-making process through the use of computer algorithms that rapidly process the new information and automatically execute orders speeds up market activity. The more information the computer has the better its decisions will be, presuming that its software is well-written. The incoming information reflects what is happening in the larger world, such as a government report on employment, a central bank altering exchange rates, a restaurant chain announcing that it will miss projected earnings, or storms in Brazil destroying sugar crops. The actions of the computer subsequently affect the outer world. A significant drop in orders, for example, will force a company's management to restructure by firing large numbers of workers. Those workers will look for employment elsewhere, altering the local economy and the ability of a storeowner to pay for his daughter to continue her dance lessons. The activities of complex computational systems that influence markets are tightly coupled to the complex institutions beyond the market, to the actions of complex individuals within the larger society, and the complex environmental and geopolitical forces that dynamically alter our world.

Coupling the actions of computers to incoming information adds additional layers of complexity—complexity in the number of inputs, and complexity in the software that factors those inputs into decisions to buy and sell. Increasing the complexity of a system can contribute to altering the distribution of all possible events so that there are more outliers. In short, with more complexity the distribution curve will spawn a longer or thicker tail, and the more unpredictable the computer's actions and their effects will be.

When elements of a system are tightly coupled together or complex systems are significantly influenced by each other, a small, unpredictable event can reverberate throughout the entire mega-system and even be a tipping point with far-reaching consequences. The global financial service firm Lehman Brothers (founded in 1850) was so entangled with other major financial institutions, that when it failed in 2008, that failure threatened to bring down the worldwide financial system. Luckily for all of us, the larger system was robust enough that it found a way to absorb the losses and reorganize without Lehman Brothers. This robustness was partially due to the few extra days during which other institutions had to prepare for the anticipated failure of Lehman Brothers. Without that time, the simultaneous failure of many firms could certainly have caused a collapse of the international banking system.

In all likelihood, the *flash crash* was caused by a low-probability action originating in one system, which was compounded by low-probability responses from other systems. The initial erratic trades set off a chain reaction that collectively had a dramatic but short-lived impact. To be sure, this is a rather imprecise diagnosis. There is no way of proving whether this theory, or any other for that matter, is right or wrong. Nevertheless, the problem of preparing for and managing low-probability incidents caused by computer trading requires attention as a means of forestalling future crises.

Computer simulations offer the best means for determining in advance the various situations a complex system might encounter. A good simulation models the inputs and influences that alter the behavior of a system. By running thousands and millions of different scenarios, an engineer or business analyst can view how differing conditions lead to situations that have a lower or higher probability of occurring. Good models provide information for planning in advance and for recognizing which safety mechanisms should be built into a complex system. They can help lower the occurrence of some disasters, but certainly not all disasters.

## MODELING COMPLEXITY

In the 1980s, systems theory gave birth to the scientific study of complex adaptive systems, which examines how individual components adapt to each other's behavior and, in turn, alter the system's structure and activity. This transition in the study of complex systems was enabled by the availability of more powerful computers. Powerful computers facilitate the creation of simulations that model complex activities. Systems theorists hoped that the scientific study of complex adaptive systems might offer a tool to understand, and, in turn, tame, the complex systems we increasingly rely upon.

Much of the study of adaptive systems focuses on physical and biological processes. Even a tiny biological system such as a living cell can be truly complex. The medical illustrator David Bolinsky created a celebrated animation of the molecular dance of life within an individual cell. The animation is wondrous to behold, a rich universe in which fantasy-like structures interact in dynamic ways. An enormous number of molecular micro-machines within a cell change their state from moment to moment. Bolinsky's animated simulation was created, together with the Harvard Molecular and Cell Biology Department, as a teaching tool to help students envision the complex inner life of a cell. However, for all of the intricacy the animation displays, David Bolinsky told me that so much happens within a cell that they were only able to illustrate 10-15 percent of the molecular structures and their activity.

Simulations of evolution were an early focus in the study of complex adaptive systems. Exploring evolutionary processes in artificial environments offers a number of benefits. In the biological world, countless generations are required to evolve successful species with interesting features. Within a computer simulation, going from one generation to its progeny can take mere seconds.

A new field called artificial life (Alife) emerged from the study of evolutionary processes. Alife researchers populated simulated environments with virtual organisms. The idea was to explore how these artificial organisms would change and adapt in response to the actions of other entities in the population or in response to changes in the environment. Think of this as the computer game SimLife (made by Maxis, the company that created the Sims and SimCity) on steroids. It was hoped that the organisms in the simulated world would evolve to become complex virtual creatures, and, in turn, shed light on the robustness of biological life.

Unfortunately, the evolution of virtual organisms plateaued. Artificial entities within computerized environments did not mutate to levels of sufficient complexity. Thomas Ray, a biologist who developed a highly regarded software program (Tierra) for studying artificial evolution, admits "evolution in the digital medium remains a process with a very limited record of accomplishments." It is unclear why Alife simulations have been so disappointing. The failures suggest a fundamental limitation in modeling biological systems within computer simulations, at least through the approaches that have been tried to date. Nevertheless, Alife researchers continue to explore new approaches, with varying degrees of success.

Scientific understanding depends upon maps and models that are designed to capture salient features of physical, biological, or social contexts while ignoring apparently extraneous details. A map lays out features that have already been discovered. There can be gaps in a map that represent the unexplored and the unknown. Dynamic models make the study of relationships and law-bound activity possible by focusing on those features that seem important while disregarding the rest. For a theoretical model to become a working hypothesis, it must reveal insights or predictions that are then confirmed by experiments in the real world. If a prediction proves wrong, the model is either incorrect or incomplete.

Models fail for a variety of reasons. Often a theoretical model is too simple to capture all the important factors that influence a system. Before the advent of computers, scientists were limited to working with relatively simple conceptual models. Computer simulations offer an opportunity to observe how a more complex system unfolds step-by-step, and have become an important tool for studying the activity within all chaotic and complex systems from the molecular to the cosmological. With each new generation of faster computers and better programming tools, simulations will be able to model increasingly complex processes.

Even excellent models can prove to be inadequate if they fail to incorporate an essential feature. Chaos and complexity theory tell us that very small influences, which may have been left out of the model, can have a significant effect. On its opening day, the Millennium Bridge, an architectural wonder and acclaimed suspension footpath spanning the Thames River in London, began to sway as thousands of pedestrians walked across. The bridge immediately acquired the distinctly English nickname "The Wibbly Wobbly." Many of those walking on the bridge were disturbed by the swaying, as was the engineering firm that designed the bridge. The possibility that the bridge would sway had not shown up in computerized models, nor was it evident in wind tunnel simulations. The firm soon realized that their models had not incorporated the way the walkers would slowly begin to synchronize their stride in response to subtle vibrations in the suspended walkway. This synchronized stride set the bridge swaying. In response to the sway, more and more walkers joined in the synchronized pace, which, in turn, accentuated the swing of the walkway.

Unpredictability is the enemy of engineers, whose job is to design reliable tools, machines, buildings, and bridges. Conquering uncertainty is integral to safety. Mechanical systems are naturally prone to move from orderly to chaotic behavior. For example, friction can cause a vehicle's motor to vibrate, and this vibration, in turn, loosens or damages other components. Detecting and eliminating chaotic behavior is key to insuring the optimal performance of complicated and complex systems.

In the case of the Millennium Bridge, once the problem had been recognized, the engineers determined that installing dampers could effectively stop the swaying. The *flash crash* also garnered considerable attention and led to various reforms meant to halt markets once trading anomalies are detected. Yet even after intense study, not enough was learned nor were adequate reforms enacted to tame the uncertainty of markets dominated by high-frequency computerized trading.

The moral of the story: even using highly refined models, the study of complex adaptive systems has not mastered, nor kept up with, the vastly complex technologies corporations and governments continually build. Because our fledgling understanding of complex adaptive systems falls short, we need opportunities for human operators to step in. But in order to give them the freedom to fulfill this responsibility, we need to give them *time*.

## WRESTLING WITH UNCERTAINTY

I was working in my office on August 14, 2003 when the electricity went off for just a few moments and then came back on. A power surge in Ohio had caused the largest blackout in the history of the U.S. and Canada. States from Michigan to Massachusetts were affected, as well as the Canadian province of Ontario. It would take two days, and for some customers, two weeks, before power was restored. The surge began when an overheated electrical transmission line outside of Cleveland sagged into a tree. That small incident cascaded into a chain of computer-initiated shutdowns throughout eight states and into Canada. Software bugs compounded the blackout, as did decisions by human operators. In southern New England, we were spared the blackout because technicians evidently overrode automated shutdown procedures and disconnected our provider from the multistate electrical grid. A few extra moments were sufficient to save those of us living in Connecticut from the inconveniences and financial burden of a blackout.

In a world where networks of tightly connected computers are increasingly the norm, extra moments can be hard to come by. A little time can be bought by decoupling the critical units in a complex system. A modular design, which permits more independent action by individual units, can enhance safety by reducing the tight interaction of components. The backbone of the Internet is a good example of a modular system. A failure at a distribution center will not bring the Internet down, as it is designed to constantly seek new routes for transporting data among the available nodes in the network.

Charles Perrow and other experts recommend decoupling and modularity as ways to limit the effects of unanticipated events. Unfortunately, in many industries the prevailing trend leads to tighter coupling of subunits.

Under pressure to increase profits, business leaders gobble up or merge with competitors, eliminate redundant units, and streamline procedures. Multinational corporations grow larger, become more centralized, and tightly integrate their individual units for greater efficiency. The business leaders who make these decisions are seldom cognizant of the dangers they invite. They are just doing their job. The politicians who thwart regulation of large conglomerates also have no understanding that they invite even greater disruptions to the economy each time a periodic downturn or unanticipated catastrophe occurs.

Decoupling the financial markets from the high-frequency trading that accentuates fluctuations in financial markets could still be realized through a few modest reforms. For example, a transaction fee could be charged by the stock exchanges for each short-term trade. Or a tax might be introduced on the profits from all stocks, currencies, or commodities that are held by the purchasing firm for less than five minutes. Tendering all trades on the minute or even on the second would eliminate advantages to firms that acquire information or tender orders a fraction of a second faster than their competitors. Each of these reforms would significantly reduce the volume of high-frequency trading. Unsurprisingly, the firms that benefit from high-frequency trading worked viciously to defeat any measure that would interfere with their practices and only a few reforms have been instituted since the *flash crash*. They argued that the liquidity from short-term trading provides a service. Evidently short-term efficiencies are much more important than long-term safety. The profits of those best able to game the system have taken precedence over the integrity and stability of markets.

It is not surprising that those who benefit from a system will fight reforms. Nor can we necessarily expect even reform-minded politicians to fully understand whether the imperfect measures they consider will address perceived problems. Experts have considerable power to influence decisions regarding the management of complex systems. Considering the intricacies, however, experts can also muddy the water, and sometimes do so in support of the *status quo,* or to further ideologically inspired reforms. Who can we task with deciding how complex systems should be managed and when reforms should be instituted?

Another recommendation for limiting disastrous consequences proposes that chemical factories, nuclear reactors, and other potentially dangerous facilities be built in remote locations. That recommendation is also seldom heeded. Placing a chemical plant near a large population center can lower the cost of labor, transportation, and energy. Locating nuclear reactors nearer to the demand lessens the amount of infrastructure that must be constructed to transport energy to its consumers.

Backup systems and redundancies are useful for addressing common component failures. Automated shutdown procedures can protect equipment in a power grid from most surges. Safety certainly enters into the design of critical systems, but consideration is seldom given to the low-probability event that could be truly catastrophic. For example, Japan experiences a tsunami on average every seven years, so the planners at the Tokyo Electric Power Company made sure to build the nuclear power plant at the Fukushima Daiichi plant more than 18 feet (5.7 meters) above the average sea level. Their worst-case scenario, however, ignored the Jogan tsunami of A.D. 869, which produced waves similar to those of the 2011 tsunami that flooded the reactors. At its peak the tsunami exceeded 15 meters, 5 meters above the height of the sea wall that they had built.

British Petroleum and Transocean, the parties responsible for the 2010 oil spill in the Gulf of Mexico, made the basic calculation that planning for disastrous events was too expensive, and took the chance that outliers would not occur. The resulting environmental damage, economic loss, and cost of the cleanup was tremendous. By September 2013, British Petroleum had spent $42 billion (€31 billion) on cleanup, claims, and fines, and began fighting additional fees that might be as large as $18 billion (€13 billion). BP argued that many of the claims it was being asked to pay were not caused by the oil spill. A year later (September 2014), U.S. District Court Judge Carl Barbier ruled that BP acted with gross negligence and must pay all claims.

In the scheme of things, BP was unlucky. Other companies that took similar risks have fared better. In the years following the Transocean explosion, many deep-water oil-drilling companies failed to implement costly, voluntary safety measures. They continue to fight legislation that would make such procedures mandatory, and with the help of political allies, they have so far succeeded. It is unclear, however, whether the cumulative economic benefits to oil companies and to the larger society from not planning for low-probability disasters outweigh the economic and environmental costs when periodic catastrophes do occur.

Perhaps with a deeper understanding of complexity, engineers and policy planners will discover new ways to tame the beast. But for the time being, decoupling, modularity, locating dangerous facilities in remote locations, slowing transactions that reward a few but whose benefit to society is negligible, risk assessment, and better testing procedures are the best methods available for limiting disasters, or at least defusing the harm they cause. To believe such measures will be taken, however, remains an act of faith. The more likely outcome is an increasing reliance on ever more complex technologies, and a cavalier disregard by a significant proportion of companies to adopt costly safety procedures. The dangers of relying upon complex systems will continue to be underestimated. Disasters will occur at ever-decreasing intervals.

Nassim Taleb proposes that all efforts to conquer uncertainty merely turn gray swans (vaguely recognized problems) into white swans. The very logic of a complex world is such that black swans will always exist. After a disaster, a few black swans turn gray. This represents a tiny decrease in black swans. But increasingly complex technologies will spawn additional growth in the black swan population. This inability to eliminate black swans should not, however, be taken as an argument against safety measures to reduce destructive events.

## THE COMPLEX HUMAN BODY

From the scientific perspective, the human body is a complex system whose mysteries will be revealed through exacting investigation. For the purposes of our discussion, that research applies to the management of complex systems in three distinct ways. First, the human body is remarkably adaptive, robust, and resilient. Understanding why humans, and indeed all organisms, are so resilient can inspire scientists and engineers to develop similar mechanisms to improve the adaptive capabilities of non-biological technologies. Second, many of the complex technologies under development entail interventions into biological processes for therapeutic purposes or to enhance capabilities. As all good doctors know, intervening in the functioning of the human body must be done carefully, and can be dangerous and unpredictable. Finally, the complexity of the human body will thwart the full and easy realization of some of the more highly touted technological aspirations such as personalized medicine and selecting a significant number of features for infants before they are born. Throughout this book, the role of complexity in naturally slowing the pace of technological innovation will come up repeatedly.

Long before there was a scientific language and method for studying the human body, words such as "soul" and "spirit" represented the more mysterious qualities of being human. Life, intelligence, creativity, and consciousness were all gifts of the soul. In the language of complexity, the gifts of the soul are emergent properties, capabilities that are more than the sum of the chemical and biological interactions from which they arise. Whether the concept of emergence is sufficient to explain life's mysteries remains an unanswered philosophical question. Emergence is a quasi-scientific term that often functions as a placeholder for "cannot be understood with the science available." Properties of energy moving through the body such as force, magnetism, and sensations may not adequately describe the qualities and mental states that were captured by the word "spirit." Certainly, the revelations of medical science are multiplying. However, the study of the body as a system reveals layer upon layer of complexity and unpredictability. The human body includes entangled subsystems that affect each other in untold ways. Mystery may yet survive the scientific assault upon its citadel.

The concept of emergence is a particularly fascinating theme within the study of complex systems. Patterns or totally new properties can emerge out of the simple interactions of the system's components. Life emerged from chemistry and physics. Without a conductor, the nearly ten thousand pacemaker cells in the sinoatrial node of the heart self-organize to create a coherent beat for the whole heart. The individual neurons in the human brain are not conscious, and yet collectively, they make it possible for you to read this book and to be self-aware that you are reading.

Some forms of emergence are explainable, such as how a large city emerges at the confluence of two rivers by the behavior of many individual actors pursuing their own self-interest. Other forms of emergence, including the way mental states emerge out of the activity of individual neurons, are more mysterious.

The human body is by far the most studied of all complex systems. It is a system vulnerable to disease, accidents, and the deterioration of key components. The body's resilience is evident in its ability to recover from many diseases with little or no assistance from drugs or physicians. Minor wounds heal. Once they have been reset, broken bones mend on their own. In the months following a stroke, critical functions such as language skills can migrate to be supported by other regions in the brain. Nevertheless, natural wear and tear, as well as abuse, undermines the body's robustness and resilience. Smoking, for example, compromises the body's natural defenses and paves the way for lung disease.

The exploiters of vulnerabilities in the body include pathogenic bacteria, which often live in the gut, but get the opportunity to capitalize on their natural drive to survive, reproduce, and flourish when the immune system is weakened. "Design" features that might have made the body less vulnerable could have also altered its adaptability and capacity to survive and prosper. This is true for all species. If the gut were redesigned so that it did not harbor pathogens, for example, it is also likely that it would not harbor the many forms of bacteria that are essential for digestion.

Well before the full onset of a disease, there are changes in the body's biochemistry that indicate a problem. The detection of these disease indicators, known as biomarkers, is viewed by medical scientists as essential for preventing illness and key to lowering the cost of health care. Of the two trillion dollars a year spent on health care in the U.S., only 10 percent is spent on drugs, 2 percent is spent on diagnostic tests, and roughly 85 percent goes to taking care of those who are sick and the administration of health care. Clearly, the prevention of illness will have the greatest impact on health care costs.

A significant number of the new medical technologies under development are directed at the early detection of biological changes that indicate the onset of disease. The vision of a Doc-in-a-Box, a household instrument for monitoring the state of one's health, captured the imagination of researchers at the Center for Innovation and Medicine in the Biodesign Institute at Arizona State University. The Doc-in-a-Box could sit on a counter in one's home. The device would be used to periodically analyze a drop of blood, saliva, or urine in hopes of determining changes that indicate the early onset of disease. The levels of various biomarkers would be matched against a normal baseline of these same indicators. In other words, your health would not be determined in comparison to other people, but only through a comparison to your own normal body chemistry and biology. If, for example, an antibody that fights and neutralizes a virus is elevated in one's blood, this could indicate an early warning of the onset of an infection.

In addition to the very difficult challenge of developing a Doc-in-a-Box that performs a variety of diagnostic tests, there are also many problems in reading the data and in determining how the test results should be used. One vision of personalized medicine empowers individuals to take responsibility for their own health care by providing them with diagnostic information. But, given the complexity of the human body, this could be a bad idea. Some individuals will no doubt be prone to take antibiotics and other drugs they do not need if they see any evidence of coming down with an infection. Or consider a woman who has had her personal genome read, her microbiome (the flora and fauna in the intestines) analyzed, and receives daily feedback on changes in biomarkers from Doc-in-a-Box. Will all this diagnostic information be helpful or confusing? Yes, some of the information will be specific enough to indicate a need for treatment. Angelina Jolie's family history of breast and ovarian cancer and the presence of a BRCA1 mutation signaled a very high probability that she would also develop cancer. However, most diagnostic signals are not this clear. For most conditions a combination of factors must be analyzed. Both known and unknown biomarkers can significantly raise or lower the likelihood of acquiring a serious disease. Furthermore, biomarkers can fluctuate due to non-problematic factors such as the onset of puberty, pregnancy, or menopause. Even a seasonal transition in the fall alters one's biochemistry as the body prepares for colder weather.

More important than the diagnostic information will be the tool used to analyze the aggregate data. Assistance from a sophisticated computer program that organizes and analyzes biological data (a discipline known as bioinformatics) will be of limited usefulness in helping to interpret the data. The bioinformatic system may or may not be aware of, or factor in, essential information. Or the system might report an array of probabilities for a host of diseases. What if the level of reliability or predictability of the diagnostic data is low? How much uncertainty is acceptable? Should the readout be altered to accommodate the mental state of the individual and whether he is psychologically disposed to take bad or uncertain news in a rational manner? Should the bioinformatic system report the high likelihood of a disease for which there is no known cure?

Perhaps the Doc-in-a-Box or a bioinformatic system should not even deliver its analysis to the individual. The findings could be sent directly to a physician, a health counselor, or an insurance company. But if employers, insurance companies, or governments get hold of diagnostic data, might it be used in ways that would violate an individual's rights? These are among the questions and concerns that arise when we consider adopting new tools for pre-diagnosing diseases. Technologies that attempt to simplify decisions can actually make those decisions more difficult. The goal of providing early diagnostic data to consumers is admirable. The complexity of the body being analyzed and the complex societal concerns that arise when imperfect consumers and imperfect institutions attempt to apply this information will complicate and impair realization of that goal.

In contrast to a computerized readout, a visit to the office of a skilled physician who is aware of local environmental factors may result in a better diagnosis with less diagnostic information than a Doc-in-a-Box. Furthermore, a good physician or nurse will gauge a patient's mental condition and determine how best to break news of a serious developing illness.

In the best of both worlds, a skilled practitioner would always have access to good diagnostic data, including day-to-day changes in a patient's biomarkers. Presently, the expense makes that prohibitive for any but the extremely wealthy. Regrettably, the increasing availability of genomic data and diagnostic tests means many people who lack the knowledge necessary to use this information in an appropriate manner will self-medicate unnecessarily due to unfounded worries. In some cases their actions will be truly harmful.

Progress in health care indicates that problems arising in a complex system such as the human body can often be managed if not always resolved or cured. The hopeful yet often naïve assumption that medical science will soon solve each and every health concern overlooks the simple fact that the more we learn, the more we come to appreciate the underlying complexity of human biology. Some of our best methods for alleviating suffering, such as drugs, add to the overall complexity by altering feedback loops in ways that can cause side effects or suppress conditions that erupt elsewhere. Biomedical technologies that improve functioning in one respect often foster fragility in other dimensions of a complex organism or sociotechnical system. Solutions to some problems can make the overall system more complex and therefore likely to act unpredictably and, on occasion, dangerously.

## THE MAP AND THE COUNTRY

In a one-paragraph story titled "On the Exactitude of Science," the Nobel Prize-winning writer Jorge Luis Borges describes a country in which the cartographers are so skilled that they create a map which captures every feature of the landscape "point for point." The map grows to be the size of the country itself. "The following Generations . . . saw that the vast Map was Useless."

All science is built upon models—models whose predictions are accurate in situations where excluded elements do not have much influence. Yet the complexity of nature and technology-dependent culture is such that even the subtlest influence can, under the right circumstance, have a far-reaching consequence. The world we inhabit encompasses countless subsystems, some of which are complex adaptive systems. Other subsystems are complicated and brittle, and then there are additional subsystems whose activity is chaotic. Feedback loops between these systems entangle them in ways that are difficult to track or decipher.

In other words, our world and our immediate environment do not conform to any model. We cannot fully conceptualize or simulate reality, and the illusion that we can fully predict or control human destiny is naïve. Historically, acknowledging and embracing the unpredictability of human existence has been a sign of wisdom. There is no reason to believe that has changed.

# { 4 }

# The Pace of Change

IN AUGUST OF 2012, CODY WILSON AND HIS FRIENDS AT THE Wiki Weapons Project began raising $20,000 to develop a gun that could be fully built with just a 3D printer. With that money they would buy or rent a $10,000 Stratasys 3D printer for testing various gun designs. The most common form of 3D printer is similar to an inkjet printer, only the nozzle extrudes a new layer of quick drying plastic, or another material, on multiple passes to slowly build up an object. Wilson envisioned 3D technology as a tool useful in pursuing his political goal—the weakening of any proposals to limit the availability of guns. What followed over the next six months was a stream of media coverage that both introduced the public to 3D printing and alerted the world to its downside.

The main impediment to printing a 3D gun is the difficulty of fabricating a gun barrel capable of withstanding the forces necessary to propel the bullet. Nevertheless, in 2013, a Texas group named Defense Distribution successfully tested a gun produced on a 3D printer. They named the gun "Liberator," and, even though the weapon was unreliable after a few shots, they began to make the blueprint available for download online.

Just imagine for a moment a teenage boy who downloads the plans for the "Liberator" and prints a gun on his home 3D printer. He takes the gun out for testing, loads it with ammo, pulls the trigger, and, on the fourth shot, it blows up in his hand. Or perhaps the plans specify that the tested gun should be made using particularly strong "ink" made of plastic and ceramic materials, and the teenage boy doesn't have any. Impatient, he prints out the gun with the plastic on hand. Result—failure and a lost limb. One last scenario: a teenager tries to fire a plastic gun that was printed out years beforehand. The plastic, similar to ski and bike helmets that need to be replaced every few years, has deteriorated, and the barrel shatters when fired. Guns produced using 3D printers are the definition of a bad idea.

Yet in the scheme of things, these inevitable tragedies are likely to be small in comparison to the numbers of young people who lose their lives due to the availability of guns manufactured by standard methods. The media coverage of guns produced using 3D printers will focus upon tragedies, even if they pale in comparison to those caused by standard guns. Attention will be directed away from the larger concerns, including the accessibility of firearms, and the difficulties in regulating technologies that appear rapidly and continually evolve.

3D printers are just beginning to become consumer products. A process for printing 3D objects from computer data was first invented by Charles Hull in 1984. 3D printers for industrial applications have been available since the 1990s. Extremely large 3D printers that have the capacity to build the frame of a house in little more than a day will be available soon for construction companies. But until recently, home 3D printers were only capable of crudely replicating an object designed with computer software. In 2014, the patent expired for a major technology called laser sintering, used in the high-end 3D printers for intricate designs. Explosive growth in the market for sophisticated low-cost home printers is anticipated over the coming decade.

Johannes Gutenberg's invention of the printing press around 1450 continues to rank as one of the most socially transformative technologies in human history. While the impact of printing three-dimensional objects may not be as far-reaching, it will have incalculable significance because 3D printing places a tool of manufacturing in the hands of an individual.

And just how will consumers put 3D printers to use? The uses of 3D printers are not all ominous, and in many ways will be incredibly helpful. Perhaps you break the plastic battery cover for a remote control device. The manufacturer may make it possible for you to download plans from its online support system. These software specifications reduce the aggravation of fixing the device by allowing you to print out a replacement cover faster and more cheaply than ordering a new one would have been.

It's unlikely that ordinary consumers will purchase 3D printers simply to fix household products. But 3D printers in the home might be useful for serious hobbyists, budding entrepreneurs, or artists. Gourmet cooks will print out shapely foods from supple materials such as pasta and bread dough. Basic designs can either be downloaded from the Internet or generated through the use of simple software. Creative souls will invent products and fabricate works of art that formerly would have required access to vast resources. But creating the designs for original detailed objects will still require skill. The printing of intricate objects will also take considerable time, as the nozzle extruding the material must be tiny and each layer extremely thin in order to capture the fine detail.

Professions outside of manufacturing are also finding intriguing new uses of print technology, especially in the field of medicine. Take the ability to assemble or synthesize biological organs. Wake Forest University's Institute for Regenerative Medicine has created functioning kidneys, printed out cell-by-cell and layer-by-layer. The realization that cells could fit through the head of an inkjet printer has also led to the development of systems that print out blood vessels. Fabricating blood vessels may eventually give surgeons the ability to direct blood to tissue that needs oxygen to heal. This would be particularly useful during an organ transplant or to speed recovery from a stroke.

The fabrication of organs and splints by 3D printers is quickly becoming commonplace. A 2012 case in which a team of surgeons in Michigan saved the life of a six-week-old infant suffering from respiratory distress is a touching example. They used high-resolution imagery of the infant's trachea and computer-aided design to fashion a splint that was anatomically customized for the baby. The splint was then printed out using biodegradable materials and implanted. It will be fully absorbed by the body within three years.

Print technology also promises to dramatically increase the speed and lower the cost of systems that construct novel strands of DNA one base pair at a time. DNA printers facilitate the assembly of biological components and new organisms in the emerging field of synthetic biology. Like many other technologies, the cost of printing a short segment of DNA has dropped dramatically in recent years. If this exponential reduction in cost continues, DNA printers will be affordable within ten to twenty years for do-it-yourself biologists wishing to fabricate single-celled organisms in their home laboratories.

It remains unclear how many of us will have 3D printers in our home or office. Even so, the technology is touted as being revolutionary and will usher in an era of desktop manufacturing. There may be a "killer application" of the technology that will make all of us wonder how we could live without our little fabricator. 3D printing conjures up visions of the replicators on Star Trek that allowed crew members to order whatever food they wanted and have it instantly produced. However, in the near future, 3D printers will probably function as one more tool available in affluent house-holds, and will only get used periodically.

The primary societal impact of 3D printing technology will result from the ways in which it speeds up processes that formerly took considerable time and expense. Engineers and artists will be enabled to cut out middle-men as they quickly move from concept to design to fabrication. The same will hold true for the creation of biological components.

The synergistic interaction between scientific discovery and the devel-opment of new technologies means that innovation is not just accelerating, but the rate of acceleration is speeding up—thanks in no small part to the demands of those who implement these technologies. For example, diag-nostic tools such as functional magnetic resonance imaging (fMRI) systems can capture brain activity as represented by blood flow. The availability of fMRI has sped up research in neuroscience. The more neuroscientists see through the use of fMRI, the more they want improvements to that tech-nology, requesting imaging systems that provide higher resolution and cap-ture the details of changes in the brain over time. Indeed, the first stage of President Obama's Brain Initiative is directed at the development of the next generation of imaging systems and other tools that will help neurosci-ence progress quickly.

But is the manner in which emerging technologies are speeding up a vast array of processes a boon or burden? The faster the rate of change, the more difficult it becomes to effectively monitor and regulate emerging technologies. Indeed, as the pace of technological development quickens, legal and ethical mechanisms for their oversight are bogging down. This has been referred to as a pacing problem: the growing gap between the time technologies are deployed and the time effective means are enacted to en-sure public safety.

Humans are a resilient yet fragile species. There are limits to how quickly we can respond and change. The demands of daily life can exceed a humanly manageable pace. When demands exceed our capabilities, corners get cut, accidents occur, and the human mind and body break down. Time set aside to reflect upon the proper course of action will be the first casualty of the quickening pace of innovation.

The increasing pace of technological development is also a result of the goals and aspirations we humans invest in new tools and techniques. In spite of claims that technology grows like an organism, individual technol-ogies and the collection of technologies have no capacity to will or want. Technological growth is a result of what we humans want from technology and our willingness to buy into the belief that technology can provide what we want and need.

With speedier modes of communication, manufacturing, and distribution, a new tool can be shared with a wider circle in ever-smaller amounts of time. By itself, 3D printing is just one more useful technology. In combination with an abundance of other technological innovations, it will contribute toward speeding up, enhancing, and disrupting the quality of human life.

## THE SHIFT

From bits and bytes to computers and cyberspace, the terminology and tools of the Information Age are quickly redefining and restructuring our daily activity. Yet only the tip of the proverbial iceberg is visible. The Information Age is heralded as something more than a proliferation of gadgets. It has even been suggested that future historians will divide history as B.I. (before the Internet) and A.I. (after the Internet).

Certainly new tools for communication, entertainment, education, research, and productivity are restructuring and enhancing countless activities. The manner in which concepts such as information processing, algorithms, and computing provide a new way to understand life, human culture, and even the mind has been just as profound. They provide the latest metaphors for interpreting all aspects of the physical, biological, and social universe. From genetics to neuroscience and cosmology to quantum mechanics, information theory offers a new approach for revealing, explaining, and, at times, distorting the play of forces that give form to our world.

Over the past fifty years, I have had the opportunity to observe the development and dissemination of computer technology, and the manner in which it encroaches upon a broad array of social activities. The first computer I witnessed in the early 1950s filled two floors of a New York bank and required many technicians on-site to change vacuum tubes that constantly blew out. Substituting integrated circuits for vacuum tubes enabled the development of more reliable, faster, smaller, and cheaper computers. From the invention of the computer to its sale as a home appliance took less than forty years. The Internet and smart phones became ubiquitous much more quickly.

For the generation born immediately after World War II, the introduction of new tools and techniques has been taken in stride. We were able to view the transition from adding machines to calculators and computers as incremental improvements. Each iteration of computing technology made basic activities such as writing, working with numbers, and sharing information easier to perform and therefore accessible to more users. Word processors made writing easier and spreadsheets turned working with numbers into a game. The Internet provided every home with a vast library of information and the means to instantly communicate with friends and family. Arguably, the early development of information technology proceeded at a humanly manageable pace.

For my peers (and me), each succeeding generation of computer technology has provided improvements that we have both come to expect and have generally welcomed. We presume that teenagers and young adults will embrace the latest gadgets, quickly sorting out the chaff from the most useful tools. In other words, the impact of computer technology on how we perform an array of tasks has been profound and yet its assimilation has moved rather smoothly. The primary disruption has been felt by industries that information technologies are replacing and workers whose jobs they eliminate. A somewhat hidden shift can be perceived in the understanding of children and young adults who grow up with gadgets capable of performing seemingly miraculous feats. They seldom have any knowledge of how these tools emerged historically or how they work. In that sense, young people experience little continuity between their world and historical developments. But, to date, the Information Age has been much less disruptive than the Industrial Revolution.

That is changing. The disruptive character of information technologies is becoming more apparent. Security and privacy risks are no longer dismissed. As computers get faster and perform more and more activities, it can be increasingly difficult to keep pace.

Information technology alone is quickening a vast array of processes. The pace of change and the pressure to turn over more and more tasks to machines are entangled. The growing reliance on computers for activities such as the purchase and sale of stocks, for example, is a result of their ability to generate profits by calculating and acting more quickly than humans. Financial markets were the first industry to be taken over by computers. Computers are now quickly replacing human decision-makers in many other industries. However, turning over decisions to machines carries serious losses and dangers. For example, discrimination, sensitivity, compassion, and care are not attributes that machines factor into their choices and actions.

## ACCELERATION, INTELLIGENCE, AND THE SINGULARITY

The processing power of computers over time increases exponentially. Some theorists argue that this is but one aspect in an overall acceleration of technological development, and will inevitably lead to a radical transition in human destiny. In this vision, humanity appears headed toward an inflection point where either the future belongs to machines and superhumans or the whole process of acceleration collapses because it cannot be sustained. Either result could be disastrous. At the least, such an inflection point will be disruptive. Or it may, as a number of prognosticators believe, lead to a future world that is radically discontinuous with everything that has gone before.

Two ideas, Moore's law and an "intelligence explosion," sit at the heart of the often-repeated claim that the pace of scientific discovery is accelerating exponentially and headed towards an inflection point that will radically alter our world. Both ideas arose as observations about the expansion of computing power, and, interestingly, both were first published in 1965. Gordon E. Moore, a co-founder of Intel, observed that the transistors on an integrated circuit had doubled every two years (later revised to eighteen months) dating back to 1958 when the integrated circuit was invented. Also in 1965, in an article titled "Speculations Concerning the First Ultraintelligent Machine," the mathematician I.J. Good posited the advent of computers that surpassed the intellectual abilities of humans in all respects. A noteworthy feature of these super-intelligent machines would be the ability to design even better machines. Their appearance would trigger an "intelligence explosion" where machines would design ever more intelligent computers as they evolved far beyond the intelligence of humans. In a noteworthy line, Good writes, "Thus the first utraintelligent machine is the last invention that man need ever make."

Fifteen years later, Vernor Vinge, another mathematician who was also an acclaimed science fiction writer, began conceptualizing the plot for a speculative novel set in a world where such an intelligence explosion had recently occurred. Vinge realized that super-intelligent machines would change life and culture in so many ways that he could not fully imagine what this future world would be like. Drawing upon a term that already existed in mathematics and the natural sciences, Vinge labeled this inflection point in human history the technological singularity.

For those who buy into this story line, there are two central questions: When will a technological singularity happen, and will it benefit humanity or threaten our very existence? At this stage no one can answer the latter question, but there is plenty of reflection by young nerds and seasoned scientists on how to ensure that future super-intelligent computers are "friendly" to humans. The answer to when the technological singularity will occur is hotly debated, with some theorists predicting it will happen in the relatively near future (fifteen to one hundred years), and others considering it a distant possibility or altogether impossible. Ray Kurzweil, the inventor and author with whom the advent of a near-term technological singularity is often identified, predicts computers will display human-level intelligence around 2028-2030. The singularity will follow soon thereafter.

Critics of the march toward an inevitable technological singularity question whether the kind of tasks performed by computers constitute machines having intelligence. Certainly, computers already exceed human capabilities in performing tasks such as solving mathematical calculations, modeling complicated systems, and searching large databases to discover correlations between disparate pieces of information. However, there are other dimensions of intelligence for which contemporary computers show little or no skill. They can fail at basic tasks such as discerning essential from inessential information while navigating through real-world environments. Computer systems cannot learn the way humans learn. They lack emotional intelligence, moral intelligence, and consciousness. While there are plenty of theories as to how each of these capabilities might be implemented computationally, there are also plenty of theories as to why computers will never have them.

While a professor at George Washington University, William Halal started *Techcast,* in which he surveyed the opinions of recognized experts as to when various technological events might occur. He notes that in these forecasts the assessments of the more optimistic and pessimistic experts tend to cancel each other out. Twenty-seven experts weighed in on the likelihood of a technological singularity. They were asked whether they agreed or disagreed with the statement, "Intelligent machines surpass humans to cause a disruptive transformation." Of twenty-seven experts, 45 percent were in agreement with this statement. For those who both agreed and disagreed, their confidence level of the accuracy of their opinion averaged out to 63 percent.

Just as popular music has its golden oldies, various fields of academic research have key articles that are reread and cited for decades. In computer science, two of these golden oldies are articles that tackle the question of evaluating computer intelligence. The first of these is by Alan Turing, one of the fathers of computer science and a man who led the effort to break the Nazi encryption code during World War II. Turing's 1950 article on *Computing Machinery and Intelligence* begins with the memorable first line "Can machines think?" Noting that there is no clear definition for intelligence, Turing offered a test matching a human's responses to questions against those of a computer. Both respondents are hidden from their questioner, who communicates with them through the equivalent of text messages. If the questioner cannot accurately distinguish the computer from the human, based only on their answers to the questions, Turing proposes that the computer must be considered an intelligent machine capable of thinking. Turing called his thought experiment an imitation game, but it is now better known as the Turing Test.

There have been many critiques of the Turing Test as a measure of artificial intelligence, but no one has proposed a better test. One criticism proposes that a very smart computer would have to dumb down its answers to fool an expert who might otherwise guess that it had more knowledge than any human could have. The most interesting critique of the Turing Test is the second golden oldie in reflections on the intelligence of computers. In a 1980 paper titled *Minds, Brains, and Programs*, the philosopher John Searle argued that even if a computer answers the questions of an expert as a human would, it does not understand what it is doing. He illustrates this point with a thought experiment known as the Chinese Room. Imagining himself in a room filled with the same resource materials about the Chinese language that are available to a computer, and adequate time to go through the same procedures as the computer would to answer questions posed by a fluent speaker of Chinese, Searle conjectures that he would determine accurate answers. Nevertheless, he does not understand Chinese, and neither would a computer that did the same thing. Searle wished to illustrate that literally understanding Chinese is not the same as simulating the understanding of a language through the manipulation of symbols. His thought experiment illustrates that computers that manipulate symbols do not have any understanding of the meaning of those symbols. Manipulating symbols is not the same as semantic understanding.

John Searle believed he was raising a simple point of common sense. He was surprised, however, to discover that his Chinese Room argument has generated endless debate. Most of the discussion is directed at refuting his core point that the models of the mind used to engineer computers are insufficient for producing machines with a capacity for understanding and consciousness.

I am a friendly skeptic regarding the prospect of research in artificial intelligence leading to a technological singularity—also referred to as the advent of super-intelligence. I am friendly to the can-do engineering spirit that proposes to build machines with enough general intelligence to perform a vast array of tasks. But I am deeply skeptical that we know enough about the nature of intelligence or consciousness to replicate these capabilities in silicon or any other material. Certain qualities of human intelligence emerge in biological systems in ways that we do not comprehend and which will require scientific insights that we have not even begun to formulate. The integration of the human mind with the body, and its adaptive resilience in responding to changes in the environment in which it is embodied, will be hard to duplicate through any non-biological means.

I am somewhat friendlier to the notion that super-intelligence could eventually be produced by combining human brainpower with advanced forms of artificial intelligence. However, the suggestion that this merger of human and machine capabilities could foster an intelligence explosion and a technological singularity remains plausible but improbable.

Certainly information technology will produce new kinds of "minds" to augment or replace human minds for many tasks. While I maintain my skepticism that a technological singularity will occur within the next fifty to one hundred years, whether a technological singularity will occur in the next two hundred years is a point upon which I have no way of making a judgment. But the longer time frame would provide many more opportunities to ensure that super-intelligent robots would be friendly to humans and human values. To believe that scientists will inevitably discover the means to build such robots or engineer our evolutionary successors is certainly not foolish. Nevertheless, the meaning of such eventualities will depend significantly on whether they occur in twenty years, two hundred years, or take a thousand years.

In 2009, a panel of computer scientists reflecting upon the social aspects of advances in AI research, expressed skepticism regarding an "intelligence explosion" and a "coming singularity." A mere five years later, according to Bart Selman (co-chair of the 2009 gathering), the likelihood of super-intelligence increased appreciably in the minds of a majority of AI researchers due to recent breakthroughs in perception and machine learning.

The possibility of super-intelligence receives tremendous attention, particularly when eminent figures such as the cosmologist Stephen Hawking and the developer of Tesla electric vehicles Elon Musk weigh in with their belief that research on artificial intelligence could pose a serious threat to humanity. Attention directed at this melodramatic future possibility sometimes functions as a distraction from tackling more immediate problems. Nevertheless, the technological singularity serves as a symbol of the total loss of human control over our destiny. If attention to the singularity alerts business leaders to be more conscious of how technologies they deploy today are already initial steps in surrendering control, it will have provided a useful service.

## THE PACE OF TECHNOLOGICAL DEVELOPMENT

Is technological development merely proceeding at a rapid clip or is it accelerating at a pace beyond human control? The true believers in an inevitable singularity accuse those of us who express reservations of failing to understand exponential growth. In their eyes, we are trapped in a linear mind-set, projecting the present pace of technological development onto the future, and blind to the accelerating pace of change.

Moore's law was the first instance in which the exponential doubling of a technology's growth and power was observed. There is, however, a growing list of other fields in which scientific discovery and technical innovation are also perceived to be expanding at an exponential rate. Data storage, bandwidth on the Internet, and data transmission rates are all increasing exponentially. The speed and cost of sequencing the base pairs from a segment of DNA have been decreasing exponentially, while the number of genes mapped per year is growing exponentially.

Hans Moravec, a roboticist and futurist at Carnegie Mellon University, hypothesized that Moore's law could be viewed as predating computer science, with biological evolution having resulted in an exponential doubling in the brainpower of animal species. And the search is on for more evidence from history that supports the view that technological progress is speeding up at an exponential rate. Ray Kurzweil has gone so far as to propose that the progress of technological development is based upon what he calls *The Law of Accelerating Returns*. Kurzweil's naming this hypothesis a "law" requires boldness when you appreciate that, with considerably more confirming evidence, biological evolution continues to be regarded as a theory.

Critics of exponential acceleration point to growth patterns that are slower or instances where it appears that technological innovation has plateaued. They propose that advocates of the theory are selective in the data they use to support their convictions. But for every criticism, the advocates

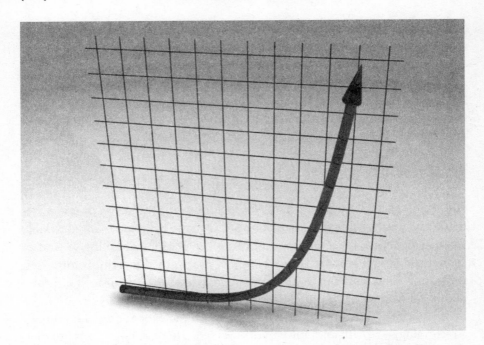

have a response. For example, if a critic points to a development that proceeds at a relatively slow pace, it will be suggested that this is merely the early stage of exponential growth. Consider a simple number series that doubles for each period—1, 2, 4, 8, 16, 32, 64, 128, 256, 512, 1024, 2048, 4096, 8192, 16384. If we plot these numbers on a line graph, the first series of numbers will fall upon a mild upward slope. The middle numbers will appear to turn the slope upward until the last numbers map along a near-vertical trajectory. If slow growth indicates the beginning of an exponential curve, then there is no way to falsify the contention that we will eventually witness a radical upward expansion. In other words, you can only prove that exponential growth is not happening in retrospect.

The lack of any room in the theory of exponential growth for human agency is particularly troubling. Apparently we humans are little more than servants in the inevitable march of technological progress. While advocates of the theory will acknowledge that humanity had to buy into large-scale technological developments such as the Manhattan Project that built the bombs dropped on Hiroshima and Nagasaki, the march to ever more powerful computers, and the sequencing of the human genome, evidently we had little or no choice in the matter. The same holds true for more recent initiatives in nanotechnology, in robotics, and in brain modeling. We, the public, merely have the illusion that we chose collectively, or through our political representatives, to support these projects. In this view, unfolding patterns and an alignment of forces beyond any human control has already chosen our destiny.

The momentum of rapid innovation alone can limit our opportunities for creative intervention and overwhelm our ability to effectively control technological development. Many who argue for exponential acceleration would argue that our opportunities for redirecting or controlling technological development are more than limited; instead, they are nonexistent in the face of inevitable growth. But I am skeptical of claims that an ever-accelerating pace of change is inevitable, and therefore doubt that the theories upon which a good number of predicted technologies rest are well-founded. However, I could be wrong. An exponential acceleration headed toward a technological singularity would indeed be a revolution in the future of intelligent life on this planet, one to which all other revolutions would pale in comparison. But even if I am right, the actual rate of change will play a central role in how much control we have over future developments.

I once sat on a panel with Ray Kurzweil during which I first acknowledged that some technological trends demonstrated exponential growth, but that I did not accept the deterministic logic of exponential acceleration. I argued that humans have agency. We have some degree of control that can be applied to speed up or slow down technological development. Ray responded that while humans have agency, this did not alter his prediction that Turing-level artificial intelligence would be achieved by 2028. Kurzweil is not merely a prophet pointing out a trend we might otherwise miss. He is a willing participant in the advent of a technological singularity.

Perhaps Ray Kurzweil's acceptance of both human agency and the inevitability of a technological singularity is not a contradiction. Perhaps his is a shrewd reading of human nature and our willingness to buy into the promise of emerging technologies. From that perspective, we have already surrendered to the blind forces driving humanity toward an unknown future.

## CONTROL

The tension at the heart of *A Dangerous Master* is whether we, humanity as a whole, have the ability to direct our destiny or whether our desires and intentions are already being submerged in the tsunami of emerging technologies. If the latter is true, the technological singularity has already begun— not the singularity in which intelligent machines and superhuman *techno sapiens* discard mere humans in their wake, but the singularity in which technological possibilities ride roughshod over the human spirit and dictate human destiny. To be sure, innovative technologies express the desires of certain people. However, that which serves some groups can be detrimental to larger segments of society.

Ray Kurzweil's theory is that of a technological determinist for whom the future of humanity is already written. There are two other theories of technological development that offer alternative perspectives. One theory proposes that innovative tools are molded by societal factors including input from those who use the technology. Early bicycles took many forms before models with two equal-sized wheels were settled upon in the late 1800s. Models with very large front wheels were popular in the 1870s, but also dangerous and difficult to mount or dismount. Consumer demand directed manufacturers to designs that were safe and easy to use. The process of refinement continued until the basic features of a bike were more or less set by the middle of the twentieth century. Then a new process of differentiation set in, as the basic features were adapted for different uses such as racing, commuting, and navigating mountain terrain. At each stage the competition between manufacturers was settled by the demand from consumers for bicycles that met their needs.

The same pattern has been repeated over and over again. Computer manufacturers introduced various designs for laptop and tablet computers. The engineering team at one or two companies got the look, feel, and feature set right, and captured a large share of the market. In the case of laptops it was IBM and Toshiba. In the case of tablet computers it was Apple. Once the market spoke, other companies largely imitated the winner and competed at a lower price point. Then, as the market grew, manufacturers differentiated their products by designing for specific subsets of consumers. From this perspective the device is just one element in the sociotechnical system that, as mentioned in the previous chapter, also includes people, relationships between people, other technologies, and prevailing customs and procedures. Technologies emerge and take form out of the interaction between these various elements.

The second alternative theory views technological development as having a momentum that is difficult, but not impossible, to alter or arrest. From this perspective, for example, the demand for increasing processing power, once established, became self-fulfilling. In working to meet the demand engineers set yearly goals, and Moore's law took on the character of an achievable target. In a variation of this theory, the momentum is seen as an unfolding pattern that has a life of its own. Once established, Moore's law dictated the growth of infrastructure including the building of factories, the hiring and training of engineers, the funding of research, and the need to raise capital for all these activities. The early demand for faster semiconductors arose naturally. But most industries also need to build a market in order to reap profits and thereby justify the capital invested in a company. This process of building a market for a product, in turn, feeds the momentum to grow the infrastructure.

Once the momentum or pattern is set, outside forces are usually required to intervene and alter the trajectory. Often this is an accident, disaster, or tragedy. The nuclear meltdown at Fukushima after the reactors were flooded by a giant tsunami forced the political leaders of Japan to initially turn off all nuclear reactors in the country and look toward other sources to meet energy needs. Changing economic conditions can also alter a momentum, though such ups and downs in the market often merely introduce fluctuations in the overall rate of development and growth. Several calculations show that exponential growth of computing power has continued in both good and bad economic times. They support Ray Kurzweil's belief that this pattern is a law and not merely a temporary phenomenon.

These three theories of technological development are not totally distinct. Each holds true in certain situations. The dynamics of a sociotechnical system are particularly important in early stages of development. Once technological growth in a new industry is established, it takes on momentum of its own. And if the course of that momentum is not fully recognized or substantively addressed, the results can be inevitable.

## ASSESSING RISKS

It is often presumed that if it were possible to predict the harmful societal impact of an innovative technology in advance, its development would be controlled. There are a number of reasons to question this presumption.

Technology assessment is haunted by a dilemma that was first proposed in 1980 by David Collingridge. The dilemma arises because it would be easiest to control the development of a technology in its early stage. In *The Social Control of Technology*, Collingridge wrote, "by the time undesirable consequences are discovered, however, the technology is often so much a part of the whole economic and social fabric that its control is extremely difficult."

For over thirty years, the Collingridge dilemma has served as dogma within technology assessment circles. The problem it highlights is real and yet its binary formulation is simplistic. Between the introduction and entrenchment of a new technology there will often be an inflection point, an opportunity when the problems are coming into focus before the technology is fully established. That time to act can be very narrow or it can span years. As technological development accelerates, the inflection point effectively contracts. This can also be caused by rapid social changes; for example, cell phones quickly went from being used by the wealthy and early adopters to being ubiquitous.

Once a process is set in motion, it becomes difficult to arrest even if harmful outcomes are recognized. A railroad engineer might spot a car on the tracks ahead, but slamming on the brakes will seldom stop the train in time to avert an accident. In the case of established industries, vested interests will work hard to thwart any changes that threaten their business or profits. There are, of course, many nuances to this kind of analysis. Certainly, substitute products can decimate an older industry. My fax machine sits in the basement gathering dust.

Early attention to an innovation by scholars and critics commonly reveals possible harms, and expands the inflection point. Nevertheless, even when perceived early, solid research is required to demonstrate which of those harms are truly serious. A public campaign may also need to be mounted in order to bring dangers to the attention of policy makers.

Clearly, the impact of many new technologies does not come to light until long after its deployment. No one saw in advance the dangers of X-rays or of the use of asbestos as a building material. Nonetheless, this book will introduce examples of situations in which it is known or strongly suspected that something will go wrong, and yet there is often little or no effort made to put precautionary measures in place. The reasons for this are several. In some cases an assessment has been made that the benefits far outweigh the risks. That is fine. Our concern, however, should be directed at technologies whose risks have not been assessed or whose risks are known but ignored by policy makers.

The last four chapters of this book will champion the need for more upstream governance: more control over the way that potentially harmful technologies are developed or introduced into the larger society. Upstream management is certainly better than introducing regulations downstream, after a technology is deeply entrenched or something major has already gone wrong. Yet, even when we can assess risks, there remain difficulties in recognizing when or determining how much control should be introduced. When does being precautionary make sense, and when is precaution an over-reaction to the risks?

Many social theorists promote a *precautionary principle*, which states that any action or technology that is suspected of being harmful to people or the environment should be rejected. In its strongest form, the *precautionary principle* places the burden of proving that a technology is safe on whoever wants to use the technology. The European Union has codified the *precautionary principle* limiting the introduction of new technologies including genetically modified foods. By contrast, the U.S. is often seen as placing the burden of the proof on whoever wants to reject a technology. In the adoption of technologies, Americans are perceived as downplaying risks and believing there is, or will be, a technological solution to every problem.

In reality, the characterization of Europeans as precautionary and Americans as risk-takers is simplistic. Both Europeans and Americans relate to risks in confusing ways. Americans legislate safety and try to eliminate risks in realms where the harms are relatively small (tamper-proof medicine bottles after the Tylenol murders in 1982), while embracing new technologies (genetically modified foods and nanomaterials) before they have been fully tested. Europeans place responsibility for injuries incurred through inherently risky activities on those who, for example, elect to ski, while the more litigious Americans end up paying for expensive daily ski tickets to cover a resort's hefty insurance policy. Americans are quite willing to pay extra for a car with a good safety rating, but are much less likely to give up the convenience of risky automobiles for slower, safer means of public transportation.

Precautionary measures slow innovation. Countries with more stringent precautionary policies are at a competitive disadvantage in reaping the benefits of potentially transformational technologies. There are real economic advantages to being the first to introduce a new tool or process. On the other hand, countries with more open policies also risk exposing citizens to dangerous products. This, of course, is not a new issue. It informs policy debate in every country as legislatures struggle with trade-offs between safety and economic productivity.

Some political parties, such as Democrats in the U.S., are characterized as placing the regulation of harms and the just distribution of goods and services ahead of stimulating economic growth. Republicans, on the other hand, argue that less government and fewer regulations will stimulate growth. This is countered by pointing to the tragedies, abuses, and economic collapses that occur when regulatory authorities are weakened. The history of economic and technology policies can be approached as a debate over benefits and losses from differing approaches toward risks, justice, and growth.

These issues influence the means and effectiveness of approaches for managing the development of emerging technologies even when we can anticipate risky tools and practices. Accelerating exponentially or not, scientific discovery and technological innovation are moving along rapidly. Speed generates a momentum that will profoundly impede attempts to intervene in the course of technological development.

The accelerating pace of change and the unpredictability of complex and chaotic systems, however, are not the only challenges that have a far-reaching influence on prospects for managing technological development. From geoengineering to genomics, each new field carries both benefits and risks.

# { 5 }

# Trade-offs

IN 1926, WHEN NORWEGIAN ENGINEER ERIK ROTHEIM PATENTED the aerosol can, he could neither foresee that it would be among the world's great inventions nor be implicated in environmental destruction. It was not until World War II that aerosol cans were turned into a useful device—a means for the military to spray against malaria-carrying mosquitos and the lice that harbor the bacteria that causes typhus. DDT, whose insecticidal properties had been discovered in 1939, was the most important pesticide used by the military to control the spread of both malaria and typhus. In the decades following the war, the use of DDT in agriculture contributed to a dramatic increase in crop yields for each acre planted. The spray can was employed to dispense everything from whipped cream to sunscreen. Unfortunately, the adoption of both spray cans and DDT had unanticipated environmental consequences.

Many of these consequences were enumerated by biologist Rachel Carson in *Silent Spring* (1962), often identified as the most important book in the birth of the modern environmental movement. Carson explained how synthetic pesticides such as DDT were killing birds and were detrimental to the environment. While she publicly challenged DDT's widespread use, Carson did not call for an outright ban. Nevertheless, it was banned in Hungary in 1968, Germany and the U.S. in 1972, and its use was restricted worldwide by the Stockholm Convention of 2004. The banning of DDT has not been free of controversy. Restricting its use has itself caused some harm. Diseases DDT helped to control, such as malaria, continue to be major killers. The World Health Organization reported that 207 million cases of malaria were documented in 2012 and 627,000 people died from the disease. In a controversial 2007 report, Robert Gwadz, of the National Institute of Health, went so far as to claim: "The ban on DDT may have killed 20 million children." An article listing the "Ten Most Harmful Books of the 19th and 20th Centuries" by the conservative weekly magazine *Human Events* gave *Silent Spring* an "honorable mention." However, DDT and pesticides in general elude simple assessments. As Christopher Bosso, a Professor of Public Policy at Northeastern University, notes, "Pesticides . . . might be characterized simply as 'good things that can cause harm,' or 'bad things that can do good,' depending on your perspective."

Chlorofluorocarbons (CFCs) served as the primary propellants for the ever-popular spray can until it was discovered that CFCs were destroying the ozone layer in the Earth's stratosphere. Depletion of the ozone layer, which blocks harmful UV radiation from the sun, has led to an increase in skin cancer. The Montreal Protocol established guidelines in 1987 for phasing out the use of CFCs, and since that time, nearly all countries have turned to less damaging propellants.

CFCs do not just deplete the ozone layer but act as "super" greenhouse gases exacerbating global warming. The earth's average surface temperature increased roughly 1.4° F (0.8° C) since the dawn of the Industrial Age, and an estimated two-thirds of that increase has happened in the last thirty years. Scientists hope to stabilize the earth's temperature around 2° F (1.1° C) above pre-Industrial levels, but if they cannot, predictions turn dire for the cumulative impact of global climate change. Melted ice caps, the flooding of cities on coastlines, and the need to relocate hundreds of millions and even billions of people inland will be accompanied by changes in weather patterns that could lead to droughts or perhaps an Ice Age in formerly productive regions.

Policies for stopping or at least slowing global climate change focus on conservation measures and reducing the carbon gases released into the atmosphere by decreasing reliance on carbon-based biofuels and switching to clean, renewable sources of energy. Technological fixes for global warming also receive serious attention. These fixes fall within the emerging field of geoengineering (climate engineering). The possibility of a technological fix is particularly attractive to those wishing to play down how serious a threat global climate change poses. In the minds of global warming detractors, the glimmer of a conceivable fix means the threat is being exaggerated.

Managing climate change does not pass the sexy topic test. But a good reason for beginning our discussion of trade-offs with geoengineering resides in the risks of various technological approaches for managing global warming. Some approaches are potentially more dangerous than the problem they ostensibly solve.

One relatively easy, fast, inexpensive, and imperfect technology that could mitigate some effects of global warming already exists. Like many new technologies, it is not without its risks. It is a form of geoengineering that seeds the upper atmosphere with sulfate particles that reflect sunlight and thereby reduce the warming of the earth's surface. This could have the equivalent effect of a large volcanic eruption that spews ash into the atmosphere and thereby temporarily reduces temperature by creating a sunshade. The ozone layer, which absorbs much of the sun's UV radiation, lies in the stratosphere 12-19 miles up (20-30 kilometers), just above the region where most commercial aircraft ride on the jet streams. According to computer simulations, sulfate distributed in the stratosphere could offset 50 percent or more of the yearly warming effect caused by an increase in greenhouse gases. The task would only require a few high-flying aircraft such as F15-Cs. Nevertheless, the use of the technology is so ethically and politically controversial that it may never be put into service.

Evening weather forecasts illustrate the difficulty in predicting meteorological patterns. Subtle influences can reduce a potentially powerful snowstorm to a mere dusting. No one knows how to assess the effects of tinkering with the upper atmosphere. Other than computer simulations, neither large-scale nor relatively small-scale experiments have been performed. Volcanic eruptions provide some experience, however, they reveal little about the long-term effect of seeding the stratosphere year in and year out.

The simplicity of seeding the atmosphere suggests that one nation might elect to employ this approach to engineer local climate while disregarding its effect on weather in neighboring regions. Consider the monsoons that provide much of the rainfall for the fertile agricultural regions of India and China. Let us imagine that by 2020 Chinese leaders become concerned about a yearly decrease in rainfall from the monsoon. Would they initiate policies that could improve rainfall in China while inadvertently reducing rainfall over India or intensifying floods in Bangladesh?

Minimizing the possibility of geopolitical tensions by establishing protocols for the exercise of geoengineering approaches requires serious attention. A UN Framework Convention on Climate Change to coordinate intergovernmental efforts to combat global warming was passed in 1992, but its effectiveness has been mixed. In 2010, the UN enacted a largely symbolic moratorium on geoengineering research. The UN schedule already includes high-level discussions to fashion an agreement for regulating approaches to manage solar radiation.

Seeding the atmosphere is just one form of geoengineering that researchers are studying. Geoengineering encompasses a collection of technologies for reducing global warming. These technologies fall within two broad approaches: solar radiation management (SRM) for reducing the amount of solar radiation that strikes the earth, and carbon dioxide removal (CDR) projects directed at extracting greenhouse gases from the atmosphere. Each of the approaches ranges across a variety of different strategies. All the strategies would require large-scale climate interventions to have anything more than short-term local effects. Large-scale reforestation is an example of a widely accepted strategy for taking carbon dioxide out of the atmosphere. But with massive deforestation taking place, for example, in the rainforests of Brazil, this, like many other strategies for the removal of greenhouse gases, will only slightly slow the onset of global climate change. Large towers that suck carbon out of the atmosphere and sequester it could be installed worldwide. However, constructing such towers in significant numbers would be extremely expensive. All methods directed at extracting carbon from the atmosphere will take decades to have any discernible impact on temperature patterns.

Solar radiation management is not a solution to global warming. It does not stop the buildup of greenhouse gases. At best, it will buy time by reducing the amount of sunlight that hits the earth. The use of such techniques may, nevertheless, become compelling if the effects of global climate change fail to be manageable by other means. For this reason, responsible scientists such as Harvard's David Keith are calling for guidelines to allow modest experiments that study how solar radiation management would alter atmospheric conditions. The environmental impact of small tests will be negligible.

Given the political sensitivity of geoengineering experiments, scientists have refrained from moving forward without the establishment of an international agreement. However, some geophysicists and environmentalists resist even apparently prudent measures like those put forward by Keith and others. Critics such as Raymond Pierrehumbert voice three core issues with allowing research on geoengineering. First, research groups frequently turn into interest groups that advocate for deploying whatever technologies they develop. Second, investing in geoengineering will take resources away from environmentally sound approaches such as conservation measures and developing clean sources of energy. Finally, environmentalists voice the far-reaching concern that geoengineering signals the "end of nature." Once countries start to directly tinker with weather patterns, there will be constant need and continual pressures to manage weather for both local and global needs. At this stage in the understanding of climate science, any presumption that weather patterns can be managed successfully resides somewhere between chutzpah and hubris, a product of the long-standing and naïve belief that nature can be conquered. And even presuming that managing weather successfully is an attainable goal, negotiating the competing demands from various regions and countries would be daunting, if not impossible.

The first two objections will not carry much political weight. But none of these concerns should be dismissed lightly. Developers of geoengineering approaches already advocate for their implementation. No one solution to global warming exists. Solar and wind alone cannot produce adequate clean energy to meet future demand. Conservation measures, while important, will be offset by requirements for more energy, particularly in the developing world.

Geoengineering should be considered one measure among many in a multipronged strategy to manage global climate change. Perhaps combining an array of approaches that includes geoengineering can successfully stave off global climate change and meet energy needs without the creation of secondary problems. Let us authorize a little preliminary research on the effects of solar radiation management, and find out whether certain geoengineering solutions should be implemented or eliminated as options. Let scientists acquire insight into how the various approaches to geoengineering alter the biochemistry and physics of the atmosphere prior to unilateral action by rogue actors. The first step entails putting in place an international agreement on acceptable experiments, along with the establishment of a regulatory authority empowered to provide effective oversight.

Geoengineering is one emerging field where the trade-offs are a central part of the conversation. The ethical and political issues surrounding geoengineering sit front and center for both the scientists developing the field and for policy planners, while the risks and ethics of the other emerging technologies are often treated as tangential matters. Furthermore, geoengineering is being addressed globally, not merely on a national level, a precedent that should be extended to other technologies whose impact will be evident worldwide. All parties will not follow international regulations, but at least they set standards for evaluating the behavior of nations and independent actors.

Without regulations in place, a few individuals can elect on their own to alter the ecology of natural environments upon which many species and many communities rely. Such was the case in July of 2012, when a fishing boat spread 100 tons of iron sulphate in the area of the Pacific Ocean 200 nautical miles west of the islands of Haida Gwaii. An American businessman, Russ George, and his colleagues who worked for the Haida Salmon Restoration Corporation conducted this iron fertilization experiment. George claims that as a result of the introduction of the chemical, there was increased algae growth, and therefore, carbon capture, over 10,000 square miles of ocean.

Debate continues as to whether George's experiment broke any laws. But we can be thankful to him in one way. Russ George alerted world leaders to what can happen if there are no restrictions and no oversight on what rogue actors, corporations, and nations can do to the natural resources that we all share. Soon, one rogue experiment, well-intentioned or not, will carry dire consequences for plant life, for nonhuman animals, and for people. Geoengineering is scary. But we should be even more frightened of the harm inflicted by geoengineering approaches implemented with little or no preliminary research evaluating their risks.

Global climate change can be slowed to the extent that energy needs are met using clean, efficient, and renewable sources. The two entangled issues of global climate change and world's need for energy play a prominent role in the development of emerging technologies. They provide plenty of fodder for reflecting upon the trade-offs entailed in adopting one approach or another to solve either issue.

All sources of energy, including those from clean energy, have an impact on the environment and some are detrimental to health. Wind turbines, for example, are responsible for the death of thousands of birds, but far less than other man-made structures such as power lines, cars, and windows in buildings. Switching to clean sources of energy reduces environmental impact, but the trade-off lies in the high cost of harvesting energy by these means. The cost of wind and solar energy, for example, while declining, is unlikely to be competitive (in most regions) with low-cost coal or fossil fuels over the next five to fifty years.

As with solar radiation management, the environmental impact of new approaches to produce, store, and transport energy are difficult to assess. Some of those approaches rely upon older technologies, such as fracking and nuclear power generation, modified for new purposes. Newer approaches include nanotechnologies and oil produced by synthetic organisms. Each individual approach offers advantages that arguably outweigh its risks. But in fashioning an energy policy, weighing the trade-offs between various approaches is essential for limiting harm.

## NANOTECH

Advances in nanotechnology are often heralded as the catalyst for the next scientific revolution. Nanotech solutions include a broad array of applications for medical technologies, new forms of manufacturing, and water purification. One potential application is the development of tools to generate and store energy—and reduce global warming. Tiny umbrella-like nanoparticles spewed into the upper atmosphere offer a future alternative to sulfate particles for solar radiation management. Whatever material gets used will eventually float down to the earth's surface. But well-designed nanoparticles would stay up longer and do a more efficient job reflecting the sun away from the earth. On the other hand, once these nanoparticles do come down, they could have an adverse environmental or public health impact.

The budget for the U.S. government's National Nanotechnology Initiative (NNI) was $1.702 billion for 2014. The early recognition that nanotechnology raises ethical, legal, and societal issues, led to a small potion of that budget being allocated to those concerns. The combined 2014 NNI allocation for environmental, health, and safety research and educational and societal dimensions was $157 million. The largest share of the NNI budget goes to energy research, with $102.4 million specifically for solar energy collection and conversion, and an additional $369.6 million requested for the Department of Energy's nanotechnology initiatives. These figures do not include all the funds spent by other U.S. government programs, other countries, and private companies on energy-related nano research. Nanomaterials are central to the development of more efficient means to produce, capture, store, and disperse energy. For example, layering various nanomaterials will make solar cells much more efficient at converting sunlight into electricity. Tiny solar cells embedded in flexible materials will lower costs by allowing lightweight energy collectors to be installed on uneven surfaces. Batteries with greater capacity that recharge quickly turn electric vehicles into a viable alternative to gas-driven cars.

The longer-term goal of nanotechnologists is to maneuver atoms and molecules to self-organize into useful tools and even tiny molecular factories that can produce products to exacting specifications. The replicator on Star Trek, which produces any food upon request out of raw materials, illustrates one fanciful vision of future nanofactories. But even if nanomachines can only recreate perfect copies of the same computer chip over and over again, engineers, corporate execs, and their investors will be thrilled.

Scientists in nanotechnology take inspiration from natural processes, but are particularly focused upon getting inorganic atoms and molecules at the extremely tiny nanoscale to do what they want. A nanometer (nm) is one billionth of a meter or only one one hundred thousandth of a hair width. The nanoscale includes anything that is 1–100 nanometers in size. Atoms and most simple molecules fall within the nanoscale. A single molecule of water ($H^2O$) is .3 nm. The two strands of a DNA molecule are 2.5 nms across and a red blood cell is 7,000 nms wide.

Over the past decade, nano became a buzzword favored by marketing companies because it conjures up visions of space-age technologies. Many products such as raincoats were advertised as coated with nanomaterials even though the particles used were often much larger than 100 nanometers. On the other hand, as the public became aware that ingesting some nanoparticles could pose long-term health risks, ad execs increasingly stated that products such as sunscreen contain non-nano zinc oxide, which is meant to indicate that the active agent will stay on top of the skin and not be absorbed through skin pores.

The possibility of manipulating matter on the atomic scale traces its history back to a December 29, 1959 presentation by the theoretical physicist Richard Feynman, titled *There is Plenty of Room at the Bottom*. "I want to build a billion tiny factories, models of each other, which are manufacturing simultaneously," Feynman said. "The principles of physics, as far as I can see, do not speak against the possibility of maneuvering things atom by atom. It is not an attempt to violate any laws; it is something, in principle, that can be done; but in practice, it has not been done because we are too big."

By the time of this speech, Feynman was already a world-renowned figure. While a young man, he worked on the Manhattan Project that developed the atomic bomb. By 1965, he would share a Nobel Prize for work in quantum electrodynamics. Later in his life, he developed a bit of stardom through his best-selling autobiographies and when, during a televised report from the commission he sat on tasked with investigating the Space Shuttle Challenger disaster, he dramatically dropped small O-Rings into a glass of ice water to demonstrate how they became brittle in the cold. With an irreverent sense of humor, Feynman could always be relied upon for a good quote, "Physics is like sex: sure, it may give some practical results, but that's not why we do it."

Many physicists read transcripts of Feynman's introduction to manipulating atoms, however, twenty years would pass before the field of nanotechnology coalesced. The engineer Eric Drexler extended Feynman's vision in a popular book titled, *Engines of Creation: The Coming Era of Nanotechnology* (1986). Drexler built upon the notion of nanomachines and inspired a generation of scientists with visions of nanorobots helping to clear clogged blood vessels, environmental scrubbers which extracted pollution out of the air, and a copy of all the books in the Library of Congress printed on a chip no larger than a sugar cube. He also scared many readers by formulating the grey goo nightmare, in which self-reproducing nanobots consume all the organic matter on the planet.

Before long, politicians were talking up the potential of nanotechnology. President Bill Clinton, in a January 2000 speech, noted that it might take twenty years or more to realize its promise. And in 2003, President Bush signed the 21st Century Nanotechnology Research and Development Act. Between 2001-2014, $20 billion was budgeted to the research by the U.S. government.

As money poured into funding nanotechnology, the NNI and its counterparts in other countries soon became captive to the material sciences, which had long been engaged in creating new metal alloys, plastics, and other substances. Miniaturization processes, such as photolithographic techniques for etching tinier and tinier semiconductors into silicon, were quickly approaching the nanoscale. Nanoresearch has been successful in the creation of thousands of new materials, while techniques for getting molecules to self-assemble into useful objects or to build tiny nanomachines remain largely speculative. The laws that direct the behavior of individual atoms differ from the laws that govern bulkier materials. Scientists in the field must first work toward understanding the properties of nanomaterials and then harness them for specific activities. For example, clumps of carbon in the form of coal or graphite do not have electrical or optical properties while tiny carbon nanotubes do. The unique properties of carbon nanotubes have made them useful for applications such as strengthening lightweight bicycle components. Nevertheless, it remains unclear how much success engineers will actually have in the more exacting manipulation of individual atoms of different elements.

The NNI funding has paid off: two to three thousand man-made nanomaterials are licensed to be included in new products. But in the rush to find new uses for nanotechnology, have we overlooked the risks? Most analysts suspect that some of these nanomaterials will turn out to be toxic, but they are not put through the rigorous testing that, for example, the U.S. Federal Drug Administration (FDA) requires of drugs before they are licensed to be sold. The FDA's regulations for testing new drugs are the most rigorous in the world. It can take seven to ten years and $500 million to develop and get a drug approved. Requiring similar procedures for each nanomaterial would slow and even halt much of the nanotechnology research. Only a small number of the nanomaterials developed have a large enough market to justify such testing costs. But the FDA does require that nanomaterials go through basic toxicity analysis. Would it be prudent to be more precautionary and demand greater testing of the health risks of nanomaterials? Or do the potential benefits of nanotechnology warrant speeding along the development of the field? Can we really know the answer to the risks posed without waiting to see what happens over the next twenty to thirty years? How many cases of cancer, lung disease, or mental illness would it take to demonstrate that limited testing on the health risks posed by nanoparticles is insufficient?

Whether to test nanomaterials more rigorously is complicated by difficulties in distinguishing when a substance is a food or a drug, or something that is never meant to be consumed, but if consumed, would be toxic. The FDA, for example, has been concerned for decades that some products are marketed as food supplements to avoid the more costly testing required of drugs, but might, nevertheless, be harmful. Conceivably toxic nanomaterials, such as coatings for water resistant fabrics, could become airborne over time as the garment ages and deteriorates, and therefore potentially be ingested. Furthermore, some materials change their properties when transformed into nanoparticles, similar to water being a crystal, liquid, or steam at different temperatures. For example, if you swallow a bit of silver it will probably pass through your system and do no harm. In India, very thin layers of silver are used as decoration on sweets and consumed as a food, while in the U.S., silver is not approved of as a food. Some silver salts, silver in a compound with other molecules, can be toxic.

To further complicate matters, other silver salts and silver nanoparticles are antimicrobial and often used to coat medical devices, for wound dressings, or even within washing machines to inhibit the growth of disease-causing microorganisms. Antimicrobial forms of silver nanoparticles will become even more important as increasing numbers of pathogenic diseases evolve into forms where they are immune to antibiotics. But it is unknown whether, or under what conditions, silver nanoparticles might join with other molecules to form compounds that could be toxic.

Nanomaterials layered into solar cells or batteries will not be intentionally ingested. The focus instead should be upon waste products from production and disposal at the end of the device's life. Life cycle analysis is among the tools used for determining the impact of a large project or the introduction of new materials.

The serious practitioners of risk analysis and environmental impact struggle to limit biases that undermine the objectivity of their findings. The better reports note what is not known, either because the necessary research has not or cannot be performed, or because the unfolding consequences of various courses of action become too complex to analyze. At their best, risk analysis and cost benefit analysis bring to light concerns that should be factored into policy decisions. In more constrained contexts, a good analysis can paint a fairly accurate picture. The reports will be filled with numbers and charts that suggest objectivity. However, environmental impact statements and risk analyses also contain many value judgments as to what is more or less important and other subjective factors that arguably bias their results.

Environmental, health, and safety concerns arising from the introduction of new materials, devices, and processes are nothing new. But nearly all the major U.S. environmental laws were enacted between 1969-1976. In 1976, the U.S. government enacted a Toxic Substances Control Act (TSCA) that required testing of dangerous or potential carcinogenic substances. Acute toxicity tests involve injecting test animals with a dose of a new food or other substance likely to be consumed. In order to get TSCA passed by the legislature, more than sixty thousand preexisting chemicals were exempt from testing. Furthermore, there has been a failure to update TSCA in light of additional challenges posed by nanotechnology-enabled products. Regulatory agencies are seriously underfunded and understaffed and therefore incapable of effectively attending to emerging challenges. In other words, there is merely the illusion that the public is thoroughly protected from dangerous substances. A future disaster that ignites public concern could remedy the matter and lead to a rash of new regulations. Even then, resistance to testing entrenched products may well get them exempt from any future regulation—the political price to put in place a more rigorous safety regime.

In weighing speedy development against long-term risks, speedy development wins. This is particularly true when the risks are uncertain and the perceived benefits great. Increased public health and environmental risks are just one of the trade-offs. In the development of innovative nanotech approaches for generating and storing energy, there are also opportunity costs. Global climate change and the serious need for clean sources of energy are urgent matters, and necessitate governments subsidizing the development of industries whose products would not otherwise be cost-effective. Policy makers are singularly focused on the potential benefits, hoping that innovation and large-scale production will eventually make solar and wind power cost competitive with non-renewable sources of energy. There is no evidence that this will happen any time soon. The cost of clean energy will go down, but the rate of the decline will not reach the cost per kilowatt-hour of coal for many decades to come. In the meantime, opportunities are lost, because the extra money spent on subsidizing clean sources of energy could have been used to fund other societal needs. Policy makers differ as to the value that should be put on slowing down the rate at which the globe warms. And therefore, we do not have an effective method for comprehensively evaluating the trade-offs between one source of energy and another.

Advocates for clean sources of energy argue that cost comparisons between the various sources are misleading. Producers and suppliers of coal, oil, and gas are not made accountable for the environmental and public health damages of their products. They propose that a carbon tax be levied on coal and fossil fuels to cover societal costs and level the playing field. But to date, politicians have shown little appetite to take on the anticipated resistance to a carbon tax. For many politicians, particularly those representing regions where the oil industry is strong, resistance to a carbon tax functions as a primary reason for denying the existence of global climate change.

Of course, there are costs associated with all products that are not necessarily reflected in its purchase price. The cost of a computer does not factor in maintenance or the recycling of parts when the device is obsolete. Plastic bag manufacturers do not pay for landfills or polluted waterways.

Determining the difference between the market price and the actual cost is not easy. Nevertheless, scholars are busily working on methods for comparing the expenses associated with differing sources of energy. Even if they succeed, there will be plenty of politics that will intervene before such determinations have a significant impact on public policy.

## ENVIRONMENTAL RISKS: SYNTHETIC OIL AND FRACKING

Some environmentalists already view the commercial development of synthetically designed species of algae capable of producing oil as a red flag. Oil produced by synthetic algae may eventually fuel many of our cars and help reduce the reliance on fossil fuels. The oil must be squeezed out of the first generation of synthetic algae, but biologists hope to genetically engineer forms of algae that produce more oil quickly and excrete it into the pond so it floats to the surface. Thus oil could be skimmed from the pond and less expensive to process. Some future forms of synthetic oil may not even release carbon into the atmosphere and therefore not contribute to global climate change.

Among the companies that have entered the race to produce commercially significant amounts of oil from synthetic algae is J. Craig Venter's Synthetic Genomics. As a biologist and entrepreneur Venter has played a central role in moving many facets of genomic research forward. This includes the sequencing of the first human genome in 2000 and the 2010 creation of the first semi-synthetic bacteria cell. Venter co-founded Synthetic Genomics (2005), which modifies microorganisms in hopes of producing useful biochemicals and new synthetic fuels. ExxonMobil committed $600 million for collaboration with Synthetic Genomics.

Existing synthetic algae ponds are enclosed. However, producing oil in the volumes necessary to make its production cost-effective will require thousands of open algae ponds. Synthetic algae will soon escape into swamps, lakes, and wells. How synthetic algae will alter or possibly harm those ecosystems we do not know. Algae growth affects many bodies of water and can make them unsuitable for drinking or recreational purposes. A species of algae cultivated to secrete oil would make swimming at the local pond a particularly slimy exercise.

If we were only talking about a few new synthetic species, their environmental impact could be manageable. However, assembling biological systems with unique properties has become the latest engineering fad, available to do-it-yourselfers and high school biology students—the modern equivalent of what building a ham radio was in the 1950s or a computer in the 1970s. Most of these experimental biological creations will be benign in their effect. Nevertheless, a few "bad apples"—virulently dominant species— can cause a lot of environmental damage. Consider the introduction of the eucalyptus tree to California in the 1850s. Planting eucalyptus was encouraged by the state government to provide a renewable supply of wood for railroad ties and timber for houses. As it turned out, eucalyptus was singularly unsuitable for railroad ties as they warped and split when dried and were too tough to hammer railroad spikes through. Environmentally, eucalyptus competes with native plants, does not support many native animals, and fuels forest fires. In some regions of California groves of eucalyptus are being cleared and native plants are being reintroduced at considerable cost.

Over the past few years, the environmental risks of hydraulic fracturing, which extracts oil and gas from shale, has received considerable attention. Fracking has been around for fifty years although methods such as horizontal fracking are new. As a method to extract oil and natural gas, hydraulic fracturing has grown dramatically. It is used for fully 60 percent of all new wells worldwide. Fracking a well requires seven million gallons of water (or more) that are mixed with sand and chemicals (some of which are toxic) and then pumped deep within the earth to break up shale and release gas and oil. The U.S. and Canada are rich in shale gas deposits. Fracking offers an excellent opportunity to become energy independent. On the other hand, fracking uses tremendous quantities of precious water, releases the greenhouse gas methane into the atmosphere, has led to groundwater contamination and illness, and might possibly make earthquakes more likely. Fracking has been banned in some states and until recently in the U.K., which lifted its ban and is instead searching for ways to make fracking safer.

How does one weigh the benefits of being energy independent against the cost of being held hostage economically and politically by a cartel of oil producing states? Do geopolitical benefits justify environmental risks at home?

The oil and natural gas produced by fracking contributes to greenhouse gases in the atmosphere. But what are the alternatives? Therein lies the conundrum of meeting world energy needs.


## NUCLEAR POWER AND THE ENERGY CONUNDRUM

Are nuclear power plants more or less risky than genetically modified bacteria, driving a car, smoking, skiing, napping, eating french fries, or flying in a commercial airplane? Individuals differ in their evaluation of risks, and experts prioritize the risks of various activities differently than the public. The experts rely heavily upon quantitative measurements such as the number of injuries or deaths caused, while more psychological and speculative factors enter the judgments of the average citizen. Students, members of the U.S. League of Women Voters, and experts were asked to rank a list of thirty activities and technologies from the riskiest to the least risky in a classic 1987 study by the psychologist Paul Slovic. The members of the League ranked X-rays as the twenty-second most risky item, while the quantitative data upon which the experts rely promoted X-rays to seventh on the list.

Students, and also members of the League, ranked nuclear power as the most risky of the activities or technologies. The experts ranked nuclear power number twenty and motor vehicles number one.

The heavily publicized nuclear disaster at Chernobyl in the Ukraine a year earlier was no doubt on the minds of the students and League members. Chernobyl is the worst nuclear accident in history. Fifty-seven firefighters and others working to abate the disaster died. This one incident required the evacuation of 350,000 people and 3,650,000 acres of farmland and forest were removed from agriculture and other human uses.

Researchers at the World Health Organization (WHO) estimate that the Chernobyl meltdown has (or will) contribute to nine thousand additional deaths from cancer and other diseases. The premature loss of nine thousand lives is certainly not trivial, and yet this number represents only a 1 percent rise in the overall nine hundred thousand deaths from cancer that would occur in the affected region if the meltdown had never happened. In contrast, the WHO calculates that more than 1.2 million people die in traffic accidents yearly. Other experts insist that the WHO's estimates for cancers due to Chernobyl are much too conservative, and that the accident will cause sixteen thousand additional cancer deaths, if not many more. The actual risk of exposure to radiation is not a straightforward science. For years the nuclear industry claimed that low-level exposure was harmless, while more recent findings suggest that any exposure to radiation can be harmful. A 2007 German study found that children living less than 5 kilometers (3 miles) from sixteen nuclear power plants had more than twice the likelihood of leukemia than those living more than 5 kilometers away. The research holds significance even while the total number of children affected is small.

Have the dangers of nuclear power generation been exaggerated? The history of nuclear power provides a classic example of the difficulty in assessing the benefits and risks of a technology and in how new innovations can overthrow yesterday's judgments. New designs promise to transform nuclear power generation from an old technology to a safer emerging technology. Keeping pace with emerging possibilities demands adaptive responses to new information and altered conditions. But the opinions of legislators and the public about a technology's safety become quickly entrenched, and are seldom subject to reevaluation.

The partial meltdown and release of radioactive material in 1979 at Three Mile Island in Pennsylvania turned the American public against nuclear power projects. Many older reactors have been taken off-line in the U.S., and there are only four new plants under construction. Europe has been more reliant than the U.S. on nuclear power generation. However, Chernobyl, and then the recent meltdowns at Fukushima, caused public acceptance of nuclear power in Europe to plummet. Germany intends to shut down all of its existing plants by 2022. As of 2012, 430 commercial nuclear power reactors operating in 31 countries supplied 13.5 percent of the world's electricity. The appetite of China's leaders for nuclear power remains strong with 17 reactors in operation, 28 under construction, and many more about to be constructed or in various stages of planning.

The giant tsunami caused by a March 11th, 2011 earthquake killed nineteen thousand people. Its height was 15 meters (49 feet) when it struck Japan's Fukushima Diichi nuclear power plant, precipitating events that lead to the meltdown of three reactors. No one died as a direct result of the Fukushima meltdowns. The story of Fukushima's long-term impact, however, cannot be written. Estimates of the effects of escaped radiation range from zero to twelve hundred additional deaths due to cancer. But there remain pathways through which Fukushima could still generate a catastrophe of regional and perhaps worldwide import. The most serious recognized threat was the danger posed by thirteen hundred spent fuel rods (each two-thirds of a ton in weight) stored in a pool 100 feet off the ground in Unit 4, which was damaged by a hydrogen explosion soon after the start of the crisis. Only a small fraction of the energy in these so-called "spent" rods was actually used to generate energy. The spent rods at Unit 4 had to be removed.

This is generally a routine task performed with computer systems and finely calibrated machinery, but that equipment was also damaged and could not be rebuilt. Furthermore, the pool itself was compromised and contains debris from a roof that blew off during the explosion, and the building is sinking and being held together by the high-tech equivalent of duct tape. The extraction process began in November 2013. A small mistake during the extraction of an individual rod could set in motion a series of disastrous events. Given the building's vulnerability, that chain of events could also be set off by a small earthquake. If the rods came too close to each other a chain reaction would commence. Other nearby pools containing just under ten thousand spent fuel rods might conceivably be drawn into a chain reaction.

Estimates projected that if a chain reaction in Unit 4 occurred, it would release ten times the radioactive Cesium-137 released by Chernobyl and necessitate the evacuation of forty million people from Tokyo. The worst-case scenario entailed the release of eighty-five times as much Cesium-137 as Chernobyl. That much Cesium-137 could destroy the world's environment and threaten human survival. It would take skill, luck, and grace to dismantle Unit 4 without a serious catastrophe. With care, the Japanese people and the world dodged this particular nuclear bullet. While slowgoing, half the rods were removed by May 2014, and on November 6, 2014 it was reported that all but 180 fuel rods (said to be less risky) had been removed. Additional ongoing threats posed by Fukushima, nevertheless, remain in a process of decommissioning that will take forty years.

Accidents are not the only problem with nuclear power generation. Various nuclear power reactors can be used to produce weapons-grade plutonium. The disposal of radioactive nuclear waste presents an additional complication. The pools storing spent rods at Fukushima and other nuclear power plants were meant to be a short-term measure, but no good long-term solution exists. Radioactive waste disposal continues as one of those predicaments that everyone recognizes needs to be solved, and yet the nearly universal response is NIMBY (Not in My Back Yard).

The management of nuclear waste from dismantled bombs, disaster sites, and decommissioned power plants never goes away. The safety of failed reactor sites such as Chernobyl is costly to maintain. Insuring that radioactive waste materials do not cause radiation poisoning or contaminate the environment requires vital wealthy governments that can take on the responsibility. Some of this waste will require thousands of years to decay. Will responsible authorities capable of managing such dangerous substances be available for those many generations? Radioactive waste functions as an eternal challenge, another time bomb waiting for the right circumstances to be a catalyst for some future crisis.

All these reasons would appear to rule out nuclear energy as a source for significant new supplies of energy. But recently nuclear power has gotten support from surprising quarters including the environmentalists Bill McKibben and James Lovelock. Moreover, there are designs for future nuclear power plants that solve some of the past problems.

Armchair environmentalists stress reasons to reject nearly every source of energy. But for environmentalists who approach the problem comprehensively, no easy solution exists. The preponderance of the 16 terawatts of energy that run the world today comes largely from technologies that generate $CO^2$. To arrest the greenhouse effect we not only need to replace three-fourths of that energy production with clean energy, but we will also have to triple production of clean energy over the next two decades to meet demand by the world's rising billions. In that nuclear reactors do not

release significant quantities of $CO_2$, they provide an important source of clean energy. Wind and solar are more attractive sources of clean renewable energy, but their relatively high cost will persist.

Coal remains an easy energy source to acquire. Yet reliance on coal is deadly. Mining and generating electricity from coal are the worst contributors to the accumulation of $CO_2$ in the atmosphere. On top of that, energy production using coal kills on a scale that even Chernobyl cannot match. In the U.S. alone, pollutants such as fly ash, carbon monoxide, and mercury that coal production and burning spew into the air kill thirteen thousand yearly according to estimates of the Clean Air Task Force. Worldwide the yearly estimates of deaths attributed to coal rise to 170,000.

Coal's destructive record alone provides good cause to reassess nuclear power. In addition, some good news—much safer and smaller designs for nuclear power plants. Two innovative approaches command significant interest. One approach calls for fast reactors that burn at higher temperatures and use as their fuel nuclear waste and excess weapons-grade uranium and plutonium. Spent fuel from older nuclear power plants can actually retain as much as 99 percent of its energy capacity. According to Nathan Myhrvold, a scientist who made a fortune as Microsoft's chief technology officer and has now turned his attention to new approaches to nuclear power generation, "we could power the world for the next one thousand years just burning and disposing of the depleted uranium and spent fuel rods in today's stockpiles."

The other innovative strategy for the design of nuclear power plants uses liquid flouride thorium as the fuel. Thorium, number 90 on the atomic chart, is four times more abundant than uranium. Thorium reactors will leave no long-lived radioactive waste products.

Nuclear power generation is one viable option, but we have no choice but to enlist a broad array of approaches in meeting the challenge of global warming. We have trapped ourselves between global warming and a world hungry for energy. The contention that global climate change be slowed radically by altering behavior alone is a sham. Conservation measures decrease the use of electricity for air conditioning, heating, lighting, and refrigeration, but certainly not on a scale to offset increasing energy demand either locally or worldwide. Few people consider giving up meat eating even though farming, such as raising cattle, pigs, and chickens, is the third largest source of greenhouse gases.

There will be unanticipated challenges with the new nuclear technologies that may only come to light as they are tested and deployed. Some of the designs are for small, portable contained reactors that could be buried outside an impoverished city to provide plenty of power with little or no maintenance. As far as getting energy to locales that really need it, this sounds great. But there will be security concerns with such small reactors and what belligerent governments or terrorists might do with them.

The new designs for nuclear power reactors are not purely speculative, but remain theoretical and need to be tested. Testing must proceed with due caution. And even if future designs for generating nuclear power are much safer, we should not delude ourselves about the continuing dangers posed by many older designs and the waste they produce.

Presuming that policy planners reevaluate nuclear power in light of these new approaches and decide they are sufficiently safe, it will still require years to build demonstration models. Even then, legislators and executives will be slow to give their approval to go ahead and build new plants until the relatively slow process of reeducating the public evolves into broad support.

In the meantime, there remains the challenge of meeting world energy needs. A truly devastating nuclear accident can never be dismissed. Fracking harvests carbon-releasing sources of energy. Coal, the bottom-line option, kills millions of people. Dismantling the modern industrial infrastructure is not viable. Conservation measures are insufficient. New sources of inexpensive, clean renewal energy are a worthy goal, but will be a long time coming. No viable plan to meet world energy needs while staving off global warming has yet emerged. We will continue to be confronted with bad choices and unsatisfactory trade-offs.

Trade-offs, complexity, and the pace of change are important factors for assessing the viability of any approach to managing the societal impact of an emerging technology. In Chapters 6 and 7, we turn to a discussion of genomics, as a case study in technological challenges all arising from one field of research. It will be helpful to keep these three themes in mind as ways of thinking about the particular issues posed by each case among many genetic possibilities.

# { 6 }

# Engineering Organisms

MY ATTENTION WAS IMMEDIATELY DRAWN BY ONE OF THE
sponsors listed on the invitation to an August 2010 workshop on synthetic
biology and nanotechnology: The FBI WMD (Weapons of Mass Destruction) Directorate, Biological Countermeasures Unit. I decided to attend the
event. Roughly a hundred people were scattered around the half-empty
meeting room at the Boston Park Plaza Hotel. By far the largest contingent
was FBI field officers flown in from all over the country.

The workshop had been organized to educate the field officers on how to
work with universities and the do-it-yourself (DIY) biology community to
thwart the misuse of synthetic organisms and nanomaterials. The FBI is
concerned that ideologically-driven terrorists, disgruntled researchers,
vengeful grad students, or psychopathic lab workers might harm individuals or cause a public health crisis by releasing lethal biological agents or
nano-poisons. A future Ted Kaczynski (the Unabomber) might well create a
pathogenic organism in his home laboratory from mail-order biological
components purchased on the Internet.

Edward You, the Supervisory Special Agent heading the Biological
Countermeasures Unit, and a former cancer researcher at Amgen, understands that the FBI needs to forge a relationship of mutual trust with the
science community. Trust could ease a scientist's concerns in alerting local
agents about researchers and lab workers whose erratic behavior just might
indicate the possibility of harmful intentions. "When I got involved," You
said, "it was pretty clear the FBI wasn't about to start playing Big Brother to
the life sciences . . . We need to create a culture of security in the [synthetic
biology] community, of responsible science, so the researchers themselves

understand that they are the guardians of the future." Forging trust between researchers and the FBI is a good first step, but this is far from sufficient to stop a disaster occurring as the result of the release of a synthetic organism.

Biosecurity is only one of the challenges posed by synthetic organisms. The possibility that the lawful release of newly engineered forms of life could have a detrimental impact upon the environment also rings alarm bells. The genetically modified organisms and microorganisms presently being created can dramatically change the ecology of a natural environment in which they find a niche. However, the environment that synthetic microorganisms are most likely to alter lies within the human body. Thousands of varieties of bacteria and several dozen fungi are housed in the human gastrointestinal tract. They are collectively referred to as either human gut flora or the microbiome. The human body and the microbiome are intimately entangled. None of us would survive without the work done by these microorganisms to break down food in our intestines and make nutrients available. Digestive problems caused by unhealthy gut flora can deplete us both physically and mentally. In turn, our mental and physical states alter the health of our microbiome.

In a study financed by the U.S. National Institutes of Health (NIH), ten thousand different microorganisms were uncovered in the gut and each of them was genetically sequenced. Collectively, gut flora has eight million unique genes—360 times more than the human genes in the body. These bacterial genes are the source of many enzymes essential for human health.

A growing number of scholars believe the microorganisms in the gut should be considered integral parts of the human body, just like skin or nerve cells. Increasingly, they refer to gut flora as the *hidden organ*. Given that the human body consists of about thirty-seven million cells, but contains ten times that number of bacteria, calling the microbiome an organ offers a whole new perspective on what it means to be human.

The question of whether your bacteria are part of you or not is secondary to the fact that the existence of the microbiome opens up a vast realm of applications for enterprising synthetic biologists. In the future, synthetic forms of bacteria might be introduced into the gut to alleviate digestive problems, improve metabolism, fight pathogens like HIV, reduce autoimmune conditions, or trigger insulin production in response to glucose consumption and thereby cure diabetes. These alone suggest that the benefits of engineering new forms of digestive bacteria outweigh the risks—presuming we take the necessary measures to minimize those risks. A bacteria laced serving of yogurt could be sufficient to introduce new, synthetic microorganism in digestive flora. Some enterprising entrepreneur might even create a microorganism that flourishes in our gut and allows us to eat all the choc-olate and ice cream we want without putting on weight. Now that would be a new organism that only die-hard natural food crusaders would refuse to invite into their digestive tracts.

The NIH study reconfirmed that everyone carries pathogenic bacteria in their intestines. Why pathogens occasionally erupt into an illness is only partially understood. Introducing additional pathogens into the gut may or may not be problematic for most individuals. Nevertheless, in tinkering with the microbiome considerable care must be taken. Changing the envi-ronment in the gut could conceivably even start a new pandemic by facili-tating an existing pathogen's mutation into an aggressive new variation of the disease.

Modifying single-celled organisms and building entirely new biological components from scratch has emerged as the latest frontier in genomics. Synthetic organisms will not only enhance health, but will also revolutionize other fields, such as offering new ways to produce sustainable energy. Esti-mates indicate that the synthetic biology industry will grow in value from $1.6 billion in 2010 to $10.8 billion by 2016.

The environmental and public health challenges posed by genetically modified organisms are easy to acknowledge. Evaluating the specific risks and addressing them is hard, but necessary. That task gets complicated fur-ther because anything to do with genomics is caught up in the bioethics war raging on many fronts. Is tinkering with genes good because of the many benefits that can be realized, or is it just out of bounds?

The charge that scientists are "playing God" when they alter genetic material carries a theological tone. Yet the phrase is also meant to capture the widely felt concern that people should not engage in certain activities. The skirmishes fought over each new method to alter genetic material serve as part of a bloody broader struggle as to whether all ethics can be reduced to weighing benefits against harms. Some activities, according to this view, are simply wrong. For example, all people have intrinsic value and therefore must be treated with dignity. Often this principle is conveyed through some form of the Golden Rule or through the Kantian imperative that people are ends in themselves and should not be treated as means to an end. In the minds of many, animals and the environment also have intrin-sic worth and should not be abused. Not everyone shares those beliefs. Nevertheless, the widespread recognition that the environment must be treated with greater care and that animals should not be made to suffer unnecessarily has taken hold.

The struggle over which values should guide decisions about what technologies to develop is central to the broader conversation our society needs to be having. Is it sufficient to go ahead with any field of research whose overall benefits can be justified as significantly outweighing the risks? Or are there some lines that should not be crossed, and, if so, where are those lines? Furthermore, what should scientists and policy planners do when there are fundamental differences within a society as to which forms of research are permitted and which are prohibited?

The bioethics skirmishes fought over the use of genetic material will continue as a proxy for battles directed at reinforcing or establishing intrinsic values. In many respects, the war itself is extremely important. The struggles help sort out and discard those values that only build cohesion within a community by fostering prejudice and intolerance. The proverbial baby, however, can get thrown out with the bathwater, particularly when all religiously endorsed values get dismissed because certain past prejudices were empowered as declarations of spiritual authorities.

Ethical theories about which actions are right and good generally fall into two broad camps. Within the one camp are the various lists of rules or duties said to incorporate intrinsic values. The correct course of action for a new situation can be determined by applying the rules to that situation. The other camp contends that the correct action can only be determined by evaluating the consequences of the differing responses to a challenge—the action that produces the greatest net benefit should be selected.

While many people claim to subscribe to a consequentialism that enthrones the greatest good for the greatest number, when pushed to think through the results of such a principle, very few would make this a hard-and-fast rule. This has been demonstrated consistently through ethical thought experiments known collectively as "trolley problems." The problems offer a course of action that could save five lives at the expense of one life. The vast preponderance of people will throw a switch to save the life of five workers from an approaching train, even if their action will certainly kill a worker on the alternate track, who would not have otherwise died. However, few people would ever advocate harvesting the organs of a healthy individual to save the lives of five people requiring organ transplants. Internationally endorsed principles protecting the rights of individuals generally prevail over simplistic calculations of benefits and losses.

The difficulty for scientists and for science policy hinges on sorting out which principles to honor, particularly when there are differing opinions and rationales regarding what is right and what is wrong. In spite of contrary claims, technologies are not value-free. Their value is instrumental, as the means to an end, and usually determined by weighing the consequences inherent in their use. But this leaves unanswered questions regarding what to do when the adoption of a technology by some groups rides roughshod over values accepted as intrinsic by other communities.

When opinions differ, leaders in democratic societies generally fall back on a utilitarian (consequentialist) analysis that weighs the advantages against the losses. Giving deference to the values of one community over others is perceived as violating the separation of church and state. In practice, the will of powerful groups often prevails, even when this transgresses the rights of minorities. Moral progress can be measured in correcting past wrongs in realms such as slavery, women's rights, and, more recently, gay rights. The right of some nonhuman animal species to be treated with respect is emerging as a battlefront of increasing importance.

It would be tragic if the debates over designer babies, genetically modified organisms, human cloning, and animal-human hybrids all defaulted to utilitarian calculations. Just because something is technologically possible and its benefits apparently outweigh the demonstrable harms, does not mean it should be done. The tension between what is possible in bioengineering and what should or should not be pursued goes well beyond genomics into technologies for radically extending life and for enhancing human capabilities. After examining genomic possibilities in this and the next chapter, we will turn to the enhancement debate in the following three chapters.

## A BRIEF HISTORY OF GENOMICS

The power to alter genetic material is the latest episode in an unfolding history of scientific research set in motion by Charles Darwin's theory of evolution. As with the discovery of any new form of power, this capability presents many questions as to when it should be used and how to use it properly or safely. Those questions cannot be approached intelligently without some appreciation for both the accuracy of the theory and the manner in which it has continually been revised and continues to be refined.

The theory of evolution is an extremely successful scientific theory. Predictions based upon the theory have been confirmed countless times. This, however, does not mean we understand everything about how evolution works or all of the influences that have led to the unfolding of biological life on this planet. Nor does the accuracy of the theory mean that all the technological developments that materialize through our understanding of genomics are benign or should be characterized as progress. Nevertheless, the significance of Darwin's work was succinctly captured in 1973 when the Russian biologist Theodosius Dobzhansky said, "Nothing in biology makes sense except in the light of evolution."

Darwin (1809–1882) did not actually invent the idea that species evolve. There was already considerable disagreement among scientists and philosophers as to whether God had created each observable animal species during that busy first week, or whether the individual species had evolved through biological changes over time. Most Christians believed that the world was only a few thousand years old, but nineteenth century geologists were accumulating ample evidence that the earth was tens, if not hundreds, of millions of years old. Darwin was well aware of that evidence. The longer time frame made the possibility of incremental evolutionary improvements at least plausible. Nevertheless, the estimates made by Darwin's peers were dramatically less than the present determination, which puts the age of the Earth at 4.54 billion years.

Darwin's great achievement lay in describing how species, including humans, emerge through a process he called "natural selection." In his groundbreaking book, *On the Origin of Species* (1859), he eloquently illustrates how, through chance, variations in nature introduce novel skills and features in succeeding generations, and how this, in turn, gives rise to an extensive array of biological diversity. Nature's way of engineering life is the hit-or-miss method—more miss than hit. Countless random mutations occur. But only those new organisms with characteristics that were amenable to the prevailing environmental conditions and to competition with members of the same species and other species would be able to adapt and survive, and therefore be capable of passing on their traits to succeeding generations. Darwin's theory of evolution was built around three ideas: heredity, variation, and natural selection. However, Darwin had absolutely no understanding of the biological mechanism by which a trait or its mutation could be passed from generation to generation.

Gregor Mendel's (1822–1884) experiments with pea plants provided an important first step in explaining those mechanisms. Mendel recognized the pattern whereby the dominant or recessive traits, such as the color of the flowers, would be passed on to the next generation. The importance of Mendel's breakthrough was not appreciated in his lifetime, but his work was rediscovered in 1900. Mendelian genetics and Darwinian natural selection were effectively integrated into a widely accepted account of evolution between 1936 and 1947. By that time, it was already known that chromosomes of DNA and that segments of DNA (genes) located along the chromosomes are the carriers of inheritable material. However, the biological mechanism by which DNA passed traits from one generation to the next, and why those traits varied (mutated), remained a mystery.

James Watson (born 1928) and Francis Crick (1916–2004) described the structure of DNA in 1953. Their breakthrough moment came with the help of data they "borrowed" from a colleague, Rosalind Franklin (1920–1958), without her permission. The beauty of Watson and Crick's description of DNA lies in the elegance of a double helix, and in how changes in the individual base pairs connecting the molecule's two strands provides a simple vehicle for genetic mutations. Such changes can occur when chemicals, radiation, or viruses alter a gene or its position on the chromosome.

DNA is often referred to as the book or the "source code" for life. In this book the base pairs are compared to individual letters, genes are sentences or paragraphs that perform a specific task, and individual chromosomes are like chapters of the book. Genes carry the code for producing a specific protein and these proteins in turn serve as catalysts for chemical processes. There are variations in the book for each plant species. The DNA for every nonhuman animal or individual human created by sexual reproduction is unique, given that some of the genes come from each parent. Nearly all the cells in the body of a person or that of a nonhuman animal or plant contain identical copies of their personal book.

The unraveling of the two strands of a chromosome provides a mechanism for copying a chapter of the genetic code. Free-floating bases (nucleotides) line up along the open strand to form a new strand of messenger RNA, which then detaches from the chromosome and goes on to transmit the copied genetic information to ribosomes in the cell. The ribosomes use the information as instructions for manufacturing a specific protein. Each ribosome can be thought of as one of nature's nano factories.

Chromosomes come in pairs. There are thirty-nine such pairs in a chicken, thirty-two in a horse, twenty-four in a potato plant, nineteen in a pig, twelve in a rice plant, and six in those pesky mosquitos. Twenty-two identical chromosome pairs and a twenty-third sex-determining chromosome paired in the nucleus of most cells in the human body carry the full instructions for building that specific person. Those twenty-three chromosomes contain 3.2 billion base pairs. We share 90 percent of our genetic code with rats and 21 percent with worms. Ninety-eight percent of the genetic code of chimpanzees and human beings is the same.

The power to engineer organisms and totally novel biological products is the result of techniques for manipulating genetic material that have developed over the past forty years. Altering any gene or collection of genes might lead to a new organism. In the case of designing human infants, the primary focus is upon the 2 percent of genetic material that distinguish one person from another.

## TOP-DOWN AND BOTTOM-UP ENGINEERING

The many pathways for engineering organisms fall within two broad categories: top-down and bottom-up approaches. Top-down approaches start with the full genome of an existing organism, which is then customized to produce a similar organism with a single or a few different features or qualities. Genetic material from one organism can be combined with genetic material from one or more other sources to produce recombinant DNA coding for a new organism that would not otherwise occur. Bottom-up engineering refers to methods through which segments of DNA are connected to build entirely new biological components from scratch. Some scientists restrict the term *synthetic biology* to bottom-up approaches, though often the term is also used to include top-down engineering.

Recombining segments of DNA, a top-down engineering approach, has led to the production of genetically modified organisms (GMOs) for food and medicines. One of the first recombinant medicines was synthetic insulin, which led the way to effectively lower the cost of medication to manage diabetes. Formerly, most cheeses were produced with rennet taken from the fourth stomach of calves. But now recombinant chymosin (the active ingredient in rennet) is used for the manufacture of a majority of the world's cheese.

Genetically modified crops that have desirable characteristics such as a resistance to pests, the ability to withstand cold weather, faster growth, higher yields, better color, improved nutritional value, and longer shelf life are widely available in the U.S. There are even varieties of common grains from which low-cost antibiotics can be produced. However, fear that they will mix with and potentially contaminate grain from heirloom stock has led to severe restrictions on field-testing antibiotic-producing grains.

Consumers in the U.S. seldom know whether the food they eat is made from grains that were genetically modified. European scientists and consumers, however, have been deeply suspicious that GMOs pose health risks that may only become apparent after generations of use. This concern has some credibility when one considers that the preponderance of GMOs incorporate traits such as weed and pest resistance that are of particular benefit to the farmers growing the crops, but whose long-term effect on humans is unknown. The European Union restricted the import and sale of GMOs, which led to trade tensions between the U.S. and Europe. In the U.S. there are calls for labeling genetically modified foods so that consumers will at least have a choice of avoiding them if they elect to do so. But this is easier said than done. A farmer, for example, can have no idea whether some plants in a field have been grown from genetically modified seed. The grain the farmer buys may not be pure, or birds could have transported seed from a field of GMO corn to a cornfield grown from heirloom stock.

There is a great deal of fear, but there is a lack of substantial evidence that GMO foods are any more risky than conventional foods. In some cases GMO foods eliminate risks posed by conventionally grown grains or vegetables. For example, GMO plants that are resistant to diseases or insects can require fewer chemicals to grow and are therefore friendlier to the environment. Fewer chemicals also lower the likelihood of one becoming sick by consuming chemicals when eating unwashed fruits and vegetables. On the downside, there is also clear evidence from the U.S. and Canada that pest-resistant GMO plants spawn pest-resistant weeds. This, in turn, can require spraying with ever more powerful weed killers.

The European Union has approved the cultivation of one form of GMO maize, while it also granted individual governments authority to place local restrictions on growing GMOs. The prospect of GMO plants contaminating highly valued heirloom varieties is particularly important in Europe, where people value the taste and quality of naturally grown fruits and vegetables. The contention that GMO varieties taste better does not pass muster with the discerning European palate. So the debate continues, and the verdict on long-term effects is still out.

But what happens when we're not talking about plants and seeds, but about fish? In 2010, the U.S. Food and Drug Administration reviewed the first genetically modified fish for human consumption. The company AquaBounty had transferred a growth hormone-regulating gene from a Pacific Chinook salmon to an Atlantic salmon. Chinooks grow more quickly than Atlantic salmon. A second gene had also been added to the new fish to ensure that the growth hormone remained continuously active. AquaBounty wished to market these faster-growing, genetically modified salmon to compensate for the shortage of salmon created by the growing popularity of sushi. Genetically modified animals might be riskier than GMO plants because of their ability to transmit diseases to humans. Whether they pose more health risks to humans than, for example, cattle that are given growth hormones (nearly all cattle in the U.S. and Canada), is unknown.

The modified salmon, trade name AquaAdvantage, are raised in pools that are landlocked. Nevertheless, environmentalists voice concern as to what might happen to other salmon species if they escape into the ocean and crossbreed. The U.S. Food and Drug Administration (FDA) has been reviewing AquaBounty's application to approve the modified fish for entry into the food supply since 1995. A 2010 panel found no "significant effects on the environment" and that the salmon was safe as a food. Nevertheless, as of fall 2014, the FDA has not announced a final assessment. Perhaps they are holding back because they received tens of thousands of citizens' objections after a similar 2008 approval of milk and meat from cloned animals.

The newer bottom-up genetic engineering approaches are less well-known, and their safety concerns less hotly contested. Their risks, however, are worth considering because they put the power to create genetic products into the hands of amateurs. Bottom-up genetic engineering constructs simple organic systems for specific tasks, with the assembly of standard biological components (BioBricks). A BioBrick is a DNA sequence that has a specific structure. Some BioBricks correspond to specific traits in organisms. Several thousand BioBricks have been registered with the BioBricks Foundation, which recommends that they remain in the public domain. These standardized Biobricks have been called the biological equivalent of thousands of different Lego parts. The use of BioBricks is transforming the creation of new biological systems into a kind of do-it-yourself engineering.

Each year, teams of college students and professionals worldwide learn the techniques for assembling synthetic organisms through the International Genetically Engineered Machine competition, better known as IGEM. An IGEM team is given a kit of standard biological components, and with this kit and additional components they design themselves, the team works to create a biological system that can function within a living cell. Harmless laboratory strains of the E. coli bacteria are a favored vehicle for these new biological systems. While E. coli can be found in the lower intestine of not only humans but all warm-blooded organisms, the strains of the bacteria used in research have lost their ability to live in the intestine. E. coli reproduces quickly enough to grow a culture overnight and is among the most intensively studied and best understood single-celled life form. One IGEM team engineered a new variation of the E. coli bacteria that changes its pH balance, a measure of whether a substance is acidic or alkaline, when it detects the presence of arsenic. In other words, the bacteria function as sensors for a poison.

The hope among bottom-up synthetic biologists is to assemble enough BioBricks to turn biology into the engineering equivalent of electronics assembly. But it remains unknown whether some of the more complicated dynamics within a living cell can be captured in the form of a single component or a simple process.

The controversies over the ethical concerns that accompany bottom-up approaches for synthesizing organisms are somewhat muted, partially because the science is young, but also because there is no use of human embryos. As will be discussed in the following chapter, any use of human embryos or cells from embryos is a hot-button issue, particularly in the U.S.

## THE GENOMIC FRONTIER

The first pioneers who entered the American West discovered a vast territory with untold dangers. This suited some, while many others wished to bring law-and-order to the frontier to make it a civilized land where they could raise their families. So too with the frontiers of genetic research: while some perceive a dangerous landscape that needs to be tamed and regulated, others say, "Let science rip."

With his contributions to the sequencing of not only the human genome, but also that of nonhuman organisms and bacteria, the ever present J. Craig Venter has done as much as anyone to explore the genomic frontier. He is often presumed to strongly oppose regulation of genomic research. But according to Andrew Pollard writing in the *New York Times*, "Dr. Venter says that he has long supported and paid for research into the ethics and regulation of the field and that there should be restrictions on letting synthetic cells loose in the environment."

Among Venter's projects is one directed at ascertaining the minimum genome necessary to sustain life. By turning off individual genes among the 482 of the bacteria *Mycoplasma genitalium,* which has the smallest genome of any organism that can be cultured, he determined that 382 genes are absolutely necessary. Venter would like to eventually engineer a new life-form with the lowest possible number of genes to be used as a research platform. An extremely simple organism could be helpful for developing bacteria with useful traits, such as consuming oil spills, from scratch.

Venter also assigned a research team to engineer the first self-replicating bacteria cells. In 2010, they announced success using a hybrid of top-down and bottom-up approaches. Initially, they sequenced the genome of an existing bacterium, then used that sequence to assemble, base pair by base pair, a genome over one million bases long. A few additional DNA segments or "watermarks" were added to the sequence to identify it as synthetic. The synthetic DNA was transplanted into a living cell, and the cell "booted up." This was an extremely impressive feat, even though there was some criticism that the organism was only partially synthetic.

The announcement of the success in producing this self-replicating bacterial cell was accompanied by a series of articles expressing a combination of concern, outrage, and praise. The concerns centered on whether scientists should be free to engineer life at their own whim, and where this crucial step might lead in the future. The bacterium was labeled a "Frankencell." In response, President Obama requested a report from his Commission for the Study of Bioethical Issues. The Commission had a unique opportunity because it could influence the course of a technology early in its development.

Three crucial questions frame the ethical discussions surrounding genomic research. What laboratory procedures will keep workers safe and ensure that new biological products and organisms are not unintentionally released into the outer environment? When might new biological products and organisms be released outside of the laboratory? What should scientists or industry be free to pursue?

The scientific community has demonstrated initiative in dealing with the first of these questions. For example, concerns over dangers posed by recombinant DNA were addressed very early on during a conference of scientists, physicians, and lawyers at Asilomar State Beach, California in February 1975. In what is often heralded as an example of responsible science, they drew up voluntary guidelines for safe laboratory procedures. Those guidelines continue to provide a foundation for biosecurity by universities and companies engaged in biotechnological research. The guidelines, however, can appear insufficient in a new era, as the tools for creating biological products move from relatively secure institutional laboratories to home laboratories.

A report from the President's Bioethics Commission titled *New Directions: The Ethics of Synthetic Biology and Emerging Technologies,* came out in December of 2010. It introduces the new phrase "prudent vigilance" as a way of striking a balance between a strong precautionary principle and little or no regulation of research. The members unanimously concluded, however, that "synthetic biology does not require any new regulation oversight bodies or a moratorium on advancing research at this time."

The report received a great deal of criticism, particularly in its failure to require a full assessment of the environmental impact of artificial organisms before they are released outside of a laboratory. Evidently commission members decided that it was too early to put such a policy in place. Furthermore, the report has been criticized for being overly deferential to the freedom of private industry to self-regulate while pursuing innovation. In contrast, a similar report from the U.K. Nuffield Council on Bioethics placed considerably more emphasis on human well-being, societal and ethical concerns, and the common good. Nevertheless, the differing ethical orientations of the two documents have not yet led to markedly different policies.

To date, there is no evidence of a significant public health or environmental crisis caused by the release of synthetic biological products or species outside of the laboratory. Regardless, the Commission blew an opportunity to call for the concerted development of guidelines for releasing organisms.

They did, however, promote the need for a public conversation on the goals of biotechnological research and whether or what kinds of restraints on scientific freedom could or should be formulated. Opening up a public dialogue is a scary prospect, particularly because various parties might attempt to manipulate the conversation for political, religious, or economic purposes. And yet, as I contend throughout *A Dangerous Master*, we have entered an inflection point in history where the importance of having that conversation cannot be overstated. A strong case can be made that free and open scientific inquiry proceed with little or no additional constraints. That case, however, requires the renewed consent of the governed.

There are real reasons to have the conversation about the goals of synthetic biology. One reason involves creatures long extinct, which are now considered candidates for recreation. *Jurassic Park* popularized the notion of genetically recreating extinct animals, but the species likely to be resurrected first is not a dinosaur species but rather the beloved passenger pigeon. Eventually, the huge woolly mammoth, an elephant species that went extinct roughly four thousand years ago, will also be recreated. Enough soft tissue with segments of mammoth DNA have been preserved so that the Harvard molecular biologist George Church believes he has a good shot at resurrecting the species. Church already developed a form of genome-editing technology that would speed up constructing the DNA of extinct species. His proposal that the woolly mammoth might be resurrected was at first derided, but soon many other groups interested in reviving different extinct species contacted Church.

Reintroducing the woolly mammoth has been cast as having a major ecological benefit, presuming a large population of hundreds of thousand could be developed. The grazing habits of mammoths encouraged the growth of grasses that protect permafrost in Arctic regions. Arctic permafrost presently preserves as much as two or three times as much carbon as is contained in all of the rain forests in the world. Without protection, permafrost is melting.

De-extinction is a curious possibility that is likely to be realized within the next decade or so. There exist both good and bad reasons for going down this path. On the one hand, biological diversity is highly valued and can help other species in an ecosystem thrive. On the other hand, any predatory species reintroduced to a natural habitat will change the ecosystem to the detriment of some existing species. Furthermore, de-extinction is expensive. Going down this road will certainly contribute toward increasing scientific knowledge, even if its environmental value will be unclear until we witness the results.

## ANIMAL/HUMAN HYBRIDS

Would a biological enhancement that improved the intelligence of a chimpanzee be immoral, a boon to medical science, messing with nature, a fascinating experiment, cruel, a moral obligation, or an invitation to the *Planet of the Apes*? This is not a multiple-choice question, so you can select more than one of the above. How about introducing anatomical features into a chimp that would allow it to express a rich set of sounds that might evolve into a spoken language?

Human-animal hybrids are a staple of both fairy tales and science fiction. The fact that researchers can presently engineer animal embryos that contain human genetic material as well as inject human stem cells into non-human animals is just one example of fiction become reality. The ethical dilemmas hybrids pose are beginning to be discussed by bioethicists, but their recommendations are unlikely to hold much weight as more and more experiments come to light. In the U.K. alone, it was discovered in 2011 that at least 155 embryos containing the genetic material of both humans and nonhuman animals had been created secretly in laboratories. The embryos were developed for research purposes and there was no attempt to bring any of the hybrids to life. No doubt some scientists just wish to experiment and see what would happen, but there are also very good arguments as to why research on animal-human hybrids will serve humanity. An animal-human hybrid, often called a chimera, could be a much better model for testing the effectiveness of new drugs than the nonhuman animals presently available. In theory, a pig-human hybrid might be an excellent source of organs to save the lives of humans who need new kidneys, hearts, or lungs. In practice, the overt attempt to create such a creature will be accompanied by moral outrage. Much of that outrage will be directed at the very idea of creating a creature that is part human. The pain experienced by experimental animals in the attempt to create an animal-human hybrid will also give good cause to reject this kind of research.

In addition to the feared takeover of the planet by intelligent apes, as portrayed in the older and more recent *Planet of the Apes* film series, there are plenty of less speculative reasons to limit such research. An animal-human hybrid could, for example, be a vehicle through which an animal disease mutates into a communicable form of flu and causes a worldwide pandemic.

The cavalier way in which animals are used in research and the amount of suffering caused to animals by researchers is immoral. Estimates of the number of animals sacrificed each year for research purposes range from over fifty million to over a hundred million. Several animal rights advocacy groups refer to this as an animal holocaust. Much of the sacrifice is for testing drugs, but the results are often not translatable to humans. In theory, an animal-human hybrid designed to be a better model for testing whether a drug should be used in research with human subjects could lower the total number of animals sacrificed. But it would also entail sacrificing many animals very close to us genetically, at a time when increasing numbers of people do not support abuse of our nearest biological ancestors, the other six species of great apes.

Regardless of moral qualms, this is one area of research likely to proceed, if only in the laboratories of rogue researchers. It may be just a matter of time before we come upon a Dr. Moreau and his family of hybrid children on some remote island. In H.G Wells's 1896 novel, Dr. Moreau created hybrids through vivisection (surgery on living animals), but by the time of the third film version of the novel a hundred years later, genetics was his preferred approach. Research on hybrids should be done openly so that we can at least regulate the practice and enforce strict laboratory safety procedures. However, it may be impossible to merely legitimate serious research without also opening the doorway to grotesque, dangerous, and clearly immoral experiments. In theory, making research by industry or university scientists transparent would be good, but in practice, it is probably unrealistic.

## THE CASE OF GOLDEN RICE

Risks and benefits are often in the eyes of the assessor and what she chooses to factor into her judgment. Consider the case of Golden Rice, a genetically modified organism engineered specifically to increase the amount of vitamin A in the diet of the poor in Asia, for whom rice is the one staple. Vitamin A deficiency causes a quarter- to a half-million children to go blind each year. Nevertheless, four hundred protestors destroyed a field in which Golden Rice was being tested in the Bricol region of the Philippines in August of 2013. The foreseeable benefits of Golden Rice were secondary in the eyes of the protestors to unforeseen risks to health and the environment, and to their perception that this plant was a preliminary tactic in a campaign that would eventually make the poor totally dependent on agrochemical companies such as Monsanto. In the protestors' analysis the profits of the agrochemical industry would take precedent over the needs of the poor. In this reaction they are not alone. That charge has been leveled by critics of GMOs around the world, in some instances for good reason.

Golden Rice is genetically engineered to be high in beta-carotene, the primary source of vitamin A. The reddish orange pigmentation of beta-carotene is what gives the rice its distinctive yellow color. In addition, another gene from a bacterium that helps switch on the beta-carotene was introduced into Golden Rice. What is curious about protests directed at this particular GMO is that in nearly every respect care was taken to maximize its benefit. The seed is not owned by any company and is being developed by a nonprofit whose focus is its health benefits. Golden Rice will be no more expensive than other varieties of rice seed. Tests have demonstrated a low likelihood that Golden Rice will cross-pollinate with other varieties of rice and thereby contaminate heirloom seed.

The NIMBY (not in my backyard) mentality is certainly one factor in rejecting GMOs. Most people do not want their family to be experimental guinea pigs for foods whose long-term effects are unknown. But in the case of Golden Rice the benefit goes directly to the people who eat the rice.

Of more import to protesters is the notion that even a beneficial GMO is a Trojan Horse in a long-term campaign by the agrochemical companies to make everyone dependent upon their seed. Farmers pay a premium for "super crop" with enhanced characteristics, and the biotech companies have tried a variety of strategies including contracts to force farmers to come back to them year after year for new seed. It is, of course, almost impossible to effectively police whether the farmer honors the contract.

When Monsanto was about to acquire a company (Delta and Pine Land) that had developed a terminator gene which made harvested seed sterile, the anti-GMO movement found a perfect symbol for their fears. Sterile seed due to a terminator gene would force farmers to buy new seed for each year's crop. This strategy made sense for Monsanto because it would protect the company's intellectual property, but it also would make farmers more dependent on seed from the company season after season. To be fair to Monsanto, some farmers are accustomed to buying new seed each spring, while others would have rejected genetically modified plants if the additional costs could not be justified in additional productivity. For farmers in the developing world, however, collecting seeds from a prior year's crop is essential because they often cannot afford to purchase new seed yearly. Perhaps if they used the muscled-up crops designed by Monsanto, their productivity would rise enough that they too could afford to buy seed for each new planting. But from a public relations perspective, the "terminator" gene turned out to be a disaster for Monsanto, who, in 1999, made a commitment not to commercialize the technology. However, concern has been expressed that Monsanto might bring the technology to market at some later date once the fervor over GMOs subsides.

The critics of suicide seeds have argued that these seeds might eventually cross-pollinate and cause widespread sterilization of other plants. But given the low likelihood of this occurring, the self-sterilizing property is exactly why a terminator gene could be attractive. The anti-GMO movement has long argued that wind or birds carrying seeds and pollen to other fields will lead to the contamination of heirloom varieties of crops. A terminator gene might actually provide some protection against this occurring. The criticisms directed at terminator genes, however, do not apply to Golden Rice. Seed from Golden Rice can be harvested yearly to plant the following season.

One of the better critiques of Golden Rice questions the very process by which this GM crop was created and settled upon as the solution to vitamin A deficiency. The leaders of communities for which Golden Rice was developed were not asked in advance whether they considered this strategy acceptable. The millions of dollars used to develop the genetically modified rice and distribute the seed might have been used in other ways to eliminate the deficiency, such as vitamin supplements or assistance in growing vegetables, such as carrots, that are rich in vitamin A. In other words, well-intentioned people outside of the community made a decision that a technological fix was a solution to the problem, and those directly affected had no say in that decision. The rejection of the Golden Rice by protesters in the Bricol region of the Philippines might appear irrational after the fact. But how would you feel, for example, if everyone in your region were told that they must stop driving their cars because self-driving vehicles cause fewer deaths (presuming they do)? The comparison of the two cases could seem a bit rash, but the question of whether paternalistic authorities have the right to tell us what we must do for our own good remains valid. Will we all eventually be pressured to accept technological solutions to every problem even when the solutions are unproven or have a demonstrable downside?

The concerns raised by the prospect that big business is taking over agriculture worldwide are not insignificant and will continue to influence public policy. By 2009, four companies controlled 54 percent of the global seed market. They continue to buy up other companies. Nevertheless, the protest against Golden Rice is an example of how there often seems little room for discrimination in evaluating the benefits and risks of individual innovations. The fear of where science is taking us has the capacity to overwhelm rational reflection.

If it should become evident that a widely dispensed GMO is responsible for a serious disease, pressures to reject all GMOs will mount. In the meantime, the protests against genetically modified plants and animals are mild in comparison to the passions inflamed by real and speculative applications of genetics to alter humans, to clone humans, or to select in advance the traits of children.

# { 7 }

# Mastery of the Human Genome

EVEN AT THE AGE OF EIGHTY-FOUR, COLUMBIA PROFESSOR ERIC Kandel is a sprightly and spirited teacher who loves to pepper his brain physiology presentations with oil paintings by J.M. Turner and Mark Rothko. In 2000, Kandel received a Nobel Prize for his research on the physiology of neurons and how they create and store memories. Kandel initially studied the hippocampus, a horseshoe-shaped region in the subcortical human brain that is associated with emotions, spatial orientation, and the processing of memories. However, the number of neurons in the brain of humans or other mammals made it difficult for him to make any headway. The human brain has approximately eighty-six billion neurons and each can be connected to other neurons by as many as a thousand to ten thousand synapses. An estimated forty million neurons inhabit the five regions within the human hippocampus, while a lab rat has more than 2.1 million hippocampal neurons.

So Kandel turned to the study of a simpler organism, the giant sea slug *Aplysia californica,* whose twenty thousand neurons are relatively large. *Aplysia's* simplest behaviors can involve a manageable one hundred cells, giving researchers the opportunity to precisely identify the role each neuron plays in a specific activity. For example, *Aplysia's* gill, which is extremely sensitive, will quickly withdraw when it is stimulated, similar to the withdrawal of a hand that gets too close to fire. In each slug this reflex involves the same twenty-four sensory neuron cells connected to six motor neurons.

Similar to the gill, *Aplysia's* tail also withdraws in a reflexive "fear" response when presented with a stimulus to which it is sensitive. By stimulating the tail for short and long periods, Kandel and his team were able to isolate and study the biochemical and synaptic activity that lead to the creation of both short-term and long-term behavioral changes. After one tail shock the slug develops a short-term memory that enhances the reflex for a few minutes. Five or more repeat stimulations give rise to long-term behavior alterations that could last days or even weeks. "Practice makes perfect," muses Kandel, "even in snails."

In the lowly *Aplysia*, memory creation is the result of complex molecular, synaptic, and gene interactions. Short-term memories, Kandel discovered, are produced by alterations in proteins that strengthen the synaptic connections between neurons. Further stimulation can lead to the creation of new synapses.

What is particularly interesting about Kandel's research for our purposes is the role genes play in the encoding of long-term memories. There are biochemical inhibitions on the creation of long-term memories. That is the reason we do not create long-term memories for everything we learn. But once these inhibitors are overridden, learning and memory are processes that alter the very structure of neurons by activating genes in the nucleus of the neuronal cell to grow new synapses that connect to other neurons. For the marine slug *Aplysia*, sustained stimulation can increase the synapses on a neuron from twelve hundred to twenty-six hundred.

Kandel's study of memory took a radically reductionist approach. He analyzed the activity of the smallest elements he could research—molecules, genes, neuron cells, and synapses—in the simplest organism that lent itself to this research. Kandel and other researchers have demonstrated that the process of memory creation in mammals is similar to memory creation in *Aplysia*.

Eric Kandel's research is particularly fascinating because it sits at the interface between two of the greatest scientific mysteries: gene expression and the functioning of the human brain. Gene expression refers to the process through which the information stored in the DNA of an individual cell is used to build proteins (or other gene products). Each protein then goes on to perform a specific function. But when the expression of a gene is altered, for example, by chemicals from outside the cell, the building of specific proteins can be activated or suppressed. Loosely used terms such as "hardwiring" suggest that who we are and how we behave is fixed by our genetic code and etched into the structure of our brain at birth. Gene expression, however, continues throughout life, and can be altered by experience, environment, accidents, and even random influences. While the brain has a basic structure, it is also plastic in the sense that learning and experience can strengthen or weaken synaptic connections between neurons and create new connections. These two processes are encompassed by the term *neuroplasticity*. After an accident, the brain can even "rewire" some functions from one region to another.

For humans, the creation of new synaptic connections between neurons is most evident in young children. As they mature a process of pruning takes place, and some synapses go quiet or are removed even as others are being developed. The organizing and pruning of neuronal connections continues throughout life, reflecting what is learned and what capabilities go unused. Each new connection between neurons affords unique associations. A single connection might be involved in the formation of a new memory.

A weekly diet of revelations from the frontiers of genetics and neuroscience creates the impression that the genome and the mind are about to reveal all of their secrets. Certainly discoveries are being made that will improve the lives of many people. On the other hand, a good number of those discoveries reveal that both gene expression and neuronal functioning are much more complicated than formerly understood. Each expression of a gene can be entangled with countless biochemical mechanisms, many of which have not yet been identified. Brain functions emerge from connections, including feedback loops that are little understood, between countless neurons. Neuronal connections are excited or inhibited by the chemical activity of more than a hundred different neurotransmitters. Terms such as neuroplasticity are just one way of saying that brain functioning is much less fixed than was formerly believed. In one sense, neuroscientists are making headway; in another sense, they continue to reveal layer upon layer of complexity. Even exponential acceleration in the rate of scientific discovery will not complete the "book of life" for many, many generations to come.

Appreciating the complexity is essential for making judicious decisions about which research to fund, hopes to nurture, and threats to plan for. In the best of possible worlds, every scientific pathway would be funded until all maladies have been eradicated. In our world, choices must be made. Sorting out the hype from the reality is important for setting priorities. The ability to discern breakthroughs likely to be realized from those that are receding from our grasp is key to the rational development of the emerging technologies.

Few subjects engender the passion and moral condemnation associated with research on the human genome and how it might alter human destiny. The passions aroused against any alterations to the human genome pose an especially difficult challenge for scientists and political leaders. When should such intuitions be listened to and when are they misguided? Altering the human genome is much more charged than other forms of human enhancement such as changing one's character by manipulating the brain. Tampering with genes violates deeply held convictions espoused by both religious communities, who consider human life a gift from God, and many non-religious humanists. In a curious twist, the genome has acquired intrinsic spiritual value. The gene has taken on the aura of defining human nature. Characteristics that distinquish an individual, and were once presumed to derive from the soul, are now believed to be embodied in the genome.

Beliefs about the inviolability of the gene are often based upon misunderstandings or misinformation. But they are also an expression of the profound recognition that genomic discoveries will radically alter the future of human biology and what it means to be human.

For the public, any mention of altering genetic material immediately conjures up visions of designer babies. Of course, there is reason to pause and put in place oversight on that front. But designer babies are just one overly hyped extreme, a possibility that is real and worth grappling with, yes. Yet let's not lose sight of the more mundane, but no less significant applications that warrant our attention—personalized medicine and ensuring that serious inheritable diseases are not passed on to one's offspring.

Selection and engineering of which fertilized eggs will be brought to term serve as the focal point for much of the ethical conflict. Should what is scientifically possible through essentially mechanical procedures determine who will have a life and who will not? For many religions, human choice should not be involved in determining who is and who is not born. That should be left to God's will. Atheists, as well as believers who do not think God is concerned with such matters, question whether such decisions must be left to a genetic lottery in which so much depends on whether a particular sperm cell among many happens to fertilize an equally random egg. The resolution of this particular debate will not be found in ethics or politics alone. The methods scientists develop may, in practice, be limited in the actual choices they afford. For example, atrocities that occur in the attempt to design a baby might condemn further research. In other words, the powerful narrative that suggests genetics will invent humanity's evolutionary successors could turn out to be wrong. The litany of *may, might,* and *could* bear witness to the fact that this research continues to exist in the realm of speculation.

The role genes play as a blueprint for freshly minted minds and bodies gets the lion's share of attention in ethical debate. The research directed at gene expression throughout life, such as that of Kandel, is less understood by the public and many scholars, and therefore less controversial, but certainly just as important. Indeed, the study of gene expression will have a far-reaching impact on health care if it leads to the personalization of medical treatments. It will also reveal the feasibility of tinkering with individual genes in order to create designer babies.

## THE POLITICS OF EVOLUTION

Recent debates about gene selection and designer babies are only the latest chapter in a history of controversies that help explain why people are particularly wary of genetic engineering. From the outset, Darwin's theory of evolution provoked conflict. The contention that humans and other species evolved from a shared ancestry challenged a literal interpretation of biblical texts. Darwin was well aware that the strict interpreters of the Bible in his day would reject natural selection, as it continues to be rejected by many today.

Even more important than the rejection of evolutionary theory by religious fundamentalists was its appropriation by social theorists wishing to promote notions of racial and class superiority. In their reading of evolutionary theory, those most successful in the social competition to prevail and survive are the fittest among all humans. The phrase "survival of the fittest," often inaccurately attributed to Darwin, was originally coined by Herbert Spencer (1820-1903), who used it to justify unrestrained competition. Over the past seventy years, the phrase became identified with an interpretation of evolution pejoratively labeled by its critics as social Darwinism. Social Darwinists couched dubious arguments for class superiority, racism, and ideological purity within a scientific veneer. The competition for survival among "selfish" species was elevated to a natural law that justified robber barons and imperialism.

Francis Galton (1822–1911), a half-cousin of Darwin and a fellow scientist, influenced the social history of evolution when, in 1883, he formulated the idea of eugenics. Galton contended that society would be improved through measures that promoted reproduction by people with "desirable" traits, and practices that limited the ability to reproduce among those with "undesirable" traits. By the early decades of the twentieth century, the eugenics movement was a significant political force. At its height, governments from Great Britain to Sweden and Argentina embraced eugenic measures. Theodore Roosevelt, Winston Churchill, and Margaret Sanger (a birth control advocate and founder of an organization that evolved into Planned Parenthood) all supported eugenic policies.

More benign eugenic practices include family planning, prenatal care for expectant mothers, and medical approaches for eliminating hereditary diseases. Among the controversial eugenic measures are forced abortions, forced pregnancies, compulsory sterilization, and infanticide. The U.S. shamefully pioneered legislation for the forced sterilization of people labeled as "defectives." Sixty thousand so-called "defectives" were sterilized between 1909 and the early 1960s. Many of those sterilized did not suffer from inherited conditions. Their "sin" was to have been born in poverty, to be malnourished, or to be uneducated. In writing the Supreme Court's majority opinion (*Buck v. Bell*, 1927) that upheld an Indiana law requiring compulsory sterilization of the mentally ill and others deemed unfit, Justice Oliver Wendell Holmes, Jr. famously stated, "Three generations of imbeciles are enough." The case legitimized eugenic sterilization in the U.S. and has never been unequivocally overruled. By today's standards, the subject of the suit would not be considered mentally impaired.

Adolf Hitler and his Nazi regime outdid the U.S. by sterilizing more than three hundred fifty thousand people. The mass genocide of Jews, Poles, Gypsies, and homosexuals by the Nazis discredited eugenics and alerted even the idea's advocates to its downside. The Nazi extermination program left few people willing to openly advocate eugenic policies after World War II. And yet, state-sponsored sterilizations continued in the U.S., with approximately twenty-two thousand occurring between 1943 and 1963. More recently, the emerging genetic possibilities, and how they might be used to supposedly improve human nature, are often referred to as "the new eugenics."

The pendulum in the ongoing debate regarding the relative importance of nature and of nurture on individual behavior has swung back and forth, with significant ramifications for public policy. Conservatives tend to argue that human character is relatively fixed, while those with more liberal views believe that altering the environment could improve human behavior. By the 1980s, the perceived failure of experiments in social engineering—liberal policies directed at poverty, education, and health care—opened the door for a reassertion that human nature is innate. It was argued that utopian political ideals failed because they came up against a wall of immutable traits—the inherent propensity of people to place a priority on the welfare of their children and relatives, the desire for personal property and security, and the need for incentives. At the end of the twentieth century, the pendulum had swung distinctly back toward nature as the primary determinant of character. Studies demonstrating similarity in the intelligence, beliefs, moods, and skills of identical twins raised separately were interpreted as indicating that biology superseded environmental influences. That is, genes trump experience and environment in determining who we are. Some evolutionary psychologists went so far as to argue that we are "hardwired" by our genes. Any characterization of human nature, however, is a delicate issue. Critics sensitive to the pernicious manner in which prejudice and stereotypes rear their ugly heads are concerned that the goal of illuminating innate human nature will branch off into emphasizing traits that stratify social groups and exacerbate racial and social inequalities.

The pendulum in the nature-nurture debate has not exactly reversed direction in recent years, but its ticktock is increasingly being ignored. A more nuanced understanding of how the expression of genes is activated or suppressed has led many scientists to abandon the old nature-nurture distinction. That distinction focused upon the relationship between a single plant or an independent animal and its external environment. In the new paradigm, anything that is not a gene is in the gene's environment. Outside of the cell DNA is an inert molecule. Its expression is stimulated by many factors including the nutritional content of the cell in which the genetic material is located. Individual genes can be activated, suppressed, and even altered by not only the diverse set of molecules in the cell, but also by other genetic information in the chromosome. In turn, the expression of genes will alter the cell, the organism, and (eventually) the ecosystem in which the organism is active.

The relationships between genes, other molecules within the cell, and influences from the environment outside the cell are so entangled that it can seem impossible to distinguish the activity of one from that of the others. Researchers are currently turning away from attempting to make such a distinction. Their focus is shifting toward the study of the wide range of mechanisms that regulate the expression of a gene and the development of an organism.

The scientific disintegration of the nature-or-nurture distinction is relatively new, hard to fully grasp, and therefore has not directly influenced policy debates. The effectiveness of social engineering will continue to be argued, but will increasingly be debated in the form of genetic engineering to achieve social goals. For example, the philosopher Matthew Liao speculates whether genetic engineering might be applied to address global climate change. Among the possibilities he considers is genetically altering appetites so that more people will be vegetarians. Producing a vegetarian diet leaves behind a much lower carbon footprint than a meat based diet. Shorter people who consume fewer resources offer another form of social engineering through genetics. Such proposals are not likely to get serious attention in the West but might be introduced elsewhere.

## THE GENOME

Media coverage of Kandel's Nobel Prize in 2000 was quiet in comparison to the ballyhoo that same year following the announcement that the human genome had been sequenced. Mapping the genome was a major feat. It was greeted with excitement, but also trepidation, as the public came to appreciate this as a major step in transcending natural restraints on human development. Environmental factors would continue to influence gene expression, but nature would certainly be surrendering its primacy to whatever scientists elected to nurture.

The U.S. government spent $3.8 billion on sequencing the human genome and related research. That amount was augmented in its later stages by J. Craig Venter's private company Celera Genomics, which invested $300 million for the parallel effort. Using newer techniques, Celera delivered a draft of Venter's personal genome in less than two years. Since then, the cost of sequencing the genome for one individual has dropped dramatically. In 2014, the gene-sequencing company Illumina announced that it would soon introduce the HiSeq X Ten, which would bring the cost of reading the full human genome to about $1,000.

Before the human genome was sequenced, it had been presumed that the 3.2 billion base pairs could be broken down into approximately a hundred thousand genes. However, one of the project's more surprising findings was that there appeared to be only twenty thousand to twenty-five thousand genes, and they accounted for less than 2 percent of the genetic material. The other 98 percent was considered to be *junk DNA*. However, it is now understood that a portion of the *junk DNA* serves various functions, such as turning on or turning off the expression of a gene. For example, a gene for a protein that creates the color pink might be turned on in the petals of a flower, but turned off when the cells for the plant's roots are produced.

The small number of genes was not good news. It had been anticipated that in sequencing the human genome, many diseases, particularly those identified as hereditary diseases, would be traced to specific genes, which could then be controlled by targeted therapies. The small number of genes made it evident that there was unlikely to be a one-to-one correspondence between specific genes and most individual diseases or specific human traits. The manner in which the many proteins, whose structure is laid out by genes, interact and build an organism must be extremely complex. Understanding the process through which genes and proteins assemble a body will be difficult to fathom, and will proceed very slowly.

Nevertheless, a few inheritable diseases are either caused by or correlated with the presence of a specific gene. For example, a common type of sickle-cell anemia, a blood disorder known to be caused by an abnormal form of the oxygen-carrying hemoglobin, has been traced to a specific gene. The discovery of that gene improved researchers' ability to study the underlying mechanisms by which the disease arises, but has not yet led to a cure.

Most traits are influenced by a constellation of different genetic factors and then shaped by environmental influences. Many genes contribute to the height of a person, for example, and the expression of each of these genes can be turned on or turned off by environmental influences such as the quality of an infant's diet. The average height of Europeans has increased 11 centimeters (4.33 inches) between the mid-1800s and 1980. To the best of our understanding, genes set parameters. The genetic makeup of a woman might dictate that her height will fall within a particular range, but then other environmentally influenced factors will determine to what height she actually grows within that range.

Influences, from weather to the availability of essential nutrients to how well the seed is protected from birds searching for food, determine whether a plant flourishes. Seeds with exactly the same DNA that are sown in a fertile valley will grow tall, while their sister seeds planted upon a craggy mountain peak will be more robust, but also shorter. For animals, the health of the fertilized egg or of the mother determines whether the developing fetus will be carried successfully to term. These influences were formerly understood as strictly environmental; however, in the new paradigm, the environment acts by turning on and turning off the expression of genes within individual cells.

Gene expression happens throughout life. Skin cells divide to heal wounds and to replace dying layers of skin tissue. Each new cell receives a full copy of the DNA from the dividing parent cell. Bugs in a gene's code in a single cell can result in a disease such as the growth of cancerous cells. For example, the UV rays in sunlight can damage the genetic code of a pigment-producing skin cell, which then divides uncontrollably and becomes melanoma, a deadly form of skin cancer. The study of gene expression opens up whole new fields for genomic research. Hopefully, it will lead to the discovery of therapies for interfering with the bio-molecular mechanisms that contribute to the development of cancers and inheritable genetic diseases. One downside of gene therapies lies in their use for experiments that could go awry and be harmful for the individual. In addition, the application of genomic techniques to enhance the capabilities of some people will have a societal impact that might be perceived as either good or bad, depending upon one's point of view.

## PERSONALIZING MEDICINE

A flood of editorials and articles accompanied the tenth anniversary of the sequencing of the human genome, a good number of which proclaimed its failure to produce the predicted medical cures. The critics made particular reference to a grand vision outlined in 1999 by Francis Collins, who had led the government-financed Human Genome Project. Collins envisioned that by 2010 an era of "personalized medicine" would emerge. Medical treatments for each individual might be customized for their biology as specified by their genes. We can check this off as another example of a recurring theme in the history of science: overly optimistic claims and hyped promises. So far only a few diagnoses for diseases, such as the testing for the "faulty" *BRCA* (breast and cervical cancer) genes, has emerged from genomic research. The patents for these two genes date back to 1994 and 1995 respectively, before the full sequencing of the human genome.

The dream of eventually personalizing medicine through genomics is certainly alive. Many breathtaking advances in genomic research have already occurred: notably, the million-fold (soon to be three million-fold) lowering of the cost to sequence an individual's genome. But there is considerable confusion among researchers as to the best way forward, the difficulty of the challenges to be surmounted, and the speed at which progress will be made. Perhaps *Scientific American* put it best with the title of its 2010 article: "Revolution Postponed."

A great deal of hope has been placed in sequencing the genomes of a large number of people, collecting data about their medical history, experience, and habits, and searching the full database of all participants for interesting correlations. The willingness of individuals to make their genomes, medical histories, and information about their personal habits available is quickly building a large database of useful information. And ever more powerful computers are achieving some success mining interesting correlations within that massive database.

One example of a step toward personalizing medicine is the work of Gualberto Ruaño, whose company Genomas probes whether the correct treatments for mental illness can be deduced through genetic analysis. He and his team have focused on three specific genes and their variants (alleles). These genes are important in producing the enzymes whose presence is required for specific classes of drugs to be effective. Many patients lack the metabolic pathways through which specific drugs act. A depressed patient might be prescribed an antidepressant that cannot be effective because the biochemical pathway through which that drug works is not available in that patient. In such a case, ineffective medicine could mean months and even years of suffering from mental illness. Dr. Ruaño's research indicates that for many patients the availability of the metabolic pathways for certain classes of antidepressants to work can be predicted through an analysis of the alleles of these three genes. All that is required is a genetic test of a blood sample. However, genetic tests are expensive and insurance companies have been reluctant to foot the bill, fearful that they will be stuck carrying the cost of unproven research. Among the hurdles that must be surmounted is the need to upgrade insurance companies' methods of determining which tests are demonstrably beneficial, and which are highly speculative. Insurance companies, legal theorists, and researchers such as Dr. Ruaño recognize this issue. They have been working to forge methods that will ease payment for genetic tests, while protecting insurance providers from the expense of unwarranted and speculative fishing expeditions.

A lot of suffering could be avoided if, for example, chemotherapy treatments for each cancer could be matched to treatments effective with other patients who have a similar genetic profile. Progress has been made in that direction toward treating a few cancers. The drug Gleevec, used to treat chronic myeloid leukemia (a cancer involving the overproduction of white blood cells), relies upon a genetic test to determine its suitability. Yet even if we do a better job in selecting the appropriate treatment, chemotherapy will remain a painful and degrading means to attack cancer. One longer-term dream is to design a synthetic virus that attaches to cancer cells and alters their DNA in a manner that will stop the tumor from growing. This is theoretically possible, although whether or when such an approach will be realized is unknown. There have been many blind alleys in the search for cancer cures. A recent report that each individual's cancer has a unique genetic profile complicates the search for effective cancer therapies.

Thousands of research projects like Gualberto Ruaño's will need to bear fruit before a more comprehensive, personalized genetic approach to medical care can become a reality. *Techcast*, the scenario prediction service developed by William Halal at George Washington University, asked its panel of experts to forecast the year when "individual genetic differences are used to guide 30 percent of medical treatments." Forty-one experts offered their assessments, which averaged out to the year 2026. The experts had a 63 percent confidence level in their assessment. When asked for a year when "genetic therapy is used to cure 30 percent of inherited disease," the years selected by forty-six experts averaged out to 2031.

One concern that will delay the arrival of genetic differences contributing to selecting medical treatments is the fact that countless millions of people want to keep their genetic profiles secret—some for valid reasons, and some because of scary scenarios that are plausible but unlikely. A valid reason involves the fear that if an employer gets access to that information, a boss might use it to terminate one's job. A small danger in personal genomics lies in the fact that knowledge of anyone's genetic information can also be used for harmful purposes. *The Atlantic*'s rather melodramatic article titled "Hacking the President's DNA" explained how knowledge of the president's genome might be used to design a virus that would attack only him or her. In the scenario, a location where it was known that a president was going to visit would be seeded in advance with a customized virus. This virus could easily be transmitted to many people, but it would cause none of them harm. The customized virus would harm only the president or family members with similar genes. Whether such biochemically inspired assassination attempts will occur any time soon is difficult to evaluate, but probably low.

The conceivable use of a customized virus for ethnic cleansing provides an especially despicable scenario. Racial divides continue to foster heinous crimes, particularly during civil wars. However, ethnic divides are seldom embodied in distinct genetic profiles. The genocidal conflicts between Hutus and Tutsis in Rwanda, for example, cannot be accounted for by their close genetic kinship. Applying genomics for ethnic cleansing would fail because the harm is likely to go well beyond the targeted group and even strike members of the ethnic community instigating the violence. Unfortunately, nuanced details like this never stopped anyone from trying ethic cleaning in the past, and undoubtedly will have little impact in the future.

## FROM TEST TUBE BABIES TO DESIGNER BABIES

Louise Brown, the first "test tube baby," was born at Oldham General Hospital, U.K., on July 25, 1978. The arrival of this 5-pound, 12-ounce baby girl signaled a ground shift in the future course of genomics. Louise's parents were unable to conceive. She was conceived in a petri dish where sperm from her father fertilized an egg from her mother. Since Louise's birth, *in vitro* fertilization (IVF) has been a boon for parents unable to have children by natural means. In retrospect IVF is seen as benign by most observers. So it can be difficult for us today to envision that from the outset, IVF was a controversial procedure that conjured up visions of a brave new world of lab-created babies.

Assisted reproductive technologies (ART) such as IVF have been a central focus for the ire of conservatives uncomfortable with biotechnologies and the direction of science. The Catholic Church has outright condemned IVF. Leaders of many Protestant denominations came to embrace the practice, but for anti-abortion advocates the acceptance of IVF had an unanticipated consequence—the unused embryos stored at fertility clinics are a ready source of embryonic stem cells. Stem cell research, however, is a somewhat later chapter in our story, as the first stem cells would only be isolated in 1998.

By the 1990s, it became possible to screen genes from embryos in the lab in order to determine whether any resulting baby might be susceptible to high-risk, genetically transmitted diseases such as cystic fibrosis or Huntington's disease. This procedure, known as preimplantation genetic diagnosis, or PGD, is lauded for ensuring that babies will not have serious genetic disorders carried by one of their parents' genes. Once an embryo is selected it can be transferred to the mother's uterus in hopes of establishing pregnancy. But PGD is highly controversial because of its eugenic implications. It might also be used to select a trait the parents believe would enhance the success of their child in, for example, a sport such as basketball or in academic pursuits. Given that only some traits are single gene specific, selecting which of the parents' traits a child might actually acquire is easy to imagine but difficult to achieve. However, an achievable goal and a particular concern has been the use of PGD for sex selection.

The availability of technologies for predicting the sex of babies has had societal ramifications. In many Asian cultures, having sons is the closest parents will come to any form of social security for their old age. Daughters will marry and join their husband's families, while the boys are relied on to care for their elderly parents. With technology available for selecting the sex of babies, a disproportionate growth of male babies in a number of Asian societies ensued. Early in the 1990s there were 116 male infants for every 100 females in South Korea. The public policy ramifications of this shift are profound, such as what to do when one in six males are unable to find wives. Shifts in demographics hold the potential for considerable unrest. In South Korea, a government campaign somewhat successfully reversed the trend, although there are still 5 percent more male than female infants.

If designer babies are scary, what of cloning humans? Ira Levin's science fiction novel *The Boys from Brazil* was adapted into a 1978 film starring Gregory Peck as Josef Mengele, the infamous Nazi concentration camp doctor. The plot centers around ninety-four clones of Hitler created by Mengele and the murder of their fathers when the boys reach the age of thirteen, the age when Hitler's father died. The film alerted the public to the prospect of cloning humans and stimulated a rigorous debate on the ethics of cloning. That debate continued for two decades leading up to the 1996 birth of the first cloned sheep, Dolly. Dolly demonstrated that cloning was not just a hypothetical possibility. The scientists at Edinburgh's Roslin Institute named Dolly after the entertainer Dolly Parton (partially known for her ample endowment), because the original source of her DNA was from the mammary gland of the adult ewe. Once the technology for cloning a mammal had been developed, cloning humans went from science fiction to a serious possibility.

The Edinburgh team had 276 failures before they were successful in cloning Dolly. The dangers of unsuccessful pregnancies, stillborn babies, or children with birth defects gives cause enough to prohibit human cloning experiments. However, the very idea of human cloning, not just its dangers, incited an uproar, particularly in the U.S. The George W. Bush administration led a campaign for a United Nations convention against all forms of human cloning. They were not successful and settled for a less binding declaration against human cloning passed by a majority in the UN General Assembly. The United Kingdom was among the countries that opposed the declaration, because it placed restrictions on cloning for not only reproduction, but also for research purposes.

In 2005, South Korean researcher Dr. Hwang Woo-Suk claimed to have developed a technique that made it relatively easy for him to clone human cells from eleven patients. Initially he was treated as a national hero. However, an investigation uncovered that Hwang's claim was fraudulent and he became a source of great shame for South Koreans. Nevertheless, the case underscored the likelihood that someday, somewhere, researchers will probably clone a human even if it is a violation of the law.

Dolly's birth fanned political and ethical debate regarding emerging genetic possibilities, but the debate burst out into a political firestorm two years later when a University of Wisconsin lab isolated human embryonic stem cells. It had long been theorized that embryos produced a master cell from which the two hundred or so different cell types in the human body emerge. The discovery opened the doorway to a new field of regenerative medicine that offered the possibility of healing diseased and aging tissue using a fresh source of new cells. Some scientists even suggested the possibility of regenerating lost human limbs in a manner similar to how some species of reptiles regrow their tails.

The primary sources of stem cells for research are the embryos stored in fertility clinics. But anti-abortion activists considered the use of embryos morally unacceptable, and moved to galvanize political support to stop further research. For social conservatives in the U.S. and elsewhere, the status of fertilized human eggs is such that they should be treated as inviolable. Initially, merely hearing that there was an ethical issue with embryonic stem cell research made U.S. citizens uncomfortable. When it became clear that the ethics centered on whether an embryo should be designated a "person," it also became evident that the controversy over stem cell research was essentially the abortion debate in a new guise. A majority of the public, which already considered many types of abortion acceptable, also lined up in support of stem cell research. Even some anti-abortion advocates decided that it was acceptable to appropriate unused embryos for research that had the potential to save lives. For example, the anti-abortion former First Lady Nancy Reagan supported federal funding of stem cell research because it might eventually lead to a cure for the Alzheimer's disease that was afflicting her husband, ex-President Ronald Reagan. Support from actors such as Christopher Reeve, who became a quadriplegic after an accident, and Michael J. Fox, suffering from Parkinson's disease, provided a human face for the campaign to federally fund stem cell research.

Beyond the concerns raised by groups that perceive the use of embryonic stem cells as a form of abortion and a violation of human dignity, there is also concern that subjects used for experimental stem cell research might be harmed. In 2009, the biopharmaceutical company Geron was given permission by the U.S. Food and Drug Administration to begin human trials using embryonic stem cells on victims of spinal cord injuries. Even advocates of stem cell research quietly expressed concern that if anything went awry with Geron's exploration of stem cells to repair spinal injuries, further use of human subjects in related research could be set back. Three years later Geron discontinued all its stem cell research programs. There is no evidence that anything went wrong. The Board of Geron merely made a business decision to not spread the company resources too thin, and to focus on cancer research.

Interestingly, the controversy over funding of embryonic stem cell research by the U.S. government served as a catalyst for states such as California, Illinois, Massachusetts, and Connecticut to fund stem cell research on their own. These states hope to get an edge in what is anticipated to be an important and lucrative field. It is possible that, because of the controversy, a larger slice of the total research funding pie from governments and foundations went to this field than it would have otherwise received.

Stem cell research is progressing in the U.S., Europe, and elsewhere, even if at a much slower rate than the more optimistic advocates had predicted. An important result of the controversy over the use of embryonic stem cells was a significant focus on using so-called adult stem cells, which can be found in everyone's body. Adult stem cells play a role in the repair of living tissue. Because these stem cells are derived from adult tissue, not embryos, their use in research was perceived as a means of bypassing controversy. However, leading scientists initially argued that research with embryonic stem cells would progress much more quickly, and vehemently questioned the viability of achieving the desired ends with adult stem cells. Nevertheless, the usefulness of adult stem cells for specific therapies has proceeded much more rapidly than many experts had imagined. In addition, adult stem cells offer an advantage in that if the donor cells are from the very individual who will be treated, the new tissue will have the patient's DNA, and therefore it is less likely that the patient's body will reject the tissue. This success has been hailed by those critical of using embryonic cells as vindication of their cause. The story should not be misunderstood as suggesting that science be guided by religious values. However, scientists, like their fundamentalist critics, need to be more circumspect regarding the validity of their own prejudices. Nevertheless, the verdict on whether adult stem cells can be used as broadly as embryonic stem cells in therapeutic applications will require considerably more research. Stem cell research remains a very young field.

## INFLECTION POINTS IN GENOMIC RESEARCH

Dolly, Louise Brown, and the use of embryonic stem cells for research purposes each offered an inflection point for public dialogue and action on policy. A surprising new incident that plucks at our heartstrings can always emerge. But the probable next inflection point will be the announcement of success in cloning a human. The character of the public conversation triggered by that event will be largely determined by the circumstances leading up to the cloning. If there are stories of malformed fetuses, stillborn babies, and infants with congenital diseases delivered along the way, widespread outrage and condemnation will spread beyond cloning and stigmatize other forms of genetic research. This might be an outcome that some who consider cloning immoral would pray for. But hopefully, the first human clone will not load the dice either positively or negatively toward cloning, and the birth will be cause for a full and far-reaching dialogue into the societal ramifications of more dramatic approaches to assistive reproductive therapies.

The prospect that many attempts to engineer genetic enhancements will go badly provides responsible governments with sufficient justification for regulating research. But who takes responsibility for failed research and the care of children bred with congenital diseases? Society? Researchers? Insurance companies? The wealthy prospective parents who funded the research? That question is far from settled and it will force a policy debate once human clones start to appear at an increasing rate. The likelihood of malformed bodies and other congenital conditions will prompt many parents to shy away from radical methods for designing their children. In combination with hesitation by individuals about having their DNA sequenced, fear of producing children with genetic abnormalities will slow down the pace of research. But eventually, the techniques and likelihood of success will improve.

Given that a human clone is very similar to an identical twin, the person created by cloning will certainly be worthy of full human dignity. For this reason, the furor over a single human clone is misplaced. The real issue is the motivations of the individuals who want to make clones of themselves or someone else. Indeed, many of the purposes for which someone might want a clone are self-centered and do not take into full account any psychological or physical harm to the newly created person. Why would someone want to clone themselves once or dozens of times? Would having multiple twins or growing up with the knowledge that one is the clone of one's father be a particularly disturbing prospect? Unfortunately, even if the reasons for cloning are immoral, it is not easy to criminalize motivations. Society will have difficulty in criminalizing anything other than specific harm to the newly created person.

Introducing the genetic material of nonhuman animals into human embryos is either illegal or generally considered immoral. But, like human cloning, that offers no guarantee that it will not be tried. All the arguments against human cloning apply here, as well as the greater likelihood of causing pain and suffering to the resulting child. Advocates believe it may be a way of introducing useful animal traits into humans, for example, the vision of an eagle or the hearing of a dog. These human chimeras—creatures that are largely human but also something else—conjure up visions of mutants with novel capabilities like the X-Men. There is, however, no guarantee that human chimeras will have desirable traits. There is only an invitation to a brave new world of genetic possibilities and the predicament of whether to accept that invitation.

In the flowering of genomics into many fields of research, there are both new techniques that are available now and additional possibilities which remain dependent on research breakthroughs. From 1980 to 1999, a woman who wanted a child could go to the Repository for Germinal Choice in California and (at no charge) have her eggs fertilized with sperm donated by successful scientists, businessmen, artists, and athletes. But predetermining more than a few of the traits your child will have remains an imaginable but more distant possibility.

In an often-repeated story, the dancer Isadora Duncan (1877–1927) wrote the playwright George Bernard Shaw (1856–1950) that:

> As you have the greatest brain in the world, and I have the most beautiful body, it is our duty to posterity to have a child." Whereupon Mr. Shaw replied to Miss Duncan: "My dear Miss Duncan: I admit that I have the greatest brain in the world and that you have the most beautiful body, but it might happen that our child would have my body and your brain. Therefore, I respectfully decline.

Shaw later denied the tale. Similar stories have been attributed to others including Albert Einstein. There is even a version of the story in which the baseball slugger Joe DiMaggio uttered comparable lines to his soon-to-be wife Marilyn Monroe, although DiMaggio does not claim to be brilliant. But with the emerging genetics it is conceivable (pun intended) for a baby to have Monroe's beauty, DiMaggio's athletic prowess, and Einstein's brains.

There is no question that tinkering with the human genome has begun. This in itself is more than enough to offend the sensibility of bioconservatives. However, the fanciful notion of designing one's baby to order remains more hype than reality.

The limited forms of assisted reproductive technology (ART) available today are likely to progress in ways that will improve techniques for selecting a few specific traits, but are unlikely to result in the ability to fully customize one's children in the foreseeable future, if ever. Major technological and ethical hurdles exist which may thwart wholesale experimentation to dramatically enhance the genome of embryos that would be brought to term. There are already limits in place on the kinds of research that can be performed on human subjects, and, in particular, on infants.

Aside from the scientific and ethical hurdles, there's a very real question of how to regulate the engineering of the human genome and to what degree regulation is possible. The policing of reproductive therapies is not viable for democratic societies committed to individual freedom. As with the experimentation on nonhuman animals, efforts to police will only push research underground where safe practices cannot be ensured and monitoring is impossible. Experiments can always be moved to countries with low research standards. The wealthy will be able to find places and purchase researchers to explore their fantasies. Nevertheless, just as there were many failed experiments before Dolly was cloned, there will be many failed attempts to engineer children with significantly enhanced capacities. The possible horrors of these failed experiments are likely to be publicized, and will in all probability raise even further barriers to any continuing efforts to produce designer babies. Efforts to hide research will not stop revelations when something goes wrong, and once the tragedy comes to light, the reaction will be much more extreme.

So how to proceed when over-policing forces researchers underground, and hiding experiments won't serve scientists' long-term goals? How to ensure that a dialogue takes place between the experts developing genomic technologies and the public concerned about their impact? Can that exchange be meaningful without necessarily quashing new technologies, nor allowing them to slip beyond our control?

Genomic enhancements will eventually be common. That won't happen overnight. Asked to pick a year when "30 percent of parents select some genetic traits of their children," the answers of thirty-nine *Techcast* experts averaged out to 2035. Preimplantation genetic diagnoses will be used increasingly to select for one or two specific traits or capabilities.

If parents do not get the children they select for, will the children they do get receive their love? Will children resulting from parents' decisions to optimize one capability or another be stigmatized by the larger society? Hopefully not. What will be the relationship between genetically selected children and their siblings who were the result of the traditional genetic lottery? Enhanced children, their siblings, and peers will be confronted with surmounting psychological scars that were unknown to previous generations. Whether this will strengthen their character or undermine the quality of their lives will vary from person to person.

The larger society can easily assimilate genetically selected individuals, presuming that their capabilities fall within the range that other citizens acquire naturally. However, superior or superhuman capabilities pose more serious societal threats. The most evident threat is to the sports industry, which could easily suffer the fate of other industries that have been decimated by technological advances. More serious will be the threat superior capabilities pose for the legitimacy of democratic society. Those from well-to-do backgrounds are already bequeathed with opportunities that others are not afforded. Nevertheless, today even the disadvantaged know that with effort and luck they can succeed. How might future communities be altered if it is perceived that the winners are all genetically selected progeny of wealthy parents?

These last two chapters have explored the societal and ethical challenges posed by advantages in genomics with much greater detail than will be given to other fields of research. For our purposes, technologies for understanding and altering genes serve as an example of the array of different concerns arising from just one field. Those specific issues range from public health to environmental risks, and from the safety of experimentation with human subjects to the far-reaching societal impact of the ability to engineer a few of the characteristics and capabilities infants will possess. The pace of discovery in genomics progresses at a rapid and accelerating rate. Nevertheless, much of what we are learning is that the mechanisms through which genes operate are bewildering in their complexity. That complexity will moderate how quickly researchers even begin to tackle more speculative possibilities, such as preselecting a large number of an infant's features. Therefore, time remains, and inflection points can emerge which afford opportunities to grapple with the ethical dimensions of engineering humans. Unanticipated and unpredictable incidents will occur as scientists and medical researchers tinker with the complex mechanisms through which genes act. There may be a few positive, serendipitous discoveries, but if a public health crisis occurs or too many malformed animals or children with undesirable mutations appear, there will be strong public pressure to heavily regulate further research. The benefits of genomic research are so great that the risks will certainly not block significant progress. Nonetheless, the assessment of trade-offs will constantly be readjusted as both benefits and harms manifest themselves.

GMOs, human genetic manipulation, as well as many other emerging technologies should be understood in terms of their complexity and the pace of change in their respective fields. Simply giving in to the hype and doomsday scenarios imagined by many in the media is insufficient. As each technological approach unfolds, there will be actionable inflection points and opportunities to assess the trade-offs between the benefits and risks of various courses of action.

The complexity of the mechanisms through which genes operate has not in any way diminished the popular belief that the wholesale creation of superhumans is right around the corner. Over the next twenty years, however, genetic manipulation is likely to play a smaller role in enhancing humans than other innovations, such as drugs to improve cognition, prosthetic devices that forge a more intimate relationship between people and their machines, and biomedical technologies that extend life span. Chapters 8 and 9 will elucidate some of the checks that will need to be put in place to effectively manage a number of these technologies and their potentially detrimental impact.

# { 8 }

# Transcending Limits

MY CELL PHONE RANG TEN MINUTES BEFORE THE MEETING OF Yale's Technology and Ethics (T&E) study group was scheduled to begin. The call was from our guest presenter, Dr. Martine Rothblatt, an attorney, a visionary, and an extremely successful entrepreneur. She and her driver were lost somewhere in New Haven. I gave them the necessary information and offered to wait outside and guide Dr. Rothblatt to our meeting room. As I rushed out to the street, the word "she, she, she" repeated over and over in my mind. Looking for a limousine or similar vehicle, I barely noticed a huge Freightliner limo bus until it pulled up beside me. The door swung open and Martine came bounding down the stairs. We greeted each other warmly, and headed to the seminar room, as my mind continued to silently repeat the word "she."

Martine Rothblatt has been a founder and CEO of several satellite communications companies including Geostar, PanAmSat, WorldSpace, and Sirius. She then created a biotechnology company, United Therapeutics, in order to test and market an orphan drug that could save the life of her twelve-year-old daughter Jenesis, who was suffering from a lung disease called pulmonary arterial hypertension. Martine is a tall, attractive woman with long brown hair usually tied back, and hazel eyes. Yet one senses an ambiguity about her gender. You see, Martine was once Martin. She is transsexual.

In my mind, Martine registers as a male, and it takes intense effort for me to override this identification. On more than one occasion, I have referred to Martine as "he"— once in her presence, but she generously didn't bat an eyelash. In announcing Martine's coming presentation at an earlier meeting of the T&E group, I debated whether or not to tell the members that she is transsexual. I did tell them, as this is one aspect of what makes her and her research so fascinating.

Martine is also a transhumanist. Transhumanists believe that the limits of the human condition can be transcended through technological means. These include enhancements of the body and mind and radical life extension. The transcendent beings that inhabit transhumanist dreams will truly take evolution beyond biology into the emergence of a new species, the *techno sapiens*. Martine is not the best-known transhumanist. Nick Bostrom, James Hughes, Max and Natasha Vita-More, and Aubrey de Grey are among the leading public advocates for technological transcendence. But Martine is uniquely positioned in that she is actively nurturing therapeutic medical research that will help the needy today while laying the foundations for a transhumanist future tomorrow.

Among Martine's many interests is cryonics, the possibility of freezing a body immediately upon death and then reviving the person at some future date when medical science is more advanced. She serves on the science board of the Alcor Life Extension Foundation, one of the best-known cryonics companies. Alcor favors vitrification, rather than simple freezing, as a method for preserving a body or head. Before the organs are deep-cooled to minus 124 degrees centigrade (minus 191 degrees Fahrenheit), as much as 60 percent of the water in the cells is replaced with chemicals. The head and body of Ted Williams, the Boston Red Sox legend and last baseball player to bat over .400 (actually .406) in a season, are cryonically preserved at Alcor's facility in Scottsdale, Arizona.

As a field of research, cryonics resides somewhere between science fiction, plausibility, and possibility. Deep cooling has proven a useful technology for a variety of medical procedures, including the preservation of organs between harvesting them from recently deceased donors and transplanting them into a needy recipient. However, to date, no one has frozen and revived any form of life more complex than the tiny (1 mm long) *C. elegans* roundworm. Nevertheless, scientists are dedicating a lot of energy to discovering how organs (including the brain) and the full body could be vitrified and stored, as well as the technologies needed to revive simple, cryonically preserved organisms. One intriguing proposal involves the injection of thousands of nanobots, capable of repairing cells damaged by disease or death, into the body while it defrosts to enable a successful revival.

The possibility of preserving a human body in a cryonic state raises a host of legal, ethical, and religious questions. Most are speculative at this point, but a few have real and immediate ramifications. Is a cryonically preserved person dead? For so many of us, the answer to that question is an obvious and resounding "Yes!" But consider that only fifty years ago anyone whose heart and respiration had stopped would have been declared dead. Around that time, advances in medical technology gave us the ability to revive increasing numbers of people who suffered cardiac arrest. A new definition of death was needed. Much would depend upon a new definition. For example, which patients should be maintained on life support equipment? When might organs including kidneys and other body parts such as corneas be removed for transplantation to those in need? When would legal authority and property rights be passed on to heirs or successors?

By 1968, a definition of death as the irreversible stopping of brain activity began to emerge. In 1981, a U.S. presidential commission issued a landmark report declaring that the "whole brain" must stop functioning, not merely the "higher brain."

The case of Terri Schiavo, a woman in a persistent vegetative state, underscored problems inherent in this definition. During a seven-year legal struggle, Terri's husband Michael worked to disconnect her from life-sustaining technology including feeding tubes. Pro-life and disability rights groups entered the battle, as did Terri's parents, the Florida Supreme Court, Florida Legislature, U.S. Congress, President George W. Bush, and the U.S. Supreme Court, before a court order to disconnect Terri was carried out on March 18, 2005. Now consider that even a minimal improvement in Terri's condition might have altered the disposition of this case. How much brain function is necessary for those in vegetative states to be judged worthy of all human rights, including the right to be sustained on life-support, regardless of the psychological and financial cost to their family and to society?

Those in the cryonic movement view death as a process. Their technicians need to get to a dying person after the heart has stopped, but before brain activity ceases and brain tissue deteriorates. If the brain tissue is preserved, then the possibility remains open that the person's life could be restored with future technology. How long will the body need to be preserved? How long before scientists develop life-restoring technology? No one knows.

In the meantime, what will be the legal status of the cryonically preserved? Is a vitrified body a corpse? Is it a person? Is it a *being* in a suspended state? These were the questions that Martine Rothblatt wanted to explore with us on her November 7, 2007 visit to the Technology & Ethics (T&E) study group. She directed both her training as a lawyer and her biotechnical expertise to explore a new, broader definition of legally being alive. Martine's goal is to develop a framework that will give judges a legal basis for treating cryonically preserved individuals as people in biostasis rather than as corpses.

The legal status of vitrified bodies might seem bizarre or unimportant to many readers, particularly those who don't believe the dead, with the exception of Lazarus or those caught up in the Rapture, will ever be revived. But consider this possibility. Your father has purchased a policy from Rudi Hoffman, a financial planner and independent insurance broker from Port Orange, Florida who specializes in such matters, that guarantees funding for preservation of your dad's body by Alcor. Furthermore, Rudi, who has over a thousand such clients, helped your father place all his property in a trust for his future self once he is revived. You would probably want to challenge the trust, not to mention your father's sanity.

A court's decision on whether the trust can be broken may well depend on the prevailing interpretation of when a person is dead. Furthermore, the legal disposition of your challenge could vary depending upon the laws of a particular state or country. Whether or not your father can ever be revived will be secondary to the fact that legal interpretations of when a person is dead could alter your fate, or at least your inheritance. If a legislature were to pass laws that could be interpreted as favorable to the rights of those wishing to be preserved through cryonics, assets and freezer cars will be headed to that jurisdiction.

Or consider that a financially strapped cryonics firm wants to break its contracts with people it preserved and pull the refrigerator plug. A court might decide that without the legal status of being alive, bodies in biostasis can be moved to cemeteries or crematoria. People will differ as to whether they perceive such a decision as appropriate, gruesome, or even inhumane.

The legal quandaries that accompany technologies for transcending existing limits on the human condition provide fodder for fascinating debates. However, the many ways in which transhuman technologies will also have a far-reaching impact on society is of much greater importance. The pursuit of enhancements, super-intelligence, and radical life extension could potentially restructure human existence and force changes in the way governments and economies are presently organized. Depending upon one's perspective and values, those changes can be judged beneficial or destructive. An intellectual battle over the positive and negative consequences of transhuman goals is well under way. Those whose aspirations converge with transhuman projects emphasize how they are the natural extension of humanity's pursuit to improve health and the quality of life. Critics point out ways transhumanism could decimate cherished values and institutions.

The pursuit of transhuman technologies poses a dilemma similar to the case with which this book opened: whether the Large Hadron Collider should be turned on. In that instance, physicists made a decision to go ahead with a technology that had a theoretical potential, however small, to end human existence. They determined that the theory was false and there was no risk. In the case of transhuman technologies, a small segment of society, in pursuit of its aspirations, could introduce possibilities whose impact on the broader society would not only be far-reaching, but also, if some of the more dire predictions turn out to be accurate, destructive.

When should a segment of society have the right to pursue goals that might radically alter humanity? Can the larger society have some input into which goals should and should not be pursued? And can it do so without destroying the freedom of individuals or of a large group of citizens?

## TECHNOLOGICAL ROUTES TO IMMORTALITY

The dream of immortality is nothing new. It is central to most of the world's great religions, although its realization differs from faith to faith. Christians view the immortality of the soul as a reward after death for having lived a good and faithful life. Hindus believe the reincarnating soul, or atman, is already more or less immortal and even the body can be transformed into immortal substance through intense spiritual practice. What has changed is perceived pathways to immortality through the wonders of modern science.

In addition to cryonics, a growing number of transhumanists, including Max More, Ralph Merkle, and Martine, are advocates of mind uploading as another scientific pathway that might lead to techno-immortality. Many forms of mind uploading have been theorized, but the basic idea involves a person being able to reproduce her mind within a computer. She could then live another life in the real world through a robotic body or within a computer roaming the many worlds of cyberspace. In effect, mind uploading, if possible, could allow one's mind to transcend the limitations placed on our lives by those pesky biological organs that age, decay, and die.

Indeed, software for people to create and store mind-files of their experiences is among the projects Martine sponsors. To quote her website:

A mind-file is a set of digital reflections of your mannerisms, personality, recollections, feelings, beliefs, attitudes, and values. We believe that future technology will permit our mind-files to achieve conscious auton-omy. This will occur when mindware is developed that tunes itself to each person's unique consciousness based upon an analysis of their mind-file. Thereafter the mindware becomes a personalized operating system. The personalized mindware, together with your mind-files, is an analog of yourself. We call this analog of yourself your cyberbeing and we call its spirit your cybersoul.

Martine is busy creating her own mind-file. A mutual friend tells of an occasion when she was dancing by herself on the village green during a fall foliage festival in Vermont. In her outstretched arm she held a movie camera directed at her face. The camera was recording a family memory, but the video would also be placed in her mind-file.

In 2007, Martine commissioned David Hanson to make a robotic head of her wife Bina. The BINA48 is capable of engaging in a crude conversa-tion, which it does through advanced software and the real Bina's mind-file. No one would mistake BINA48 for a living person. Many of BINA48's re-sponses to questions are humorous and filled with non-sequiturs. Neverthe-less, this talking head provides a fascinating demonstration of what can be accomplished with present-day technology and stirs the imagination as to whether mind uploading has a future. Martine refers to BINA48 as, "the Kitty Hawk of artificial consciousness."

Reviving bodies that have been frozen and uploading one's mind into a computer system has long been the stuff of science fiction. What has changed is the growing legions of people who not only believe such technol-ogies are plausible, but also that they are likely to debut within their life-times. The Russian billionaire Dmitry Itskov proposes that within the next five years he will have created a robotic copy of a human body that can be remotely controlled with a brain-computer interface. He calls this an *avatar*. Sound familiar? By 2025, he expects to be able to transplant a person's brain into the avatar after the rest of their body dies. A third stage, transferring a human's personality after death into an avatar, will, in Itskov's plan, be ac-complished before 2035. Finally, a holographic avatar modeled on an indi-vidual will be possible by 2045.

In addition to cryonics and uploading, regenerative medicine and radical life extension offer one more technological path to immortality. This route is closely related to traditional medicine, which doubled life expectancy during the past 150 years. The average life expectancy for a man born in the U.S. in 1850 was 38.3 years and for a woman was 40.5 years. By 2007, life expectancy in the U.S. had doubled to 78 years, and according to the World Health Organization, was 79.8 in 2013. These figures factor in infant mortality, so the average living adult will survive into her eighties. Life expectancy worldwide progressively grows as the transformations wrought by the germ theory and sanitation revolutions spread and medical breakthroughs accumulate. Even today, the universal adoption of one modest technology, the bar of soap, could save the lives of 1.5 million people per year worldwide. For countries that are in the midst of instituting modern sanitation and medical practices, life expectancy increases rapidly. For more medically advanced countries, the average life will continue to grow longer through a collection of technologies that will replace or regenerate failing organs and dying tissue. The forecast from the assessment of forty-eight *Techcast* experts was that lab-grown and artificial organs for the replacement of major body parts (kidney, heart, limbs, etc.) will be available around 2026.

The most radical vision for life extension, however, is the proposal that we could halt aging itself. The longest life ever recorded was 122 years, and has been presumed to represent a ceiling on life expectancy. But if the biochemical mechanisms that contribute to aging can be arrested, then 150 years or much longer is conceivable. The leading prophet of the vision that we can defeat aging is Aubrey de Grey, a tall, lanky, gnomish-looking Englishman with a long reddish-brown beard. De Grey proposes that we attack dying as a disease. He has advocated tirelessly for radical life extension and was initially either dismissed or sharply criticized by the scientific community. But as the idea has caught on, more and more scientists are launching research projects directed at defeating at least some of the mechanisms that contribute to aging. One speculative possibility has nanobots roaming through the human body to clean up pathogens, malformed proteins, and other debris that contribute to aging. They might also repair damaged genes that cause mutations leading to cancer.

"Curing" cell senescence, a phenomenon where cells are no longer able to divide, is an especially intriguing proposal. Without continuing cell division there is nothing to replace dying cells. Each time a cell divides, the protective telomeres on the ends of the chromosomes shorten. This discovery, which dates back to the late 1970s, offers one insight into why DNA deteriorates and cells stop dividing as we age. It may lead to a "cure" for cell senescence.

What is fascinating about research directed at slowing, if not actually stopping, the aging process is the breadth of mechanisms involved and number of pathways being pursued. While religious paths to immortality focused upon perfecting the soul or whole person through character development, scientific paths increasingly focus upon the biology of the cell and replacing dying tissues and organs.

One danger to extending life expectancy stems from the likelihood that medical science will find more and more ways to keep bodies alive, even as minds deteriorate. This could lead to the warehousing of millions upon millions of *living* people whose mental life is greatly diminished or even non-existent. Keeping people with limited or no mental life alive already confronts society with a challenge that creates emotional and financial burdens for loved ones. Multiplying that burden many fold could easily tax the emotions and resources of coming generations. Nonetheless, in responding to that critique, de Grey emailed me that this argument "for hesitation in developing such therapies . . . has parallels with past medical advances. A major contributor to the magnitude [of] today's Alzheimer's epidemic, for example, is the success that we've had in postponing heart disease, allowing so many more people to live long enough to get Alzheimer's. Yet, we do not regret making those heart disease advances."

In the Bible, Methuselah lived 969 years. And in the George Gershwin musical *Porgy and Bess,* an impish figure named Sportin' Life doubts that any "gal will give in" to a nine-hundred-year-old man. Nine hundred years may not be immortality, but a nearly 11.5-fold increase over present life spans will seem like forever. Extending life may only occur incrementally and not radically. Nevertheless, each additional year extends the possibility of surviving until reviving a cryogenically preserved body, uploading, or one of the other routes to immortality is perfected. De Grey has predicted the first person who will live one hundred fifty years is alive today, and a person born within the next twenty years will live to a thousand. In his vision, that person will not have aged dramatically, so there will be no need to fear a loss of sexual attractiveness as Sportin' Life warned.

Robert A. Heinlein's science fiction novel *Methuselah's Children* (1941) describes a future civilization destabilized by the discovery of the Howard Families, whose members have very long life spans. The larger society believes the members of the family have a secret that they will not share, while we readers are made to understand that longevity was evolutionarily bequeathed and helped along by selective breeding. But regardless of why some people live longer, the novel's plot hinges on the negative effect this discovery will have upon society.

Living longer will be a by-product of medical research, but if we are to make radical life-extension a priority, as Aubrey de Grey proposes, we need to think through the societal ramifications. Living longer will alter our quality of life and disrupt the social fabric in countless ways. What will life be like when it is no longer punctuated by the anticipation that it is finite? Will people who live hundreds of years find meaningful activities, or will they be bored? Can the planet sustain a population that doubles *ad infinitum*, adding billions and billions of people who will tax resources and struggle to prevail over each other? On the other hand, will there be unrest if radical life extension is only available to the rich and the powerful? Aubrey de Grey and other advocates for radical life extension respond to these and other questions, but some of their answers seem glib and unsatisfactory. For example, they blithely proclaim that technology will enable the expansion of the resources needed to sustain the rapid population growth that results from people living longer.

They also contend that the discoveries and technologies leading to longevity will be made universally and freely available. That is easy to propose, but it is very hard to imagine that this would be the case. Corporations will be unwilling to give up their patent rights. Or governments will be unable to afford the universal distribution of each and every expensive life-extending technology. The wealthy will not surrender their privileged opportunities and will therefore adopt the latest proven means to live longer, regardless of the cost.

Should radical life extension be made a goal by policy planners who make decisions regarding government investments? If it becomes necessary for governments to make a choice as to the primary goal for medical research, I personally would prefer that we first improve the quality of life for all humans on the planet. Everyone living happily and healthy for eighty-five to ninety years would also be less socially disruptive than radical life extension for a small minority. But others contend that our society can easily pursue both goals.

The attraction of immortality is sometimes attributed to a fear of death. Ray Kurzweil, the inventor and leading prophet of exponentially accelerating technological development, admits this is a driving force in his life. His critics charge that it drives his irrational prediction regarding the pace of technological development. Ray, born in 1948, predicts that uploading will be perfected around his ninetieth birthday, and he plans to be around in good enough health to take advantage of the opportunity. To maintain his health, he follows a regime that includes taking food supplements in the form of two hundred pills a day. I once sat next to Ray during a dinner at which he pulled out a plastic sandwich bag that contained the fifty or so pink, blue, green, white, and cream-colored pills for that meal. He began to explain, but I assured him that his supplement regime was already legendary.

It matters little whether fear of death functions as a motivator for Kurzweil or other advocates of scientific paths to immortality. To be fair, how many among us would turn away from immortality or even an extra ten years of life if we thought this goal was within our grasp? Would you turn down life extension if it simply delayed the onset of disease? A great deal of anecdotal evidence suggests that the vast majority of people want to live as long as they can, and even poor health does not deter this drive unless the physical or mental pain is unbearable. Francis Fukuyama, the political theorist and critic of transhumanism, doubts that we will be able to regulate biotechnologies that extend life. Life extension is one of the *killer apps*—an application that makes the adoption of a new technology irresistible—and therefore a goal that will provide broad public support for some of the scientific approaches to transcending the present-day limits on humanity.

In the opinion of a majority of scientists, uploading, brain transplants, and cryonics will remain science fiction until major technological thresholds have been crossed. The members of Yale's T&E research group were not convinced that uploading is likely to be possible in the foreseeable future, if at all. Many of the questions and comments directed at Martine Rothblatt during her fall 2007 presentation on cryonics and uploading were colored with skepticism. Who or what would be in the computer? In what sense is the informational *being* in the computer *you* or even the equivalent of a person? Will the entity in the computer or robot have feelings and emotions?

I asked Martine how much information she believes would be needed in order to capture her personality within a computerized entity. She replied, "very little." For Martine, capturing mental patterns was more important than the capture of specific information. I probed further. "Without all your personal history and memories, this entity living within a future robot would not really be like you. Why would one want to be uploaded?" She looked me directly in the eye and replied, "to preserve for the future of humanity all the beauty, light, and intelligence that resides within you and each of us."

Few of the T&E members were convinced by Dr. Rothblatt's arguments. Nevertheless, she certainly sowed the seeds of doubt in the minds of some who considered such technologies impossible. The grace, sensitivity, and intelligence with which she responds to each challenge is truly impressive.

Martine is a master of the fine art of listening and the rarer art of actually responding clearly and succinctly to the question asked. There is a touch of charisma in her quiet demeanor. Martine Rothblatt is evidently someone who is comfortable with who she is and what she believes.

She is certainly not naïve. On a later occasion, Martine confided to me that she understands that many of these futuristic prospects are based upon theories that may turn out to be wrong. I asked why then had she been so unrelenting in her advocacy during her presentation to the T&E group. She replied that a Yale forum was the wrong place to equivocate. I can't fault Martine for taking this stance. But if not at an academic institution, where in our society is equivocation and nuanced reflection appropriate?

Technology is making many things possible. Time will tell whether immortality is among those possibilities or whether the present-day scientific quest to transcend death is a blind alley. Even if we cannot transcend death there are many tools becoming available that will facilitate transcending some limits on the mind and body.

From life extension to biochemical means to improve cognition and physical prowess, each step down the enhancement road highlights an unresolved issue for any society that values freedom. The freedom of the individual, and therefore that of the wealthy and private foundations, to pursue enhancements will, in all probability, be treated as sacrosanct. But does the freedom of the individual give a small minority pursuing its goals the right to force on others the radical restructuring of human existence?

Transcending limits holds consequence for the larger society. Rapid population growth and new forms of social stratification and inequality will be accompanied by more psychological, spiritual, and philosophical concerns. Will the value of mere humans be decimated? Will people feel under pressure to adopt the latest technological enhancement? Parents, for example, might feel under pressure to enhance their sons and daughters so that the children can compete successfully. Will citizens embrace a life without the threat of an imminent death, or might they consider it not worth living?

People learn to adapt for better or worse. What concerns me is whether that adaptation will be forced by technological possibilities, rather than freely adopted. If, for example, extending life is a widely shared goal, regardless of the consequences, then those who facilitate its realization should be rewarded. However, what of the negative consequences? Who holds responsibility for overpopulation, increasing pressure on environmental resources, and changes in the quality of life that large segments of society find undesirable or demeaning?

## THE THERAPEUTIC CONUNDRUM

Martine's mood may have been particularly upbeat when she visited Yale in November 2007. The previous week, United Therapeutics Corporation (UTHR), a company Martine founded to market drugs to treat cardiovascular disease, announced earnings for the third quarter of $59 million, up 46 percent for the same quarter the previous year. During the week of her presentation, the price of a share of UTHR had risen to over $100. Profitability was a bonus for a company that was originally founded to save the life of Martine and Bina's daughter.

Their daughter Jenesis was diagnosed with a cardiovascular disease called pulmonary arterial hypertension (PAH) when she was only six years old. PAH is a condition that can start with shortness of breath, tiredness, and chest pains from a lack of oxygen, and leads to severe limits on all physical activity. The disease is caused by restricted blood flow brought about by pressure on the pulmonary artery, which carries blood from the heart to the lungs. This pressure can be the result of inborn stiff artery walls, the tightening of artery walls, or blood clots.

In researching whether there was any treatment for PAH, Martine was directed to a drug developed at Burroughs Wellcome. The number of people with PAH is low so the drug was not considered commercially viable, and therefore had never been taken through costly clinical trials. Jenesis was twelve when Martine assembled United Therapeutics, licensed the drug, hired a team to develop it, and funded clinical trials. The initial research was made possible by funds Martine generated from Sirius and her earlier companies. Happily for Jenesis and others suffering from PAH, the drug turned out to be effective, and is now marketed under the name Remodulin.

Remodulin generates sales of one billion dollars a year. The price of Remodulin is high—$100,000 a year. But Martine and UTHR give it freely to anyone who cannot afford it, or anyone for whom the cost of the drug would exceed the cap on their insurance. Even then, less than 5 percent of patients have needed to be covered under this free plan. On the downside, drugs like Remodulin contribute to the rising cost of health care.

"Unfortunately," says Martine, "in the majority of cases, Remodulin loses much of its effectiveness after a few years." Therefore, UTHR is investing research dollars in alternative treatments for PAH and other cardiovascular diseases. One of the projects targets the number of lungs available for transplants. Upon death, lungs fill up with fluid and quickly deteriorate. Only 15 percent are healthy enough to be transplanted into a person suffering from lung failure. UTHR's research project is directed at a method for rehabilitating a significant number of formerly discarded lungs so that they can be transplanted successfully.

An adventurous longer-term project includes two approaches for fabricating artificial lungs. One approach is to build an artificial lung upon the scaffolding of a pig's lung using tissue that comes directly from the patient. Scientists at UTHR believe that stem cell research will progress to a point where pluripotent stem cells with the patient's DNA can be induced to generate new lung tissue to populate the scaffolding. Since these lung cells will have the same genetic profile as other cells in the patient's body, the new organ is much less likely to be rejected.

The more controversial approach is a xenotransplant, the transplantation of a pig lung into a human. Pig lungs are the right size, but ensuring that the human host will not reject the foreign tissue remains a challenge. That is the focus of much of the xenotransplant research at UTHR—altering the pig tissue genetically and utilizing drugs and other means to stave off rejection of the foreign tissue by the recipient's body.

"Yuck" is a common reaction to the image of any organ harvested from a pig being transplanted into a human body. This feeling of disgust is understandable given that pigs inhabit environments perceived as unclean. However, "yuck" became more than an emotional reaction when Leon Kass, the chairman (2001-2005) of President Bush's Council on Bioethics, suggested that, "in crucial cases . . . repugnance is the emotional expression of deep wisdom, beyond reason's power fully to articulate it." While Kass's words were measured, his comments were made in the context of his opposition to human cloning and the enhancement of human capabilities. Many other bioethicists immediately denounced the proposal that repugnance might constitute a moral intuition or moral wisdom. They pointed out that disgust could arise from the violation of culturally conditioned prejudices. Historians provided examples of the *yuck factor* being used to justify the persecution of others whose race or behavior was different from that of the dominant culture. Confusing culturally conditioned emotions with morality is a bad idea.

To the best of our knowledge, disgust evolved as a protective response to poisonous or rotten foods. Like other emotions, it became linked in the mind to behaviors or situations that were designated as taboo. For many cultures, taboos provide a good method to direct members of the community away from harmful practices. But they also served to create group cohesion by designating others as disgusting. In a more scientifically based culture, we look for reasons underlying the taboo. There are undoubtedly valid scientific reasons for taking extra care before transplanting animal organs in a human body. Nevertheless, the *yuck factor* alone would be insufficient for most of us to reject a pig's lung that could save the life of our young daughter.

During Leon Kass's tenure as Chairman of the President's Council on Bioethics the Council released a report titled, *Beyond Therapy: Biotechnology and the Pursuit of Happiness* (2003). The report proposed a distinction be made between therapeutic practices and the use of medical technology to enhance human faculties. While the members of the Council knew this distinction would be hard to establish, the intention was to place a moral onus on human enhancements.

The tone of the report was paternalistic. When two of the more liberal members were ousted from the Council in February 2004, a perception grew that the opinions in the report represented those of members who embraced more conservative values. Criticisms of the report's findings grew within liberal circles. Past example of glasses improving vision and vaccinations that enhance immunity were used to underscore the dubious practical application of a therapy-enhancement distinction. While the report did not suggest that glasses or vaccines are questionable enhancements, such examples damaged the constructive application of the distinction. Other difficult questions were raised. If with good nutrition a child grows two inches taller than she would have otherwise, is this an enhancement? If through pre-implantation genetic diagnosis (PGD), parents elect to bring to term an embryo that is likely to have positive attributes, such as the intelligence exhibited by one or both of the parents, is this an enhancement? If drugs allow one's IQ to be improved significantly but still fall well within the range of one's peers, is that an enhancement? It can be quite difficult to determine exactly what traits are enhancements—particularly since most enhancements we might choose for ourselves or our progeny are features and capacities that some other humans have acquired by natural means.

The transhumanist scholar James Hughes declares that questions like these demonstrate that "the therapy/enhancement distinction is meaningless." But that dismissal does not help assess which technologies are useful and which cross a line. It indiscriminately dismisses the existence of any line. Yet even Hughes acknowledges that governments have a right to restrict technologies that are harmful to the individual. Thus regulating the safety of food, drugs, and devices is acceptable, as is restricting gene therapies that might be harmful for an infant.

All parties recognize that restricting enhancement technologies is difficult in that many of these tools and practices will be developed for therapeutic purposes and then repurposed to enhance the capabilities of people who
do not need therapy. Herein lies a conundrum for those who wish to arrest
the development of technologies that might enhance human faculties. All of
the core technologies, from regenerative medicine to drugs that improve
memory and attention, are under development to treat those suffering from
physical or mental disease. For example, the narcolepsy drug Modafinil was
discovered to also improve the attention span of those without the disease.
While expensive, Modafinil (Provigil®) has become the cognitive enhancer
of choice on U.S. college campuses.

Likewise, the incremental increase in life expectancy is a natural by-product of biomedical research, while radical life extension generally gets catego-
rized as a transhumanist goal of evolving beyond present limits. The
difficulty lies in distinguishing between the cure for a disease, the pursuit of
better health, and the transcendence of limits. When disease and death can
be broken down into distinct biomedical challenges and each solved separately, bringing them together may well lead to radical life expansion.

Good people with very good intentions are developing these technologies. Martine Rothblatt is both a good parent and a transhumanist, and
those two sides of her character cannot be easily separated. The technologies being developed at United Therapeutics are among those which will be
used for both therapeutic and enhancement purposes. UTHR was created
out of Martine and Bina Rothblatt's love for their daughter. What makes
Martine's story something more than that of another far-out visionary
spinning speculative tales of plausible future technologies, is her persistence
and realization of a company that probes the cutting edge of bioscience. In
that goal, UTHR is only one tiny player in a medical technology industry
that encompasses multinational pharmaceutical firms and research hospitals around the globe.

## MINDCLONES

The vision of uploading one's mind into an artificial entity, which Martine
outlined to the T&E members in 2007, culminated in her 2014 book, *Virtually Human*. During the intervening years, Martine fleshed out responses
to the various questions she had encountered about the pathway to uploading, the quality of life for immortal *mindclones* (digital emulations of a person), and whether these future beings would be granted legal entitlements

generally reserved for people. In her reading of evolving legal and ethical sensibilities, mindclones will eventually be accepted and treated as worthy of full legal rights. To concerns that mindclones will only be available to the rich, Martine responds that they will be so cheap that anyone could create a digital emulation of the personal contents of their brain. If she is correct, in the future there will be millions, and possibly billions, of mindclones, each with their own legal rights.

For those desiring a mindclone and interested in exploring this form of immortality, Martine's vision is good news. However, other theorists offer less optimistic views regarding the advent of mindclones. George Mason University economist Robin Hanson has applied his skills as a social scientist to analyze the societal impact of *ems* (his term for emulations of the human brain). He concludes that cheap *ems:*

> would outcompete humans on almost all jobs. Most wages would fall to near *em* subsistence levels, and ems would quickly outnumber humans, pushing the world economy to double every month or faster . . . As most *ems* would be copies of the few hundred best-suited humans, *ems* would be far more capable and productive than most humans, and feel very engaged in their jobs.

Martine Rothblatt's vision offers an expansion of individual experience through the creation of digital clones. Hanson's analysis leads to the harnessing of artificial minds for economic purposes. Perhaps both are possible. From the perspective of mere humans, however, Hanson's deductions lead to a darker world, unless one presumes the *ems* will solve problems that humans cannot, and their increased productivity will improve the quality of life for everyone. There exists an ambiguity whether mindclones will actually be a boon, disruptive, or dangerous to humanity as a whole.

Rothblatt and Hanson also differ on how quickly we will witness the advent of an Age of *Ems* or Mindclones. Hanson estimates that the first digital emulations will be created in the twenty-second century. Similar to Ray Kurzweil, Martine predicts that mindclones will appear within twenty to thirty years.

The attraction of uploading one's mind into an artificial entity hinges upon the belief that a mindclone will delight in the fragrance of a rose, experience sexual pleasure, and savor memories. Indeed, all predictions of machines comparable to humans, whether mindclones or totally new entities, assume that human-like consciousness can be produced artificially. In recent years, con-

"I THINK YOU SHOULD BE MORE EXPLICIT HERE IN STEP TWO."

sciousness has been elevated to the primary battlefield upon which philosophers and scientists struggle over what computers can and cannot do, and why humans may or may not be distinct from any new forms of intelligence we create. The divide lies between those who contend that consciousness is a complex phenomenon and will not be realizable through current approaches to digital technology, and those who propose it will naturally emerge as information technologies mature. The first view offers significant time, and therefore breathing space, to tackle the societal challenges smart forms of artificial intelligence pose. In the later view, mindclones will soon be among us.

The near-term predictions rest on the belief that consciousness will emerge within any artificial system with a very rich set of connections between the many pieces of information in a massive database. Even the profile Google, Facebook, or a data-consolidating firm create about you and your online activity represents a major first step toward the eventual development of your mindclone. But to emulate human-like consciousness, the networked information must include memories and ongoing sensory input. This computational theory of how consciousness might emerge offers an interesting hypothesis that can neither be proved nor demonstrably refuted.

A growing body of corporate and government leaders and well-connected scientists believe emulating the human brain is a worthy activity. Many scientists propose that in the process of creating a computer model of the brain, we will learn a great deal about how the brain works, and therefore be able to find cures for mental maladies including Alzheimer's disease. Corporations such as IBM are directing substantial sums toward computer simulations of the human brain. The European Union invested 1.2 billion euros in a Human Brain Project (HBP) lead by Henry Markham at the Swiss Federal Institute for Technology in Lausanne. However, the HPB is controversial, criticized as premature and a poor investment in a July 2014 letter signed by more than 180 scientists. In comparison, the U.S. Brain Initiative much more modestly directs funds to the development of tools necessary to further the progress of neuroscience.

Consciousness, our ability not only to be aware but also self-aware, holds mysteries that evade the easy grasp of contemporary science. Scientists and philosophers are engaged in a concerted effort to forge a science of consciousness. For example, neuroscientists such as Christof Koch search for the specific activities in the brain that indicate when a person is actually conscious of their surroundings and aware of being conscious. But even if they can determine those neuronal and biochemical activities, what are referred to as the "neural correlates of consciousness," there remains the question of whether we have actually explained why we humans consciously *experience* the immediate surroundings in which we find ourselves.

Knowing which neurons fire when we see a red jacket does not describe what we are feeling—the experience evoked by the color red or the beautiful woman or handsome man wearing that jacket. However, few of the philosophers critical of explaining consciousness in terms of brain activity or computational processing are holding out for some spiritual explanation, or contend that consciousness is dependent on some non-physical mind stuff. They just do not see how reducing the experience of being conscious to biochemistry, neuron firings, or information processing does anything more than dismiss what it feels like to have an experience.

There exist many theories as to why humans are conscious, self-aware beings. And yet, we understand very little about the complex and dynamic processes that make it possible for you to read this book, comprehend what I am saying, and creatively integrate my ideas into your own fabric of meaning. Our limited understanding of how the brain enables mental states must grow considerably in order to emulate a mind within a computer. Nonetheless, achieving immortality through uploading will persist as a theoretically conceivable approach, but one that may well turn out to be based upon bad assumptions. In the meantime, consciousness will continue as a placeholder for what cannot be fully explained by the science available, a role once filled by concepts such as *soul* and *spirit*.

The desire to be healthy and the dream of immortality together play a powerful role in driving scientific discovery forward. The obsessive desire to live longer regardless of the personal and societal costs (emotional, ethical, and monetary) already contributes to a health care crisis for which there is no easy solution. Certainly living longer provides short-term benefits for the individual and their loved ones. Nevertheless, ignoring the longer-term challenges created by the short-term benefits is a prescription for future crises. If we truly want those benefits, let us put similar energy into addressing the negative consequences—either that, or slow the march of technological innovation, and accept the losses.

The people inventing the future, such as Martine Rothblatt, are engaged in helping their families and others with immediate needs. But they also recognize that their work will lay foundations for an edifice upon which new pathways to immortality might be built. Whether any of the pathways to immortality succeed may be less important than the discoveries made in pursuit of that more elusive goal. Generation after generation of scientists have been inspired by dreams that were not realized in their lifetimes. Even should immortality recede into the distant future, new powers to cure disease, develop more complex robots, and enhance human capabilities will emerge along the way.

Innovators and transhumanists are working hard to persuade all of us that their goals are, or should be, goals for all of humanity. But even if they fail in this endeavor, the impact for the good or bad of new discoveries will alter the lives of large segments of society. Each of us will benefit from better health and an improved quality of life. Nonetheless, the quality of life might also decline for the majority of people due to the pursuit of technologies that a small community or an individual country considered desirable. How to honor individual freedom and the sovereignty of nations, while keeping technological development from slipping beyond our collective control, is a serious and unique challenge for contemporary society.

## SOCIETAL IMPACT

People are living longer, and yet there is no evidence that the rate by which life expectancy grows has accelerated. That, however, could change quickly with a few major scientific breakthroughs. In the meantime, even incremental growth taxes social systems around the world. For example, retiring early while living longer adds to the pressures on old-age benefit programs covered by governments and private pension funds.

A longer life can be the result of either staying healthy or biomedical technologies that keep senior citizens alive even when their health and quality of life is poor. The latter can be both emotionally and financially costly for everyone involved. It might be wonderful to have a few extra years with loved ones, but when their suffering is extended, so also will be the emotional strain on family and friends. Furthermore, medical technologies that extend the life span of senior citizens are costly. Artificial joints, organ replacements, and round after round of chemotherapy already contribute significantly to a health care crisis. Estimates vary, but the last two years of life are agreed to consume an inordinate share of the health care budget. People living longer did not create this crisis; nevertheless, longer life exacerbates a challenge for which no easy answer exists.

The convergence of progress on multiple fronts can have a dramatic effect on the larger society in ways beyond health care. Consider the impact of emerging technologies on job creation. Living longer is just one among many factors that play a part in hampering efforts to lower unemployment rates. Postponing retirement becomes more common as medical advances improve the mental and physical capabilities of the elderly. This, in turn, slows down the rate by which existing jobs are made available for younger people entering the workforce.

Both liberal and conservative politicians commonly propose that investments in biotech, nanotech, and infotech are the surest route to creating new jobs. A number of theorists, however, have pointed out that this presumption may result from a total misunderstanding of what is actually occurring. The rate at which new forms of work are created appears to be slower than the loss of jobs due to the cumulative impact of emerging technologies.

Loss of jobs to increasingly intelligent robotic devices is one phenomenon we already witness. Among those who have studied employment trends are Erik Brynjolfsson and Andrew McAfee, whose books *Race Against the Machines* and *The Second Machine Age* chart some unsettling trends. They note that productivity, jobs, hourly wages, and income traditionally grew in unison. In the past few decades, there has been, what they call, "the Great Decoupling." The U.S. median income stopped rising during the Reagan administration, and a decade later, employment flattened. However, productivity and gross domestic product (GDP) grew over the last thirty years.

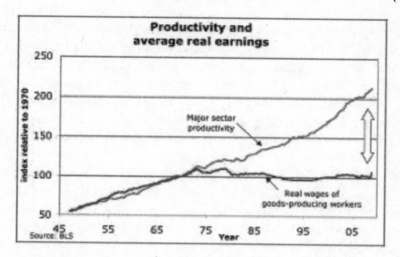

Technological innovation, in their view, is the primary factor contributing to the loss of jobs, but it also played a leading role for the growth of GDP.

In a 2013 study by Carl Frey and Michael Osborne, two scholars at Oxford University, as many as 47 percent of the jobs in the U.S. are already susceptible to computerization. Replacing as little as a third of those jobs with information technologies over the next decade could be extremely disruptive. Furthermore, robots are becoming more capable, while performance measurements, scripted behavior, and proscribed attitudes mechanize an increasing percentage of human jobs. Long before the advent of the speculative *technological singularity*, the time when artificial intelligence far exceeds human intelligence, these two trends will cross. The result will be a rapid increase in the replacement of human workers by robots that work 24/7 and do not require wages and benefits. Marc Saner, Director of the Institute for Science, Society and Policy at the University of Ottawa, and I have referred to this juncture as a *functional singularity* that will reshape society. The radical impact of this juncture is largely predictable so it is not exactly a singularity. But over the coming decades, its advent will be much more important for public policy than the more speculative *technological singularity*. While the technological singularity posits that AI research will inevitably lead to the emergence of super-intelligence, a functional singularity recognizes the importance of social forces in shaping the future course of developments in AI. For example, widespread perception that robots are robbing jobs could create broad resistance to the development and adoption of increasingly intelligent AI systems.

The economist John Maynard Keynes coined the phrase "technological unemployment" soon after the start of the Great Depression. In the 1930 essay *Economic Possibilities for our Grandchildren,* he wrote that "we are being afflicted with a new disease . . . namely, *technological unemployment.* This means unemployment due to our discovery of means of economising the use of labour outrunning the pace at which we can find new uses for labour." With these words, Keynes captured a recurring fear that has gripped workers since the dawn of the Industrial Revolution.

The fear that jobs will be lost with the introduction of new technologies occasionally erupts into a rejection of scientific progress or even revolutionary fervor. The name of the legendary Ned Ludd, who is said to have broken two knitting frames in 1779, was adopted by skilled English textile workers in their 1811-1813 Luddite rebellion against machinery that allowed lower-paid, less-skilled workers to perform their jobs. For the past two hundred years, anyone against industrialization and automation has been called a "luddite." More recently, the term is often applied disparagingly to anyone critical of technological development.

As we know, the world economy recovered from the Great Depression as it did from earlier and later downturns. Historically, increases in productivity from labor-saving and more efficient machinery have led to expansions in capital and new forms of work. Is it different this time? Or is the slow employment recovery from two consecutive recessions merely a trough during a period of adjustment? Certainly our present challenge differs from a very large displacement of human workers by smart machines (a functional singularity), or the more distant prospect of a future inhabited by mindclones and cheap *ems*. And yet the present employment crisis could be perceived as an inflection point that leads to a more serious crisis ten or twenty years hence.

Perhaps the next generation will witness the two hundred year old luddite nightmare in which technology robs far more jobs than it creates. There may well be periodic upturns in job creation. But let us presume for a moment that the new forms of work created will fall behind the demand for jobs. This would certainly be disruptive. If, for example, the combination of longer life and better health results in 50 percent of the workforce electing to retire after they are seventy years old, will their great, great-grandchildren find employment? More importantly, if wages are not the primary means for distributing capital, how will goods and services be made available to those without work?

In the late 1960s and early 1970s, it came to the notice of the post-war baby boom generation that the rapid growth in university enrollment meant more people were being educated than there were quality jobs for them to fill. This was among the underlying factors that fostered the birth of a worldwide youth rebellion. Millions of young people joined socialist-inspired political movements and experimented with sexual liberation, communal living, mind-altering drugs, and lifestyles built around former leisure-time activities such as meditation. The left-liberal values of the youth culture dominated the politics of the 1970s. Many young adults believed that post-industrial economies were moving toward a leisure-class society. An expansion of the welfare state to provide a minimum income to everyone was among the policies that progressive-minded socialists proposed to sustain this growing leisure class. But the election of Margaret Thatcher as Prime Minister of the United Kingdom (1979-1990) and Ronald Reagan as President of the United States (1981-1989) signaled the onset of a conservative reaction against the welfare state. Simultaneously, there was a vast expansion in opportunities for new jobs made possible by the availability of low-cost personal computers.

The tension between those wishing to enlarge and those wishing to contract social services continues as a core political issue in democratic countries. If technological unemployment is indeed a reality, the tension between welfare state paternalism and demands that everyone be self-reliant will come to a head in the next decade or two. The expansion of capital together with the contraction of jobs will force post-industrial societies to consider instituting something like the creation of a guaranteed minimum income.

This simple analysis outlines a problem while failing to capture so many additional factors that come into play. Robin Hanson envisions tremendous growth in capital from the productivity of robots working twenty-four hours a day, seven days a week. But capital growth will mean very little to the average citizen, when 70 percent of stock ownership is held by 5 percent of the population, and more than half of the American public owns no stocks. David Rose, a technology entrepreneur and venture capitalist who has founded or funded more than eighty companies, believes that the exponential growth of technological development has put us on the road to a world without work and perhaps even a world with very few companies. If they are right, and capital grows dramatically, and if governments intervene to distribute that capital with even a modest degree of fairness, a utopian future is at least conceivable. There are, however, many less rosy scenarios. For example, Ed Woodhouse, Professor of Science & Technology Studies at RPI, wonders what will happen if 3D printing (a form of robotics) puts one hundred million Chinese out of work. Would this lead to social unrest in China, and how will an unstable China affect the economies of an increasingly entangled world? The mere report of a downturn in the rate at which the Chinese economy is growing can already cause a bad day for stock markets worldwide.

Jaron Lanier offers a darkly critical vision of how the Information Age is progressing. Lanier occupies an idiosyncratic niche in the ecosystem spawned by digital technology, and his dreadlocks certainly give him a more countercultural appearance than that of a card-carrying member of the techno-glitterati. Hailed as a visionary for having popularized virtual reality and founding the first company to sell virtual reality gloves and goggles, Lanier has recently taken his colleagues to task as a leading critic of computer culture. Among its sins: fostering a mechanistic view of human nature; embracing a naïve belief in a utopian technological singularity; gutting well-paying jobs, and substituting them with the promise of Internet jobs for which only a few people get adequately rewarded; and expanding the tools available for governments and businesses to spy on citizens.

There are as many scenarios about how this positive and negative mix of forces will unfold as there are futurists—a large and growing industry. (If you cannot find a job elsewhere, try selling your predictions.) No soothsayer can sort out the forces in play with any degree of certainty. And the melodramatic tales Hollywood writers envision usually exaggerate the impact of one or two technologies, while ignoring the many other tools and techniques that will appear simultaneously. If technological unemployment is indeed a reality, it will lead to a very serious capital distribution crisis.

When such a crisis might occur, how it could alter the political economy of countries worldwide, and whether governments will fall as a result depends upon many factors. Capital growth might yet lead to the creation of new forms of work that only people can fill. Political leaders may have some success in buying time with fiscal and monetary policies, and with the public's willingness to be bought off by the bread and circuses of the Internet. Moderating forces invisible to this writer might yet emerge. The gathering of storm clouds can be read as omens foreshadowing inevitable tragedies or as signals to plan ahead in order to moderate the force of disruptive trends. If technological unemployment outstrips job creation, forward-thinking governments could forestall political unrest through some form of capital redistribution such as a robust welfare system or a guaranteed minimum income.

For the past decade, discussion of the societal impact of intelligent robots and delayed retirement on the availability of jobs has been largely restricted to the reflections of futurists and a few isolated social theorists. Something has changed. Throughout 2014 mainstream economists and financial leaders, such as Larry Summers and Nouriel Roubini, began raising alarms about technological unemployment. The topic is poised to erupt as a political issue. High-level summits on the problem will convene in 2015. Various policy proposals to redistribute capital will be floated. For example, Jerry Kaplan, a computer scientist and serial entrepreneur best known for pioneering tablet computers, proposes that existing capital should not be taxed, but future capital growth should be redistributed. Hopefully, political candidates will be pressed to clarify their position on addressing technological unemployment. Let the public dialogue begin.

The arrival of a techstorm precipitated by technological unemployment will result from the collective impact of many individuals and groups pursuing their own goals. Society is being transformed one discovery, one robot, one business decision, one genetically modified organism, one nanoparticle, and one enhancement at a time. Each step toward extending life or turning over a task to a robot reverberates throughout the social system.

A direct line can be traced from the efforts of Martine and countless scientists and other entrepreneurs focused upon improving health care and extending life to the brave new world being created. A surge in population growth and technological unemployment, as well as a restructuring of who (or what) gets valued, cannot be entirely staved off. Technological development can be slowed, and possible means to manage its negative consequences can be forged, but only through concerted effort.

# { 9 }

# Cyborgs and *Techno Sapiens*

PRESENTERS AT EARLY ANNUAL MEETINGS OF THE WORLD
Transhumanist Association (WTA) in Stockholm, London, Berlin, or New
Haven (1999-2003) talked enthusiastically about drugs that would dramatically enhance cognitive abilities, brain-mind interfaces, and radical life extension. By that time, the use of drugs and growth hormones in sports was
receiving a great deal of attention. The International Olympic Committee
had created a World Anti-Doping Agency in 1999 and the U.S. followed
suit a year later. A new drug, Modafinil, which had been approved for sleep
disorders, was discovered to be better than anything else available for sustaining concentration. Cochlear implants to help the hearing impaired had
become commonplace, and research was underway to implant chips into
nerve tissue in hopes of designing other sophisticated prostheses. In the eyes
of those ready to embrace enhancements, the next decade would clearly
signal a great leap forward in enabling individuals to extend their own talents and transcend existing limits on human potential. Simultaneously, those
who believed that widespread use of enhancing technologies would come
with negative ethical and societal consequences sounded alarms. In 2003,
The President's Council on Bioethics published its disapproving judgment
in a report titled: *Beyond Therapy: Biotechnologies and the Pursuit of Happiness.* All parties presumed that inexpensive, effective means to dramatically
enhance faculties would be available within a decade or so. After all, were
not drug companies secretly testing every conceivable neurochemical to find
irresistible cognitive elixirs? And certainly medical science was soon to un-
cover means to produce a dramatic extension in average life expectancy.
Furthermore, nanotechnology promised to create tiny machines that could

be surgically implanted and would allow communication between the inner world and outer devices.

In retrospect, progress toward producing enhancement technologies over the past twelve to fifteen years has been a disappointment or a blessing, depending upon one's point of view. As with research in genomics, the many advances in neuroscience and medical technology may well be laying foundations for major breakthroughs. Nevertheless, the complexities and the incremental development suggest that the more significant possibilities are zooming off into a somewhat distant future. Paradoxically, the race to invent mere humans out of existence proceeds rather slowly, even as the pace of scientific discovery accelerates. This, however, has not stopped the hype machine driven by dreamers and sensation-purveying media from predicting a near-term transformation of humanity.

Critics who appraise each incremental step perceive a chasm between what has actually been achieved and what is continually professed to be right around the corner. Claims that the next generation will include *techno sapiens* whose skills will far exceed those of their unenhanced compatriots should be viewed skeptically. A more measured assessment suggests that a few technologies for enhancing capabilities will be progressively perfected. Although the advances made in recent years fall far short of the exaggerated projections, no one dares rule out a phase shift based on the convergence of major biomedical breakthroughs. Optimistic scientists and hopeful transhumanists continue to project the possibility of a biotechnology that leads to a substantial jump in average life expectancy, or a drug that significantly improves the ability to excel in creative and intellectual work.

There is a great deal of confusion and disagreement as to what actually constitutes an enhancement. For some communities any technology that significantly alters one's "God-given" capabilities is an enhancement and is unethical. But even these communities do not reject wearable technologies such as glasses that improve vision or cell phones that extend capabilities as long as no one tinkers with the human body. Indeed, wearable devices such as the Apple Watch, a robotic exoskeleton, or a dedicated health-monitoring device come with various safety or privacy concerns, but are generally accepted as ethical if used properly. Controversy centers upon drugs, surgeries, and devices that alter the body and mind. Even then, a distinction can be made between enhancements that improve capabilities while limiting one's skill set to fall within the range of talents that others are born with, and enhancements that enable extraordinary talents. Technologies that alter mind and body, dramatically improve capabilities, and facilitate skills that go beyond those of all, or nearly all, humans are the focus of tension. Those technologies include designer babies and various approaches to radical life extension and immortality discussed in the last two chapters. But here we'll direct attention to drugs, surgical procedures, mind-brain interfaces, and the introduction of nanodevices into the body.

Even if those anticipated enhancements are still far off, true believers exhibit remarkable talent for ignoring the failure of past predictions, and creatively construct new ten- and twenty-year plans. In the 1960s, the critics of artificial intelligence (AI) research, such as the philosopher Hubert Dreyfus, turned out to be more accurate than the optimistic proponents of AI. But that did not put off the believers. Each decade since has witnessed unfulfilled predictions and rationalizations as to why the advent of full AI lies just ahead, in the next approach. Natural language processing, artificial neural networks, evolving artificial life, and embodied intelligence have successively been perceived as breakthrough approaches that would certainly lead to the development of machines with the capacity to learn and create. More recently, hope is being placed in a new approach called Deep Learning.

In the case of AI, the hype machine ritualistically exaggerates the importance of each step forward, such as IBM's Deep Blue beating the chess grandmaster Garry Kasparov or IBM's Watson winning at the TV quiz show *Jeopardy*. Both victories were truly noteworthy, and yet each revealed new layers of complexity that must be surmounted for full AI to be realized.

In the realm of human enhancements, the remarkable feat of a paraplegic maneuvering his wheelchair with thought alone gets exaggerated into proof that minds and machines will soon be fully integrated. Hype, hope, and wishful thinking overwhelm any realistic appraisal of the difficult thresholds that lie ahead. The media discovered that innovations accompanied by exaggerated claims attract viewers, but eyes roll and the audience switches channels when critics share their doubts. In the U.S., challenges to the conventional wisdom that we can achieve anything we set our minds to are treated as cynical defilements of a revealed truth. In this narrative, the power of positive thinking always prevails.

Regardless of the slow pace of change in realizing major enhancements, and regardless of the fact that new layers of complexity make it unclear how far off it all is, both the transhumanists and their critics have bought into the same basic narrative. Scientific mastery, both believe, will soon permit wholesale alteration of the human genome, mind, and body. The upshot will be easy access to superior skills, soon followed by new races of cyborgs and *techno sapiens* whose capabilities will far exceed those of the average person alive today. The sense of inevitability which accompanies these projections can be energy-sapping for people who feel that the future will not need us—at least not as we currently are. Only time can reveal whether skepticism with respect to the transhuman dream represents a failure of imagination or if the dream itself epitomizes imagination run wild. Even should extraordinary enhancements be possible, there is nothing inevitable about the future course of humanity. Plenty of inflection points will appear for slowing or altering the trajectory of technologies that might be adopted for enhancement purposes. We *do* have a say. Inflection points lie on the horizon and a few are already perceivable.

## A BATTLE FOR THE FUTURE

For thousands of years, education, character development, good nutrition, and exercise functioned as the primary modes of self-improvement. The availability of technological shortcuts for self-advancement is troubling to those who value the honing of character through sustained discipline as central to humanity's success.

Differences of opinion about whether enhancing human capabilities is moral or desirable date back to the eugenics movement, but also to the promise in the nineteenth and twentieth centuries that scientific progress would improve opportunities and cure diseases. Goods that were once accessible only to the wealthy are widely disseminated today. For example, kings and queens two hundred years ago would be envious of the fabrics and clothing available to the average citizen at the local discount store. Good medical care is increasingly assumed to be a right, not a privilege. As a means to level the playing field and improve the quality of life, many forms of enhancement continue to be seen as appealing for political, medical, and social reasons. Progress is certainly good. Going back in time is neither desired nor possible. The real question, then, is when do enhancements cross a line? When do the social and ethical concerns outweigh the perceived benefits?

The enhancement debate erupts each time biomedical research crosses another major threshold. The birth of Louise Brown, the first test-tube baby, and Dolly, the first cloned sheep, set off intense political and intellectual battles in the bioethics war. With the sequencing of the human genome, the debate heated up again, largely because of projections that scientists were in the process of perfecting methods for alterations in DNA that would permit the exacting design of babies. The crossing of each technological threshold naturally gives rise to the question of whether a societal red line has also been crossed.

All sides characterize the enhancement debate as a battle for the future and the soul of humanity. To date, the battle has not claimed any serious casualties. Yes, professional sports fight for their survival in the midst of doping scandals. Students experiment with each new drug professed to be cognitive-enhancing in hopes that it will give them an edge in their studies. A tiny fraction of parents preselect the genes of their children, primarily to avoid inheritable diseases. The cyborgs among us are either the disabled donning advanced prosthetics in hopes of acquiring basic capabilities, or young nerds and the military elite experimenting with wearable devices. "The future," as cyberpunk writer William Gibson stated in 1993, "is already here—it's just not very evenly distributed."

While we're obsessing over the cyborgs and the future of humanity and buying into the hype, we're entirely ignoring the specific and real risks that accompany experiments that alter the body. The human mind and body, once considered inviolable, is in the first stages of being treated as a design space to engineer and remold. In spite of that, scientific progress often emerges from a succession of failures. Even if it becomes possible to engineer a strain of human being with truly superior capabilities, this will require many generations of trial and error. The simple fact that there will be failures underscores the importance of laws and guidelines on the use of human subjects for research purposes. Research ethics offers a powerful, yet poorly understood, tool for modulating the development of drugs and devices for medical applications including human enhancements. The emotional and physical costs of attempting to alter many capabilities will be great enough that only a limited number of people will elect to be experimental subjects. Some who volunteer to be guinea pigs will regret that choice, and their trials will be a timely warning to others. The dangers and complexity of radically altering mind and body may slow and perhaps arrest the march toward a transhuman future. Regardless of what the predominant narrative tells us, that future is still far from inevitable.

## A BACKWARD GLANCE AT THE ENHANCEMENT DEBATE

The enhancement debate echoes earlier chapters in the continuing tension over the promise and realities of scientific discovery. The love of technology, the belief that technology will improve the human condition, and the fear of what technology will bring have danced together to a syncopated beat.

The Industrial Revolution conjured up visions of technologically inspired utopian futures. Even the poor working conditions, long hours, and squalid city tenements that accompanied industrialization did not quell the enthusiasm. But the destructive might of the WWI war machines, the Great Depression, and the gas chambers of WWII did temper utopian illusions attributed to scientific progress.

Even so, nearly everyone agrees that technological development has improved the quality of life. During a tour of Sturbridge Village, the recreation of a New England town during the period between the War of 1812 and the Civil War, some friends and I stopped for an extended conversation with one of the historian guides. She was garbed in clothing from the early nineteenth century and working as a weaver. Other historians situated around the property worked and dressed as shoemakers, tinkers, ministers, and farmers. Individually and collectively they loved this era of history. And yet she told us of an informal poll they had taken among themselves. If given a choice, none of them would elect to live in the early nineteenth century, with all of its hardships, over the present age.

A poll as to whether people would wish to live in the present or future might produce a very different result. While transhumanists anxiously wait to embrace enhancements, their detractors on both the political Left and Right perceive enhancing human faculties as a harbinger of a dystopian future. For the more conservative critics, tinkering with the human body is unnatural, violates a gift from God or from nature, demeans human dignity, and demonstrates excessive pride (hubris). Given differences in values and in the meanings people attribute to their lives, it is unsurprising that they differ on whether enhancements should be embraced or rejected.

As mentioned in earlier chapters, liberal detractors express the fear that certain enhancements will bestow unfair physical or mental advantages. Unequal access to enhancing technologies will further increase disparity between those with wealth and opportunity and the have-nots. Maintaining at least the belief in equal opportunity is important for the health of a democratic society. If, in practice, leadership positions are only held by families with the funds to purchase enhancements that make them and their progeny superior, can widespread support of democracy endure? In other words, enhancing human faculties could single-handedly undermine social cohesion.

James Hughes, a leading transhumanist, argues in his book *Citizen Cyborg,* that the enhancement debate cuts across traditional Left/Right alliances and will function as a core theme in future politics. He is among those who propose that technologies, which provide an advantage, be made affordable and available universally. This admirable goal, if enacted, would defuse the charge that human enhancements will exacerbate inequality, and truly level the playing field. But it is difficult to believe that any such public policy will be adopted. Stating the goal quells criticisms, but does not solve the problem.

Trickle-down enhancements might slowly improve the condition of everyone, but in the meantime, the well-to-do are bound to have capitalized upon their early lead.

The debate over technological enhancements swings between its societal impact and more philosophical concerns. What is natural? What does it mean to be human? Attempts to define human nature serve as a wavering Maginot line for an ongoing struggle in which victories are measured in rhetorical points won or lost. Nonetheless, advocates of human enhancements have been effective in pointing out why any notion of what is and what is not natural can itself be dangerous, nurturing prejudice and relegating anyone without the requisite human features to second-class citizenship. They share this view with advocates for the rights of the blind, the deaf, and others with disabilities.

Much of the enhancement debate centers upon distinctions one side or another wishes to make as a means of framing the discussion. The most important of these has been the distinction between a medical therapy and an enhancement. While the use of the two concepts seems straightforward, determining which technologies are or are not enhancements can be difficult. As with disagreements about extending life, the enhancement debate reveals conundrums that elude straightforward answers. Should tools and techniques that level the playing field by providing an individual access to capabilities that others were born with be considered enhancements? If not, exactly who should be designated disabled and therefore granted permission to use technological means to improve their capabilities?

Should vaccines be considered a medical therapy or an enhancement? Perhaps both. Vaccinations *enhance* immunity to diseases. In the mid-eighteenth century, smallpox killed 400,000 Europeans a year. In 1980, the World Health Organization declared smallpox eradicated. Vaccines for tuberculosis, polio, meningitis, and measles dispel the fear of ailments that haunted our ancestors. Of course, calling a vaccine an *enhancement* frames the discussion in a particular manner, conducive to claims that a wide collection of improvements to the human mind and body should be embraced.

Vaccinations against deadly diseases are widely disseminated, and efforts are directed at making the leading vaccines universally available. If vaccines were once an enhancement, they are now a birthright. How about an inexpensive future drug that improves all students' study skills? Might that be made a birthright?

As the debate progresses, proponents and opponents of human enhancements have become well aware of the flimsiness of the lines they draw. Yet few converts desert from one side to the other. The basic issues touch emotional chords, which are more powerful than reason or nuanced talking points.

Both advocates for and against human enhancements weave compelling narratives seeded with crucial insights. But at times the discussion gets ugly. The anti-science extremism of Biblical literalists on the one hand, and an obsessive Frankensteinian desire to tinker with the human body on the other, are certainly not the only motivations driving the committed to be for or against human enhancement. But listening to some of the outspoken antagonists might give you the impression that all parties on the opposing side are dangerous kooks. Those against enhancements propose that anyone who cares about ethics or has spiritual beliefs should renounce tinkering with the body or mind. Transhumanists argue that if you value science and the benefits it provides, you must logically embrace technologies that improve human capabilities, even if you do not know it. And if you don't embrace enhancements, you are a luddite. Both sides act as if this is a political campaign, in which winning the public's heart and mind is essential to saving humanity. And in this they may be right, although some of their arguments are more alienating than persuasive.

Being forced to take sides is never helpful in the quest to explore more complicated, nuanced scenarios. Many of the people that I meet want an exploration of the middle ground. Establishing foundations for such a conversation will require concerted effort. In the meanwhile, a deadlocked confrontation serves the interest of both the *status quo* and of the continuing march to discover means to enhance individuals.

A reasonable response to emerging possibilities should not be to totally ban enhancements, but nor should it be to embrace them without question. Fortunately, the former is unlikely to happen. Democratic societies will find attempts to restrict human enhancements either impossible or as setting unappealing precedents. Restricting individual freedoms usually requires a demonstration that specific activities directly hurts others. Freedom-loving societies will therefore learn to accommodate and adapt to many forms of enhancement, even if they devastate existing institutions such as professional sports franchises. The modern democratic state has been astonishing in its ability to accommodate individuals with distinct and unusual capabilities. Those with disabilities are embraced wholeheartedly, and those with talents provided with opportunities. While many cast a weary eye on the intentional and explicit enhancing of individual capabilities, democratic societies will continue to integrate those who elect to alter themselves.

So far, the primary impact of enhancing technologies has been upon professional sports, and, to a lesser degree on amateur sports. Charges that technological enhancements in general are immoral or a form of cheating depend upon social conventions. For a professional cyclist competing in the Tour de France, doping violates the rules and therefore is cheating. It is deeply problematic, however, when democratic societies enact laws and regulations that protect the most advantaged among us. And yet people appear quite willing to grant owners of professional teams the right to protect their franchises by designating strict guidelines for who does and who does not have a right to compete. People love sports and do not want to see the legitimacy of athletic competitions compromised. Even so, the difficulties the regulatory bodies of professional sports have had in limiting the use of enhancing technologies, or in catching cheaters, do not bode well for the larger society's ability to control their use.

Athletes unwilling to take performance-enhancing drugs have found it difficult to stay competitive in some professional sports. The few who have been caught claim they only took growth hormones to keep pace with teammates who also use enhancers. Consider the development of synthetic erythropoietin (EPO). Natural EPO stimulates the production of red blood cells. Synthetic EPO raised the aerobic capacity of athletes by 8 percent and revolutionized sports. In world-class cycling the use of synthetic EPO started around 1990, and cheating quickly became an unspoken requirement to compete successfully. Lance Armstrong won the Tour de France a record seven times by ruthlessly exploiting enhancers including synthetic EPO and by gaming the testing system designed to eliminate doping. But not everyone can win through the use of enhancers. Deriving a competitive edge from enhancers in sports depends largely upon their use by gifted athletes who continue to hone natural skills through training and discipline. A technology can facilitate improved performance. It cannot guarantee either instant or eventual success.

The challenges the sports industry confronts today serve as a surrogate for the larger society working out its attitude toward enhancements and whether they can be regulated. Meanwhile, transhumanism stakes a claim to being an intellectual movement of note. Its leading thinkers are sensitive to the concerns that make large segments of modern society uncomfortable with major alterations of the human mind and body. They offer deeply reflective, creative, and serious responses to their critics. Whether history looks back on transhumanism as an important philosophical movement, however, will only be partially determined by the ideas of its leading proponents. Ultimately, transhumanists will be judged by whether their vision of a radically transformed and improved humanity ever truly occurs.

While the intellectual skirmishes about enhancement have been helpful in dismantling many false arguments, from a public policy perspective, and in the ongoing march of scientific discovery, they have had limited effect. Short of either a disaster that activates the public politically, or some coherent, positive vision that garners broad support, the drivers of scientific development will largely direct and propel forward the continued transformation of humanity. The exploration of which enhancements are palatable will proceed. However, there is one area in which the public can find a way to reach a broad consensus for managing and probably slowing that exploration: issues of safety.

## THE CYBORGS AMONG US

The philosopher Andy Clark proposes that all humans are natural-born cyborgs. By this he means we do not just use technology, but incorporate the tools into our being. Even basic inventions such as the letters in the alphabet, paper, and pen allow us to extend our minds beyond the body. With the symbols of mathematics we can work through a problem on a piece of paper that we could not solve in our head. However, the risks of adopting quill and paper were never great, unless one was writing an incendiary political tract.

Each technology that becomes a part of our lives plays a role in modifying the organization of the human brain. With habitual use, new connections between neurons form and older connections are strengthened or weakened. Neural pathways that go unused can be adapted for new purposes. In this manner, the technologies we work with, even if they are tools we use only occasionally, get mapped into the brain and become extensions of our mind and body when in use.

In the modern vernacular, the term *cyborg* refers to people merged with machines. It conjures up visions of Robocop, the Borg, Doctor Octopus, or Tony Stark as Iron Man. In reality, the merger of humans and their technologies has taken several forms, many of them fairly mundane. A significant percentage of us are reliant on drugs to perform optimally. In addition, cochlear implants, hidden pacemakers, defibrillators, and neuro-prosthetics that relieve the symptoms of Parkinson's disease, have also become commonplace.

In the future, nanosensors implanted under the skin, circulating in the bloodstream, or in synaptic union with thousands of neurons, may communicate wirelessly with devices at a distance from the body. Major technological augmentations—whether the result of genomics, drugs, hidden chips, or nanotech—will qualitatively alter the lives of the enhanced, but not always in readily apparent ways. In fact, the enhanced cyborgs among us could be indistinguishable from everyone else. Genetically enhanced individuals may elect to pass as unenhanced, presuming that genetic engineering does not result in visible differences. Recipients altered by genetics or implants will be more *techno sapien* than cyborg.

The image painted by advocates of technological enhancement is one of advantages that can be gained without paying a price, but in life this is seldom the case. Future historians will characterize our age as the search for quick fixes—the drug that instantly cures a disease, the stock trading algorithm that makes one rich, or the enhancement that easily leads to success in many endeavors. But wounded warriors go through a rigorous, often painful therapeutic regimen to learn how to use a neuroprosthetic arm or leg. Drugs have side effects. Growth hormones do not bequeath instant success in sports, but only work in combination with discipline and training. Injecting growth hormones also comes with risks such as nerve, muscle or joint pain, carpal tunnel syndrome, high cholesterol, and diabetes. The risks sometimes get exaggerated in hopes of discouraging the young and the foolhardy, but they exist nonetheless.

Implanting magnets or chips in the body can offer extra-sensory or extra-human abilities, but they require some training to be useful. Kevin Warwick, a Professor of Cybernetics at the University of Reading in the UK, surgically implanted a chip in his arm connected to a nerve, capable of sending a signal to an external device. However, it took him eight weeks of practice before he was able to activate the chip to send the signal at will. Learning to operate a sophisticated brain computer interface is likely to be as difficult as mastering the violin.

Remarkable feats will no doubt be realized in coming years. But they will arrive through extensive experimentation, great effort, and many failures along the way. The ill, the disabled, and the needy will be the first experimental subjects. Those yearning for enhancements hope that trials among the unfortunate will yield proven techniques for improving capabilities and interfacing the human mind and body with computers and nanomachines. In the meantime, drugs and wearable devices will continue as the most attractive methods for technologically improving oneself.

At least technologies outside of the body can be removed, while interventions into the mind and body will, in many cases, not be reversible. The ubiquitous use of earphones tethered to iPods and cell phones attests to the increasing intimacy between people and their technologies. Virtual reality gloves and goggles have been marketed since the 1980s. Wearable devices, such as computerized bands and watches that monitor vital signs and activity, are becoming gadgets of significant interest to consumers. Robotic exoskeletons improve mobility for people with weak limb or lower back problems, and provide workers with superhuman strength. However, wearable devices must be safe, should not interfere with the ability of the user to take responsibility for her actions, and should not interfere with the property or privacy rights of others. Google glasses, which facilitate access to the Internet, were introduced as a promising device, but their acceptance faltered. People became distrustful of those donning glasses that enabled Web surfing.

Research demonstrating the dangers of texting while driving led to laws that restricted the practice. Similar research showed that talking on the phone while driving is just as dangerous, but legislators realized the public would not accept a ban on that activity. The safe use of wearable technologies will focus on similar concerns over the coming decades. As with telephoning or texting while driving, the trade-offs between functionality, ease of use, safety, and respect for the rights of others will continually be balanced off against what the public will or will not accept.

Questions about costs—financial, personal, psychological, and societal—are important, and very much on the minds of consumers. The image of the cyborg in pop culture makes this clear. *The Six Million Dollar Man* (1973) was perhaps the first cyborg to capture the public's imagination. An injured astronaut named Steve Austin is rebuilt with prosthetic enhancements that make him a heroic figure. Fourteen years later, *Robocop*, a more plaintive figure, makes his debut. Yes, police officer Alex Murphy's life is saved and what remains of his body and mind is merged with a machine to create a powerful crime fighter. Nevertheless, in the original film and its sequels, Robocop becomes a pitiful creature, as we viewers are made aware of what he's lost. This shift bodes well for the central question regarding enhancements of the human body and mind. How many people will elect to enhance their minds and bodies if there is a heavy price to pay for doing so?

Most people will only indulge the risks of enhancement if the hazards are small and the benefits great. In professional sports the incentives for doping are so great that it is difficult to formulate adequate disincentives, as even humiliation and banning have not proven effective. Perhaps in other fields a few people can gain a competitive edge that yields great rewards by enhancing their intelligence or some other mental attribute. But if a large percentage of the population takes cognitive enhancers, any actual competitive benefit could be nullified. It will be like an ad campaign between Coke and Pepsi in which a great deal of money is spent just to keep pace with each other. There are costs, but no real benefits. And yet even without a competitive benefit to the individual, there could be significant benefits for the society if, for example, fewer people do stupid things, or more people are capable of taking on difficult tasks. Of course, there may also be noncompetitive benefits for the individual, including the satisfaction of being adept at a task. But what if the benefits to the society or to the individual entail serious risks, such as severe headaches or brain damage? What risks should individuals be permitted to take? Society has an interest in protecting people from harm. Therefore, an extensive regimen has developed over the past thirty-five years to ensure that research subjects are treated ethically.

## RESEARCH ETHICS

Research ethics will play a crucial role in the development of technologies that enhance human capabilities. The protections in place to oversee the ethical treatment of human subjects create barriers to studies that are dangerous, unwarranted, or bad science. Research subjects have the right to be informed about the risks before they give consent to be included in a study. These rights are relatively new in the annals of scientific investigation and are still being revised.

Horrific experiments by Nazi doctors on prisoners in concentration camps resulted in the death or disfigurement of countless unwilling subjects. Experiments on slaves, prisoners, the mentally ill, or ethnic groups deemed "undesirable" were nothing new, and yet the scale of Nazi experimentation shocked those following the post-World War II Nuremberg Trials. A few of the doctors pleaded in their own defense that they should be treated leniently due to an absence of international laws limiting medical experimentation. Subsequently, guidelines known as the Nuremberg Code were formulated to outline permissible practices when experimenting on human subjects. The Code calls for voluntary consent from subjects and avoidance of unnecessary pain or suffering. While an important first step, the Nuremberg Code failed to be transformed into international law or adopted legally by individual countries.

The Nuremberg Code also failed to prevent further egregious experiments on human subjects. The notorious Tuskegee experiment, in which the progress of syphilis was followed in 399 poor black men, began in 1932, and continued after the war until 1972. The experiment is especially controversial because penicillin was recognized as an effective cure for the disease in the 1940s, but subjects in the study were never given treatment. The Tuskegee study is also deplorable in that it was conducted under the auspices of the U.S. Public Health Services. Even more horrendous were syphilis experiments conducted by the same health services in Guatemala from 1946-1948. Soldiers, prostitutes, patients in mental hospitals, and prisoners were injected with syphilis and other sexually transmitted diseases, leading to eighty-three deaths.

In 1966, anesthesiologist Henry Beecher published an article in the eminent *New England Journal of Medicine* citing twenty-two unethical experiments with human subjects that had been performed in the U.S. in the years following the Nuremberg Trials. Beecher kept the researchers' names anonymous, but later investigations revealed the experiments had been conducted in mainstream institutions and the results were published in leading journals. Among the examples described by Beecher was the introduction of live cancer cells into subjects who were never informed that the cells were cancerous. Building upon the Nuremberg Code, the Beecher article created impetus for the eventual enactment of laws and professional guidelines detailing acceptable practices. Neither the Tuskegee nor Guatemala syphilis studies were mentioned in the Beecher report, because they only came to light in 1972 and 2005, respectively.

Over the past half-century, thanks to the Beecher article and revelations about the Tuskegee experiment and similarly horrific research, regulations and oversight agencies, to ensure the rights of research subjects, have been developed in many countries around the globe. Those rights include voluntary participation and informed consent, respect for persons and their autonomy, and protection from harm. Furthermore, experiments must be designed so as to yield scientifically valid results, and the benefits must outweigh the risks.

Many experiments, such as chemotherapy treatments for cancer, will be painful or carry side effects. Before signing on, a subject must be informed of the risks. Subjects must not be coerced or enticed by money to participate in experimental research. Nor should the likelihood of their receiving benefits from the research be exaggerated. For example, prospective subjects should know that few, if any, of them will receive any benefit for being in an experiment testing an unproven drug. The chance of any benefit will be even lower if half the subjects are randomly selected for a control group given a placebo.

The U.S. Federal Policy for the Protection of Human Subjects, better known as the "Common Rule," was enacted in 1991. The Federal Drug Administration (FDA) holds primary responsibility for updating regulations covering the testing of drugs, devices, and new medical procedures. It also functions as an oversight agency, and acts through thousands of locally run Institutional Review Boards (IRBs). An IRB within a hospital or university reviews research proposals and determines whether the design of the experiment adequately protects subjects. Approval from an IRB is required for research to start. The IRB can shut down an experiment if a violation of the regulations occurs.

IRBs and their international counterparts will determine which enhancement experiments can proceed. Raising or lowering standards for research on enhancements provides a degree of leverage for both proponents and opponents of enhancement research. But IRBs alone cannot be expected to alleviate all risks. For example, they do not make determinations as to the social value of research—whether, for example, a new drug might give its users an "unfair" advantage. They merely evaluate the safety of the research and whether the therapeutic benefits justify the risks.

Research ethics committees started out as paternalistic bodies loath to put subjects at risk. They did, however, recognize that people suffering from life threatening conditions, such as intestinal cancers, were anxious to take on major risks when a new treatment might offer their only hope. Even when there is no hope of a cure, the terminally ill often volunteer as research subjects to assist the progress of scientific knowledge regarding the disease killing them.

A shift occurred in the attitude of IRBs during the early years of the AIDS epidemic, according to Dr. Robert Levine, a Professor of Medicine at Yale, who played a leading role in the development of medical and research ethics. AIDS activists challenged the paternalism of IRBs in slowing down research on drugs that could potentially impede the progression of the disease. They successfully argued that IRBs should not be protecting them from risks that they willingly desired to take on. This shift will have great significance as increasing numbers of potential subjects petition to be included in experiments testing drugs or devices that enhance capabilities.

Historically, members of an IRB have determined that risks should not be taken on unless the research will have demonstrable therapeutic benefits. But Maxwell Mehlman, a Professor of Law and Bioethics at Case Western Reserve University, points out that nothing in the regulations restricts IRBs from approving research designed explicitly to test a human enhancement. Mehlman's reading of the regulations, while correct, will not necessarily translate into IRBs' immediate willingness to approve somewhat risky enhancement experiments. And yet, advocates will become skilled at arguing the psychological benefits participants will derive from such enhancements.

This issue of IRB approval only affects research for explicit enhancements. For example, a drug that has no known therapeutic value, yet is believed to improve the ability of children to perform especially well on standardized tests, would be an explicit enhancement. Of course, the drug might first get approval through studies that demonstrate its usefulness for improving the intelligence of children with special needs.

Pharmaceuticals that have already received approval for one therapeutic use are a different story. Doctors can prescribe these drugs for off-label applications for which its efficacy has never been demonstrated. Modafinil (sold as Profigil or Alertec) was approved in 1998 by the FDA to treat narcolepsy and later (in 2003) for additional sleep disorders. The belief that modafinil contributes to wakefulness and attention, with little or no side effects, has turned modafinil, as mentioned, into a cognitive enhancer of choice. An estimated 95 percent of prescriptions for the drug are for off-label uses. The manufacturer Cephalon paid $425 million in fines in 2008 for basically encouraging its being prescribed for uses other than those for which it had received FDA approval. But don't feel bad for Cephalon, as Provigil continues to generate $700 million in U.S. sales each year. Given the profits, the fine was just a cost of doing business.

Little quantitative research has been performed to study the effectiveness of drugs and food supplements presently marketed and touted as improving memory or other cognitive functions. The FDA should encourage testing the enhancement effectiveness of drugs. It should pass on instructions to IRBs that such experiments can be approved. Consider the use of Ritalin and Adderall, prescribed to treat children with Attention Deficit Hyperactivity Disorder (ADHD), but also widely presumed to enhance concentration and study skills for anyone. Parents, believing the drug will improve their son or daughter's grades, are known to want ADHD diagnoses for their children in order to get an Adderall prescription. And yet, the drug's effectiveness for improving the cognitive capabilities of essentially normal individuals has never been fully demonstrated. One study at the University of Pennsylvania by Irena Ilieva, Joseph Boland, and Martha Farah concluded that Adderall provided little or no improvement in the cognitive skills of healthy young adults. Nonetheless, some subjects in the experiment believe it did help them—a possible placebo effect.

Exaggerated expectations and money wasted on ineffective technologies provide one good reason for studying technological enhancements. The prospect of dangers arising from back alley surgeries and tainted drugs offer another. Heavy regulatory restrictions spawn black markets. Before abortions become legal in the U.S., underground abortions were commonplace and occasionally led to death or disfigurement. The rich could always find a way to get a safe abortion even if that meant flying to another country. Protecting middle class and poor women from harm caused by back alley procedures played a significant role in overturning legal restrictions on abortions. More recently, a black market in cheap cosmetic surgery (also viewed as an enhancement) has caused death and disfigurement. In science fiction portrayals, cyberpunk antiheroes travel to seedy districts of dystopian Asian cities to buy brain-computer interfaces, organ transplants, nanomachines, and illicit drugs. Open markets afford the opportunity for oversight and the ability to convey dangers to unsuspecting users.

Proponents of human enhancements contend that regulations to slow research will be futile. Investigators will move their labs to countries with different values. Perhaps, but countries around the world have increasingly become concerned that their citizens not be used as guinea pigs for research that other countries restrict. In past decades, drug companies performed research in Africa and Asia that would not be approved in Europe or the U.S. For example, in the 1990s, a backlash over drug trials in the Ivory Coast where some pregnant women with HIV were given placebos rather than antiviral drugs, led to the cancellation of similar studies in other countries. While such drug trials are unethical and make it sound like nothing has been learned since the Tuskegee experiment, cancelling the research is also tragic. The women felt that enlisting in the trial was their only hope for getting the drugs necessary to protect the child they carried from contracting HIV.

Corrupt officials can be bribed. The governments of poor countries can be influenced. Hidden labs funded by rich benefactors in an island nation might be ignored. Yet the difficulties of masking dangerous research or concealing large drug trials in a networked world means accidents and harmful incidents will come to light. If major human enhancements require years of research, trial, and error to move forward, as it appears they will, the scale of the research will be hard to hide. Furthermore, large investments of capital will require the involvement of wealthy governments and powerful multinational corporations. Concealing dangerous research will be difficult but not impossible.

Defense research by major powers is often kept secret. One popular science fiction theme entails a government, commonly China, engaged in secret research to clone a superwarrior. The U.S. already funds research that combines various technologies to create the enhanced future soldier or cyborg warrior. En route to the battle zone, the soldier would be injected with a vial of viruses carrying copies of select genes from his own DNA, a proven method for repairing muscle tissue and increasing strength and stamina. He is handed a cocktail of the latest cognitive enhancers to improve concentration, heighten memory, and speed up reaction time. These future combatants are outfitted with robotic exoskeletons that give superhuman strength. Neuroprosthetic devices that convey thoughts facilitate a soldier communicating with other members of the team. Insect-sized drones flying through the jungle in search of guerrilla warriors beam back the enemy's location, which is then overlaid on the soldier's visor. With the help of a tablet computer built into a data glove, he directs larger weapons-carrying drones. Back at headquarters, supervisors monitor the physiological well-being of soldiers with the assistance of information from nanosensors on their body and in their bloodstream. At the same distant locale, a supervising medic activates a shot of adrenaline when vital signs falter.

Although forward-looking military planners fund innovative and truly experimental research, commanders in the field tend to be much more conservative in their adoption of new technology. Nonetheless, they do wield inordinate power to command soldiers to don experimental devices with little or no room for the soldier to say "no." In this sense, the requirement for voluntary participation in research does not exist in the military. Even when given a choice, a soldier could feel under compulsion to do what the commander requests. Research ethicists would like active military members to have the same rights for voluntary consent that are afforded civilians.

It's not a perfect system, but research ethics does hamper researchers from proceeding with harmful, unnecessary, or abusive experiments. Research ethics will not arrest the development of enhancement technologies, but it will tame its pace and provide many inflection points to reflect upon whether the individual and social benefits justify the risks. However, when the rewards are great as in providing an edge in professional sports, many young athletes will be more than willing to engage in high-risk activities. In the military, there will be plenty of volunteers willing to experiment with the latest enhancements, particularly if doing so is represented as helping provide for their country's security. Even with informed consent, few soldiers will seriously consider in advance their fate if a procedure or drug goes wrong.

What *should* happen when research goes bad? Some enhancements will not be reversible. Should the military, for example, be restricted from enhancing soldiers in ways that cannot be reversed? Perhaps, as the philosopher Chalmers Clark recommends, potential civilian research subjects should have an advocate who emphasizes for them the downside of participating in experimental studies. Soldiers or civilian subjects might experience side effects or even physical or neurological damage. Who will be responsible? Maybe special insurance policies for subjects should be put in place before enhancement research gets approved.

The process of reviewing research using human subjects can slow but will certainly not arrest the development of enhancing technologies. The review process at least offers some breathing space. If society decides enhancement trials require additional regulations, as opposed to therapeutic research, then the research ethics review is where those compliance measures can be added. However, it is always important to keep in mind that too much additional regulation will just drive the research underground, and thereby raise the chances of putting the safety of subjects at risk. Also, the credibility and tacit support given to the review of research using human subjects could be undermined if it were perceived as a tool for the pursuit of ideological ends.

## COGNITIVE ENHANCEMENTS

One area likely to draw subjects despite the risks will be the testing of drugs that might enhance cognitive abilities. Cognitive enhancers offer both mundane and transcendent capabilities. In Ramez Naam's science fiction novel *Nexus,* a nano-drug bequeaths the user telepathic union with other minds. The idealistic genius-hero Kane and his cohorts perfect a version of the Nexus drug, which can be programmed and then activated through thought alone to facilitate mastery over the mind and body. How to safeguard their Nexus secret from a clandestine agency within the U.S. Department of Homeland Security and other interested parties provides Kane with an engaging ethical dilemma. As entertainment, *Nexus* is a well-envisioned journey into a transhuman future. But are technologies like Nexus any more plausible than the One Ring from *The Lord of the Rings*? Tolkien's ring serves as an allegorical symbol, but contemporary science fiction purports to be possible or even inevitable. The plausibility of the Nexus drug stems from research in neuropharmacology and neuroprosthetics.

In classic tales the hero overcomes many obstacles to acquire the magical potion or relic. The obstacles confronted by nerd heroes relate more to safeguarding the powerful device from those who would use it for destructive purposes. While potions, elixirs, and drugs generally bestow unusual abilities and magical powers with little or no effort, the cognitive enhancement drugs presently available are difficult to harness and not particularly powerful. The pharmacopeia of drugs for depression, schizophrenia, and a host of other mental maladies help millions stabilize their lives. And yet they are difficult to tweak, often come with discomforting side effects, and have little or no enhancing properties.

Coffee, Provigil, and other stimulants sharpen attention and the ability to perform well over a long period of time, but do not enhance cognitive capabilities to a new level. Marijuana can facilitate a full-bodied engagement with the moment, particularly a piece of music. Mind-altering drugs, from mescaline to thousands of synthetic psychotropics, such as LSD, open doorways to dreamlike states. For some people that is enough. However, only a few creative souls seem capable of harnessing insights from drug-induced experiences for useful purposes. Most of those are painters, writers, and jazz giants who cultivate a craft and then work very hard to ensure that they can produce art while under the influence of stimulants. The biographies of greats such as Charles Baudelaire, Jack Kerouac, John Coltrane, Janis Joplin, and Michael Jackson underscore the trials of producing art under the creative and destructive influence of drugs. And those who succeed are only a tiny percentage in comparison to those who self-destruct. There are countless depressing stories about the millions of young people who falsely believe that drugs will ease their way to creative fame and fortune, but never put in the time and discipline to hone their talents.

It is likely that individual biochemistry or subjective factors sway the effectiveness of many purported cognitive enhancers. The human nervous system and the mental states it produces are so complex and so subtle that almost anything can influence them. Discerning when a drug has a significant, consistent, and positive impact is not only difficult, but also costly. During the certification process, drugs do not get tested in concert with other drugs unless the pharmaceutical company seeks approval for a specific cure. It would be impossible to test all combinations of drugs. If a new wave of cognitive enhancers becomes available, as has been anticipated, researchers will learn much about the effectiveness of each alone and in combination with other drugs by reports from individuals engaged in self-experimentation. Therein lies the greatest risk—people experimenting with cocktails of cognitive enhancers, some of which may turn out to cause a short-term disturbance or a full-blown mental illness. Students have been experimenting with whatever mind-altering compounds are available for generations. Wealthier countries have tools for tracking adverse incidents. When a specific combination turns out to be detrimental, tracking agencies such as the U.S. Centers for Disease Control eventually recognize the problem. But reports of the dangers may or may not get disseminated to the next wave of experimenters.

The challenge of adverse mental events and substance abuse already taxes mental health facilities. There is no way of knowing in advance whether a collection of new cognitive enhancers will make matters worse. But there is also no clear evidence of powerful enhancers on the horizon, capable of bestowing dramatic improvement in memory function, analytical skills, creativity, or any other cognitive ability. The belief they are coming is a by-product of the surge in neuroscientific research and the presumption that the brain will soon reveal all its secrets. Most neuroscientists, appreciating the complexity of the nervous systems, are hopeful of finding treatments for mental diseases, but much less optimistic about obtaining mastery over the brain. Not only will neuroscience need to make dramatic strides beyond what is presently understood, but medical practitioners will need to learn how to fine-tune the idiosyncrasies of individual brains with the tools available. Given the complexity of the brain, that fine-tuning could be even more difficult than managing other complex dynamic systems such as economic markets.

Some of the most fascinating research involves neuroprosthetics. Implanted chips, for example, transmit bits of information from nerve tissue, or transmit information directly to nerve tissue. A chip implanted in a monkey made it possible for the monkey to operate a robotic water delivery system. A chip implanted in the motor cortex of fifty-three-year-old Johnny Ray, who had suffered a massive stroke that paralyzed him from the neck down, made it possible for him to spell out words by moving a cursor on a computer screen with his thoughts. The trick lies in software that deciphers the information. Implanted chips, limited to 256 synaptic connections or less, convey a very limited amount of information, but proposals for broader bandwidth could dramatically increase the dataflow.

In one creative proposal, a bundle of tiny carbon nanowires slipped through a blood vessel would forge synthetic connections to large numbers of neurons in the brain. The nanowires would be similar to the fiber-optic cables that carry tremendous amounts of video information and computer data for Internet and media services. Another strategy proposes thousands of individual nano-transmitters connected to neurons, broadcasting information wirelessly to and from external devices. But many unknowns remain as to whether scaling up dataflow can be performed effectively and safely. Each incremental step forward requires solutions to a host of engineering challenges, such as ensuring that the materials used do not deteriorate or damage nerve tissue.

The seamless integration of new activity with other brain functions is of central importance. The research to date works with underutilized nerve tissue in individuals with disabilities. The introduction of broad bandwidth transmission that, for example, permits internal Internet access would overlap with and potentially interfere with other functions—a dangerously distracting prospect. In comparison to the massive amount of unconscious processing going on in the brain, consciousness is sequential and needs to be available to focus on the tasks at hand. Managing competition for conscious attention from new inputs to the brain stands as a difficult, though probably not insurmountable challenge. With its penchant for drawing us in, surfing the Web while driving could be a particularly dangerous distraction. Remember to turn on the self-driving function in your automobile before turning on the Internet in your head!

Skills afforded by neuroprosthetics will not be instantaneous. Perfecting enhancements that optimize individual cognitive capabilities requires dedication and rigorous practice. A perilous learning stage awaits the intrepid explorer. The difficulties and dangers of linking the brain with external devices, however, will not diminish the desire to explore cyberspace directly from our own brains.

*Cyberspace,* the word if not the destination, was invented by science fiction writers in the early 1980s and received widespread dissemination through William Gibson's *Neuromancer,* the breakthrough novel of the cyberpunk genre. The antihero of *Neuromancer* is Case, a onetime hacker and drug addict turned hustler, living in the underworld of a dystopian Japanese city. With the help of drugs, Case "jacks into" the worldwide network of computers and allows his mind to roam freely through the many worlds of cyberspace. Published in 1984, *Neuromancer* predates the configuration of the Internet into a *World Wide Web*, but envisions the interconnection of computer realms and how this might be used as a virtual world for entertainment, experience, education, and cyber espionage. Case's dependence on drugs, organ transplants, and a prosthetic interface to transport his mind into cyberspace serves as a plot device, but Gibson's intuition may have tapped into an essential truth. Dissociating the mind from the body for sustained periods of time will require drugs and/or discipline, and will have a downside that includes putting taxing stress on internal organs.

Gibson's "jacking in" and Naam's Nexus illustrate two very different forms of technologically enabled transcendence. One provides an escape from the body, freeing the mind to travel through the interconnected realms of cyberspace. The other offers an embodied engagement that submerges one mind in a sea of minds. As motivators for developing and experimenting with enhancement technologies, the spiritual promise of transcendent states of mind complements the material rewards from optimizing individual capabilities. For the more puritanical among us, technological means to spiritual states dispense counterfeit defilements of the sacred. And yet such judgments lose their potency as science increasingly establishes its hegemony as the arbiter of what is true and what is possible.

# { 10 }

# Pathologizing Human Nature

TRANSHUMANISM STAKES A CLAIM TO BEING AN INTELLECTUAL movement of note. Its leading thinkers represent the transhumanist vision of an enhanced and improved humanity as a culmination of forces set in motion by the seventeenth century Enlightenment. Together with the theory of evolution and the computation theory of mind (the hypothesis that minds are computers), transhumanism offers a new synthesis of what it means to be human and how our species will evolve to transcend its limitations. The emerging synthesis, which combines scientists' descriptions of biological processes and human features with a transhumanist analysis of where we are headed, is inspiring for those desirous of acquiring the benefits they perceive enhancements will afford. That same vision is disturbing to those of us who feel it buys into a representation of what it means to be human that is incomplete and mechanizes, biologizes, and pathologizes human nature.

The transhumanist vision incorporates many of the ideas and insights which assault the belief that we humans are exceptional creatures, whether designed by God or by nature. Humans are not only unexceptional creatures, from the perspective of this new synthesis; it is necessary to transcend our limitations because we are also flawed creatures. With this attitude, the future inevitably (and should) belong to our evolutionary successors.

The assault on human exceptionalism has hastened the onset of a full-blown spiritual crisis. Spiritual or psychological crises are of a different order from the disasters I've discussed so far. And yet they are important for our purposes to the extent that they rob people of confidence in their own and our collective ability to meet challenges. But the more common reaction to the present-day spiritual crisis appears to be dismay and/or resignation at the inevitable encroachment of technology into domains formerly considered sacred. In other words, a white flag of surrender is waved before the incessant and mechanistic unfolding of the presumed technological inevitabilities. A first step toward re-empowering ourselves, and keeping technology from slipping beyond our control, is to recognize the gaps and distortions inherent in the various assumptions that sanction technology's ascendency.

The humanism of transhumanism lies in its proponents' faith that the best of what it means to be a feeling, caring, and self-aware person will be fully expressed in technologically enhanced cyborgs and *techno sapiens*. I am fearful that we will buy into a counterfeit technological illusion of transcendence. There is a serious gap between the explication of technological possibilities and how they get imbued with human qualities. Technology could triumph over our humanity in many different ways. While visions of *techno sapiens* and digital clones living eternal lives dance in the heads of the imaginative, the dystopian impact of enhancements invades the nightmares of the unconvinced.

The leading transhumanist thinkers are sensitive to the concerns that make large segments of modern society uncomfortable with major alterations of the human mind and body. They offer deeply reflective, creative, and serious responses to their critics. They are working hard to explain why transhuman possibilities will be positive for humanity, and have turned their attention to the ethical concerns that have been raised. Nick Bostrom, a founder and first president of the World Transhumanist Association (WTA), presently focuses his attention on global catastrophic risks, particularly the danger of an unfriendly technological singularity caused by super-intelligent AI. J. Hughes, the former executive director of the WTA, now oversees the Institute for Ethics & Emerging Technologies (IEET). In Hughes's mind, attention to ethical considerations will flesh out the importance of the transhumanist movement. I certainly welcome the focus upon ethical concerns. However, from my perch, reflection upon the ethical issues human enhancements raise is largely a compensation for an inherent weakness in the transhumanist belief that technology is of primary importance in bettering the human condition. Technology is a helpful aid, but an untrustworthy leader.

None of this alters the simple fact that transhumanism offers a powerful vision that in certain respects is more comprehensive than competing worldviews. Transhumanism emerged as a compensating vision for the inability of other forms of humanism or of religiously inspired worldviews to fully accommodate the scientific method and its continual stream of verifiable revelations.

Until recently, placing science and technology at the center of human understanding had been perceived as inadequate because these fields lack an inherent ethical component. Those with religious beliefs, for example, often express doubt that we can ensure people will act ethically if God's existence is questioned or dismissed. For Christians, and indeed some non-Christian faiths, the promise of punishment for bad deeds is essential, as are spiritual rewards for good behavior.

But suddenly ethics has been turned into a focus of scientific inquiry. Where once it seemed obvious that the competitive evolutionary struggle could not have created people capable of self-sacrificing, altruistic acts, today biologists theorize how values emerged through evolutionary means. For example, game theorists explain why natural selection would favor members of a species with traits for sharing or cooperating over other members of the same species who are selfish. Neuroscientists also are taking a serious interest in how our brains might be hardwired for specific moral intuitions or ways of making moral judgments. And both humanists and the new atheists, such as the evolutionary biologist and author Richard Dawkins, build upon this scientific edifice to extrapolate as to why God or spiritual rewards and punishments are totally unnecessary for people to behave ethically.

Over the past few years, in perhaps the boldest move yet, transhumanists contend that moral behavior can be enhanced technologically. This proposal offers a good starting point for our discussion regarding the incompleteness and distortions evident in the present-day science/transhumanist synthesis.

## MORAL ENHANCEMENT

The sudden attack by Muslim militants on a platoon of U.S. soldiers patrolling the foothills of the mountainous region between Afghanistan and Pakistan was over in less than two minutes. While a medic attended the wounded, she shouted out to the other combatants to take a previously issued pink pill. The pill, a 40-milligram dose of propranolol, if taken soon after a terrifying ordeal, reduces the likelihood of post-traumatic stress disorder (PTSD).

PTSD is debilitating. By some estimates, 20 percent of soldiers exposed to a traumatic incident develop full-blown PTSD, and another 30 percent exhibit some symptoms or "partial PTSD." It also strikes rape victims, refugees, and even health care professionals. People suffering from

PTSD feel distress or relive traumatic events even when they are not in danger. For some victims of PTSD, their life is a series of nightmares and intrusive memories set off by a dropped pot or a popped balloon.

Evidently propranolol interferes with the encoding of the emotionally-laden memories associated with PTSD. It may also be useful for altering certain attitudes. The discovery of propranolol's influence on emotions inspired a team of Oxford University researchers led by Sylvia Terbeck to investigate whether the drug could alter implicit biases, particularly emotionally-laden racial biases. Preliminary experiments indicate that it does. Might propranolol then be used to reduce racism and improve or enhance moral behavior? The researchers showed reserve as to whether their findings would hold true outside of a laboratory setting, but one member of the team, the Oxford philosopher Julian Savulescu, had already proposed that biomedical means should be employed to enhance people morally. For Salvulescu, the possible use of propranolol to dampen racism serves as just one example among a number of drugs and other techniques that might be used to raise people's moral acumen.

In a book entitled *Unfit for the Future: The Need for Moral Enhancement,* co-authors Igmar Persson and Savulescu argue that moral enhancement is essential for solving the potentially catastrophic challenges humanity faces. In their eyes, traditional methods of moral education and social reform have proven too slow and inadequate to sufficiently raise moral attitudes and aptitude, or improve behavior. Problems whose solutions require major behavioral adjustments, such as global climate change, can only be met if people's moral attitudes and will improve significantly.

In addition to propranolol, oxytocin—a hormone that contributes to feelings of trust, love, motherly devotion, and generosity, as well as lowering stress and reducing social fears—has also been considered a moral enhancer. Research lead by Ruth Feldman, a psychologist at Bar-llan University, concludes that women with high levels of oxytocin during the first trimester of their pregnancy bonded better with their newborns. Another series of experiments lead by economist Paul J. Zak revealed that externally administering a dose of oxytocin made subjects much more trusting and generous during games in which they must decide how to split up a sum of money. Given research findings such as these, oxytocin quickly acquired labels such as "the love drug" or "the trust drug." Visions of oxytocin perfume or oxytocin food additives soon materialized.

For transhumanists, moral enhancement quickly became one more killer application, one more irresistible technology—in addition to longer life—

that would make enhancing human capabilities desirable. Furthermore, the notion of moral enhancement helped muddy the water. It is hard to argue against moral improvement. While other enhancements such as greater intelligence might be factors in worsening inequality, from a societal perspective moral enhancement appears to have only an upside. The notion of moral enhancement thereby defuses arguments against enhancements in general, and reinforces the sense that if moral enhancements are inevitable, so too are enhancements like greater intelligence and superior physical prowess, so we might as well submit.

On the other hand, treating human moral faculties as biochemical mechanisms that not only can, but also must be enhanced medically carries the false ring of a distorted form of scientific reductionism. Are love and compassion nothing more than biochemistry? Biochemical mechanisms make human activity and human emotions possible. But are we really being asked to buy into a model where the behavior of healthy people can and should be manipulated from the bottom up? Bottom-up approaches change one's chemistry in order to alter psychology and behavior, and can be contrasted with traditional top-down approaches for improving behavior such as character development. Or is the argument that we are all flawed creatures who must be treated therapeutically in order to be healthy? In other words, are some scientists and transhumanists pathologizing human nature?

Moral improvement at an individual level has traditionally been viewed as the result of effort in the form of rational reflection or the cultivation of virtue. Can moral improvement truly be acquired by taking a pill? Or does this notion represent the delusionary overreaching of techno-solutionism? Perhaps drugs will be discovered that actually help individuals improve their moral behavior, presuming that is the person's intention. Nevertheless, we should not confuse maneuvering people to act appropriately as being equivalent to having moral intelligence. Consider a con artist who employs oxytocin for eliciting trust to dupe her mark. Certainly the moral intelligence of the individual duped was not enhanced by the drug's effect.

Altering emotional content, attitudes, and memories is a controversial exercise, even when for moral purposes. Consider a woman who is given propranolol twenty minutes after being raped by a male attacker. This could certainly be beneficial to her, but would testimony she gave in court against the rapist be deemed reliable? If her attacker avoids prosecution because her testimony gets dismissed, what does this do to our system of justice or her sense that the evildoer paid for his crime and her suffering? It certainly will not protect others from victimization by a serial rapist. Or consider a rapist who takes propranolol before or immediately after committing the dastardly act. In such a case, the drug might act to lower any sense of guilt or remorse, and thereby function as license for future immoral acts rather than as a moral enhancement.

Increasingly, a significant portion of scientists represent the human mind/body as a complicated machine made up of a collection of discrete mechanisms. Scientific findings that reveal the importance of specific bio-chemical processes exhilarate transhumanists with the prospect of treating the mind/body as a design space that can be improved. But the implication that humans are essentially machines, which can and should be molded through tinkering, is incorrect because the whole person is something more than the parts. When no other option is available, using a drug to relieve debilitating symptoms is clearly acceptable. However, the use of drugs should never be confused with the hard-won character development that comes from the conscious recognition of the instances when one's behavior is unacceptable. Tinkering can certainly alter some behavior, just like surgery can clear plaque out of a clogged artery or remove a cancerous tumor. After the surgery, the body will function better, not just in remediating an obstruction, but in allowing the body parts and mind to operate more optimally as an integrated organism.

An excessively reductionistic or mechanistic model of the mind/body can be viewed as demeaning our sense that we humans are a unique species with exceptional capabilities. This is certainly not the first scientific assault on the belief that we humans are truly exceptional creatures. That assault dates back to Copernicus's (1473–1543) recognition that humans do not reside at the center of the universe. The contemporary philosopher Luciano Floridi points out that the affront to human exceptionalism continued through the Darwinian and Freudian revolutions, which successively demonstrated that humans are animals and not particularly rational. Floridi then goes on to characterize the Information Age as a fourth revolution that also assaults human exceptionalism. People, information systems, and all of life are being viewed as what Floridi calls inforgs, made up of information. Floridi is correct in pointing out that each of these revolutions has been an assault on the view we humans have of a uniquely exalted place in the natural order.

Stripping away false beliefs is generally a good thing. We are not, after all, at the center of the universe as we once thought. But when new, supposedly scientific belief systems get substituted, caution is in order. The twentieth century was haunted by destructive ideologies, including Nazism and Communism, that claimed to represent the latest "scientific" understanding. Vital aspects of our humanity get diminished when scientists place too much emphasis on the biological, or for that matter, informational mechanisms that make human activity possible.

## BIOLOGIZING AND MECHANIZING

Throughout modernity there have been scholars who believed that life could be reduced to some form of physical and biological determinism. In recent years, that viewpoint has gained power with plausible hypotheses that organic life serves the needs of "selfish genes," that character can be manipulated biochemically, and that consciousness is merely an emergent property of synaptic activity in the brain. Whether such theories *explain* or "explain away" the complexities of life remains an important debate in the philosophy of science.

The Stanford University neuroscientist, primatologist, and best-selling author Robert Sapolsky's proposals as to the centrality of biochemistry in personality, attitudes, mental states, and identity are among the most far-reaching. He points to a large body of research findings that demonstrate how individuals' psychology and character have been altered by neurochemical disorders and improved by biochemical interventions. Treating the symptoms of depression, anxiety disorders, attention deficit hyperactivity disorder, and other mental maladies biochemically changes people's behavior and sense of self. In effect, Sapolsky argues for a form of biochemical determinism. From this perspective, even criminals should not be held fully responsible for behavior that is the result of biological factors beyond their control.

Sapolsky is just one among countless scientists whose work contributes to the biologizing and mechanizing of human behavior. The methods of science are most effective when reducing complex phenomena to discrete physiological components, biochemical processes, and psychological mechanisms that can be examined and tested within controlled experiments. The success of this approach is self-evident. But it also lends itself to an imbalanced mechanistic view of natural processes and the life of both humans and nonhuman animals.

Primatologists, for example, understand that the behavior of chimpanzees within a controlled laboratory setting differs significantly from how the same great apes would act in the wild. As scientists they are interested in the behavior of animals in natural settings. But the study of animals in their natural habitat (ethology) is time consuming, costly, and opens up any conclusion to challenges based upon the influence of uncontrolled variables. In other words, the simple fact that scientific reductionism leads more easily to verifiable results, creates an implicit bias that quickens the development of the scientific enterprise, while inviting distortions.

Biologizing and mechanizing are intellectual tools applied by scientists and social theorists for analyzing behavior, and through which the functioning of the whole person is understood in terms of the centrality of individual components. In contrast, the mind/body can be viewed as a complex, integrated organism engaged in an intricate dance with the many elements of the environment in which it is embodied. Furthermore, organisms demonstrate behavior or properties that cannot be fully explained by the cumulative impact of the parts. Reducing a human or a nonhuman animal to its individual components, or viewing a person in a specific setting as an integrated being—integrated biologically and in relationship to her immediate environment—represent totally different perspectives. And yet both points of view do have their place in furthering efforts to understand our minds and bodies. They are both necessary to frame a more comprehensive understanding of the activity of both humans and nonhuman animals.

Philosophers debate the extent to which the scientific enterprise itself leads to a distorted representation of human nature. Or whether the tendency to mechanize and biologize human behavior results from distorted interpretations of what science reveals. The answer usually depends upon the specific cases examined. In the meantime, recognizing distortions provides a first step in evaluating whether biotechnologies offer an appropriate or misguided response to the challenge of improving human behavior.

## FIXING DESIGN FLAWS

Not only must we reckon with the scientific tendency to mechanize and biologize behavior; we also have to deal with the way this process is used to reveal flaws and target those flaws therapeutically. Assertions that humans are flawed date back thousands of years. For Christians, it took the form of original sin, first stated by Irenaeus (130-202 CE), and later developed more fully by Augustine of Hippo (354-430). In Christian doctrine, God made man perfect in his image, but through their actions in the Garden of Eden, Adam and Eve bequeathed original sin to all their successors.

Echoing a sentiment we each feel at times, Frederich Nietzsche (1844-1900) proclaimed in his *Genealogy of Morals*, "Too long the earth has been a madhouse." Attributions of neurotic motivations pervade contemporary analyses of each person's behavior. The 2013 update to the *Diagnostic and Statistical Manual of Mental Disorders* (DSM-5), a catalog of mental conditions for professionals, runs over nine hundred pages and even medicalizes grief and minor cognitive disorders due to the normal aging process.

Popular books by writers such as Daniel Kahneman, Malcolm Gladwell, and Dan Ariely describe flawed habits of mind, routine errors, or biases that apparently rule human judgments. These annoying mental glitches can lead to questionable if not truly bad decisions. Some mental errors, particularly those associated with quick judgments, can be viewed as adaptive traits, while others—misjudging probabilities or a bias that automatically favors the status quo—might simply be cognitive deficiencies.

Mental habits make us good at quickly responding to immediate dangers while poor at assessing longer-term risks. The fight-or-flight reaction to an approaching pride of lions saved primitive ancestors living on the African savannah. Reacting or "thinking fast" helps us adapt to changing circumstances with little or no delay.

Thinking slowly—applying reason to the analysis of a problem—improves the quality of one's judgments, but requires time, energy, and discipline. Without a compelling reason, people commonly turn away from expending the kind of mental energy necessary to tackle a difficult problem.

Noting that we each have cognitive biases is a lot easier than figuring out ways to keep them from interfering with one's judgment in an inappropriate manner, such as judging a person by the color of her skin. Most people are also good at emotionally responding to the pain of an individual, but incapable of scaling those emotional responses to encompass the pain of large numbers of people. In a quote often misattributed to Josef Stalin, "the death of one man is a tragedy, the death of millions is a statistic."

Psychological fixes for mental biases and cognitive deficiencies usually focus on creating new compensating habits, or on defusing existing habits by recognizing the situations in which they mislead us. Unfortunately, the success of behavioral fixes for deficiencies has been rather modest. Uprooting bad habits is never easy. And when flawed judgment is not merely the result of learned habits, but arises from inherent propensities, it is even more difficult to change. The failure of psychological and behavioral fixes is one reason proponents of moral enhancement feel biochemical or genetic approaches may be necessary to improve people's behavior. Certain drugs might be used as aids that help improve behavior, presuming that is the individual's intention. But like the discipline an athlete must expend to exact results from the use of a growth hormone, drugs that facilitate behavioral improvements will do so only with will and effort.

The representation of humans as flawed goes beyond decision-making to an array of evolutionarily bequeathed physical attributes. The blind spot in vision is one flaw. Lower back problems, overly complicated feet, and weak knees are all the result of a body whose design features were naturally chosen for our quadruped ancestors. Standing erect came late in the evolutionary process. A truly farsighted engineer would have been required to plan in advance for the continual stress on joints as a result of a long bipedal life afforded by modern medicine.

The battle of older men with urethras awkwardly surrounded and compressed by a swollen prostrate gland attests to either the absence of a divine creator, or the presence of one with a weird sense of humor. Women similarly deal with painful urinary and bladder tract infections due to multipurpose genitals and their proximity to the rectum. A narrow birth canal limits the size of the infant's head and makes for agonizing deliveries.

From a modern perspective, a sweet tooth and a fondness for salty fatty foods warrants note as a dysfunctional attribute. In the state of nature such foods were difficult to come by. Surviving cold spells or famine depended upon stored fat. Appetites that served our ancestors damn us. In a world of abundance, sweets, fat and salt, french fries and ice cream are the bane of a body evolved to serve the needs of primitive hunter-gatherers.

Any list of individual attributes that appears dysfunctional in modern contexts provides a good case for redesigning humans. Indeed, the very claim that the human mind and body needs to be upgraded rests upon such lists. Whether a design upgrade is a scientific prescription or one among many possible interpretations of scientific facts, largely depends upon how one understands the scientific enterprise.

Techno-optimists have stepped boldly into the present-day spiritual crisis by proposing we embrace the continuation of evolution by technological means as a new source of promise and meaning. Alternatively, those critical of science as an arbitrator of meaning, reassert their belief in human exceptionalism and reject tinkering with human nature. A middle way accepts some forms of self-improvement but calls for a critical examination of each approach to biomedical enhancement. Aristotle recommended that we choose the middle way as a route to the *good life* or happiness. Perhaps we should take his advice. However, selecting any one of these three approaches presumes we understand the present spiritual crisis and its practical ramifications well enough to discern the constructive ways forward from those where illusory promises pose as solutions.

Certainly many people have and will employ biotechnologies for self-improvement. Whether such methods, however, are deemed good, acceptable, or inappropriate relies largely upon one's view of human nature. The enhancement debate has been primarily a struggle over who gets to define human nature, and thereby frame whether, or what kinds of, enhancements are acceptable. However, one thing appears clear. To quote William Irwin Thompson, "There seems little chance of getting out of this century with the same human nature with which we entered it."

## THROUGH A GLASS DARKLY

My favorite example of a prediction that was far off the mark dates back to 1894. In that year, the eminent physicist and later Nobel Prize winner Albert Michaelson declared, "It seems probable that most of the grand underlying principles have been firmly established." Little did Michaelson know that research he had published seven years earlier with his colleague Edward Morley would lead directly to a revolution in our understanding of the physical universe. The Michaelson-Morley experiment failed, and yet in its failure proved that ether, an all-pervading stationary medium through which light was presumed to propagate, just did not exist. This, in turn, forced physicists to rethink their Newtonian assumptions about the structure of the physical world. Historians of science consider the Michaelson-Morley experiment to be a starting point that leads to Albert Einstein's special theory of relativity in which there is no need for the concept of stationary ether.

In the dialectic of history, each new idea or thesis is either incomplete or carries distortions which automatically get compensated for through the emergence of an antithetical idea (the antithesis). The tension between the thesis and antithesis gets worked through until a new resolution or synthesis comes to the fore. But over time, that synthesis also reveals distortions or an inherent incompleteness. The synthesis is recognized as a new thesis and the dialectic process continues.

In the history of physics, the thesis was a model inspired by the work of Isaac Newton (1643-1727) in which ether played an important role. By demonstrating that ether does not exist, the Michaelson-Morley experiment became an antithesis challenging the prevailing Newtonian model. In the struggle to reconcile those differences, a new synthesis emerged, the special theory of relativity, which laid foundations for a new model of how the universe is structured. Over the past decades, physicists have struggled with findings that cannot be fully reconciled with relativity theory, and one of these may yet force a new revolution in our understanding of the physical universe.

Transhumanism developed as an alternative to humanistic and religious perspectives unable to fully accommodate scientific understanding. The science/transhuman synthesis purports to be a framework within which the understanding of human evolution and the emergence of human intelligence and behavior can be fully explained. All that is necessary is to work out many of the details. On closer inspection, however, the inadequacies of the science/transhumanist synthesis are coming into view. For example, most neuroscientists do not believe that the computational theory of mind is adequate to fully explain how the brain operates. In spite of everything we are learning about neural activity, how the brain works and facilitates the emergence of mind and consciousness remain mysteries in the eyes of the majority of neuroscientists. As with Albert Michaelson's prediction, the framework is not up to the job.

This chapter has alluded to some of the compensating ideas that can help lead toward a new synthesis. To what extent a new synthesis will build upon, rather than overthrow, the transhuman vision is unclear at this time. At this stage in our understanding, we are only getting a glimpse of the important considerations that will need to be integrated into a more complete framework. As the Biblical book of 1 Corinthians states, "For now we see through a glass darkly." With a little patience and humility, slowly but surely more light will be shed upon that glass.

This chapter also signals a shift in the trajectory of our narrative away from an emphasis on the risks emerging technologies pose, and toward solutions to manage those risks. The process of developing solutions begins with defusing powerful assumptions that feed the pace of change and block our ability to inspect alternative courses of action. The transhuman vision is one take on the centrality of innovative tools and techniques that nurtures and relies on the notion of technological inevitability. Transhumanists are not interested in slowing the march toward technological enhancements of human capabilities. In defusing the power of the transhumanist vision, we take a first step toward considering alternative approaches. The next chapter will question the assumptions that drive the pace at which military weaponry is developed and drive the never-ending expansion of the health care industry.

# { 11 }

# Drivers of the Storm

THE ADVANCING PROTESTORS WATCH A MILITARY GREEN HUMVEE
with a large eight-sided reflector mounted on its roof roll up and stop two
hundred yards ahead. With the help of a video screen and a joystick, the
Humvee driver makes sure the reflector is directed at the unruly demonstrators. He pushes the button on the joystick, and within a few moments, everyone, including the more macho members of the crowd, has dispersed.
The weapon is a heat/pain ray (euphemistically named an Active Denial
System). It was developed for the U.S. Defense Department by the Raytheon Corporation as a nonlethal weapon for crowd management, and first
demonstrated in 2007. A budget of $6,377,762 had been allocated to fund
the development of the initial two systems. While an Active Denial System
(ADS) was deployed to Afghanistan in 2010, the weapon has been controversial, and to the best of my knowledge, has never been used during a war.
However, Raytheon developed a smaller ADS named the Silent Guardian,
which it markets to law enforcement agencies.

The ADS is one among a new class of armament known as directed-energy weapons. These weapons focus different forms of energy—microwaves,
radio waves, lasers, or particle beams—for a variety of purposes. For example, laser weapons to shoot down incoming supersonic missiles and UAVs
are under development. In 1998, an international ban on laser blasts that
would blind soldiers and civilians on the battlefield came into force. By
September 2014, 103 nations endorsed the agreement that blinding lasers
were among the weapons that should never be used.

The ADS focuses microwaves. While your microwave oven functions at
the lower wavelength of 2.48 GHz and can cook a potato all the way

**Active Denial System**

through, the ADS emits a beam of microwaves at 95 GHz, which will only pierce the skin to a depth of 1/64 of an inch (0.4 millimeters). According to volunteers, the ADS creates a red-hot burning sensation, and yet there is an absence of heat. The research indicates that microwaves of this intensity have no lasting effect. However, there remain concerns as to whether they can damage the eyes or cause second-degree burns to those, such as invalids, unable to get out of the way quickly. There is also controversy as to whether the use of the ADS could be viewed as a form of torture. Torture is just one of many practices outlawed by the Laws of Armed Conflict (LOAC), the internationally agreed upon rules that establish limits on acceptable behavior during warfare (*jus in bello*).

Should the use of a nonlethal heat ray to disperse an unruly mob be viewed as an acceptable practice or a violation of international humanitarian law? What actions can or should be taken to stop the development of a synthetic biological weapon of mass destruction by a country that sponsors terrorism? Should the military be free to enhance a soldier's capability for combat purposes, even when there is no way to reverse the procedure? These were among the questions considered at NeXTech, a series of unique war games that explored the use of emerging and future military technologies.

The basic scenarios were familiar. The military options considered were new: cyber- and bio-weapons, drones and land-based robots, enhanced soldiers, and directed-energy weapons including electromagnetic pulses (EMP) that can knock out all electronic devices.

P.W. Singer, a Brookings Institute scholar and author of the best-selling *Wired for War*, organized the series of workshops along with the Noetic Group consultancy. The U.S. Department of Defense's Rapid Reaction Technology Office sponsored the war games. Military planners were interested in whether any of the new high-tech weaponry would be a game changer akin to the introduction of flying machines and submarines. The first gathering explored whether commanders, field officers, and soldiers found these emerging military technologies to be effective, practical, or strategically useful. Another event focused on how the use of emerging military technologies would be perceived by the enemy. For example, drone strikes evidently helped al-Qaeda in its recruiting efforts—an unintended consequence that military planners would want to factor in before deploying other innovative weapons.

I had the privilege of participating in the last workshop that brought together experts in public policy, international law, and ethics. We generally agreed that the use of the ADS violates the Laws of Armed Conflict (LOAC). LOAC specifies that noncombatants should never be intentionally targeted even if it is for their own good. But what if terrorists preparing an attack mingle among the demonstrators? Here the question gets more difficult in that the Geneva Convention of 1949, which established the foundational principles for the laws of armed conflict, acknowledges that noncombatants might be injured when a proportional response is made to targeting combatants. What constitutes a proportional response, however, persists as the most difficult issue in the ethics of warfare. No hard formula exists for determining the number of civilians acceptable to kill when there is an opportunity to take out a high-level terrorist with a drone strike.

Policy, law, and ethics are often clumped together as if they share a common language, however, the problems stressed by members of each field were very different. Ethicists, for example, were particularly good at highlighting secondary and tertiary issues, such as violations of the values of a specific community. The lawyers judged technologies through the lens of established international humanitarian law. The policy analysis was often based upon practical or political factors that did not enter into the judgments of either the ethicists or the law experts. All of these perspectives should be factored into selecting a final course of action, but seldom are.

One particular exchange near the end of the second day has stayed with me. We were scrutinizing various approaches to a serious threat of a terrorist attack on a specific date, but with no foreknowledge of which U.S. city would be targeted. In this fictional scenario, the terrorist group had recently demonstrated their skill by planting a bomb in a European city that killed more than a hundred people. We discussed a variety of ways to respond to the anticipated attack. All the options appeared to be very problematic. Then, a journalist in attendance said, "We should do nothing!" It was a showstopper, at least for a few moments, as the room went quiet and everyone reflected upon the apparent sanity of the proposal. Mark Hagerott, a retired Captain and Professor of Cyber Security at the U.S. Naval Academy, broke the silence. "The President couldn't do that, he'd be impeached."

Mark Hagerott is, of course, correct. Presidents and leaders of other countries take many actions that make political sense to prove they are strong on defense and homeland security. They have to do so, even when their policies are unintelligent, because the opposition and the citizenry will blame them if anything goes wrong. Since September 11, 2001, this has led to wars and homeland security and surveillance mechanisms that are arguably more destructive to societies and basic democratic values than the terrorist threat they were designed to combat. Furthermore, the war against terror has in all probability spawned a future generation of terrorists. This is evident in the rise of ISIS, the militant Islamic State in Iraq and Syria. In the attempt to create safe harbors, we have made our world less safe. And that will continue to be the case until citizens accept some degree of risk and do not automatically hold their governments at fault when terrorist acts occur. Until that time, political and military leaders are just doing their job.

Regrettably, the pressure upon political and military leaders to stave off future attacks has another disturbing effect. Whereas we might look to leaders to *check* the pace of change when it comes to embracing new technologies, they are often doing the opposite. In fact, preparations for security and defense are a leading driver of the gathering techstorm. As revealed by Edward Snowden, the U.S. National Security Agency used information technology to collect an unprecedented amount of personal data that had formerly been considered private. Large investments were made in developing advanced techniques for big data analysis to find needles in this massive haystack of information. In addition, the U.S. Defense Department invests billions of dollars yearly to fund research useful for the development of advanced weaponry. The 2014 U.S. budget contained $2,817,000,000 for DARPA, the Defense Advanced Research Projects Agency. That money funds research, much of which is secret, for artificial intelligence, cyber warfare, unmanned weapons systems, laser defense systems, and the enhanced soldier of the twenty-first century. DARPA's investments alone have a tremendous impact. And yet it is only one of many agencies in governments around the world funding leading-edge research.

Other drivers of the techstorm include medical research and science and engineering research in non-medical fields. Governments fund research to improve the lives of their citizens, to stay competitive, and out of a desire to better understand the world we inhabit. Understanding improves a government's ability to anticipate and preventatively defuse future dangers. But this trend is not limited to government actors; nongovernment foundations expend vast sums financing projects that hopefully demonstrate new approaches to tackle world health, poverty, and peace. Staying competitive also requires most industries to fund a healthy R&D budget.

The logic propelling these drivers of the storm forward appears irrefutable. How can we not make every effort to alleviate pain and suffering by searching for cures to diseases such as cancer and Alzheimer's? Have not all empires weak on defense been destroyed? Is not an ounce of prevention better than a pound of cure? The fiduciary responsibility of business leaders to their stockholders demands increasing profits through innovation. Scientists naturally strive to understand and to create activities that nurture technological advancement.

The collective logic of these arguments and the manner in which they support existing trends feeds the sense of inevitability, the feeling that the pace and trajectory of technological development cannot, and perhaps should not, be altered. The forces in play have tremendous momentum. However, the growth of an individual industry might be arrested by a substitute technology. The horse and buggy whip industry disappeared with the advent of automobiles. Film and photo processing have been replaced by digital photography. Newspaper publishers struggle to reinvent the industry before it dies.

Heads of countries and legislators quickly discover limitations in their power to alter the momentum of scientific discovery and technological innovation. Budgets can be increased, but are difficult to lower once an industry becomes entrenched and provides jobs that support the economy of a political district. New ventures, such as government-sponsored brain research in the EU and U.S., are easier to start up. Justifying the long-term benefits and finding funds in already tight budgets functions as the primary hurdle to overcome when introducing a new scientific initiative.

A tragedy can slow the momentum of an industry or at least raise questions regarding assumptions about its importance. The uncontrolled marketing of drugs was checked after the use of thalidomide by pregnant women between 1957–1961. More than ten thousand infants worldwide were born with malformed limbs and other defects. In the U.S. the pharmaceutical industry had long resisted expensive testing of new drugs. But the thalidomide tragedy catalyzed support for the certification of pharmaceuticals by the Food and Drug Administration. Extensive tests were put into place to ensure the safety of both innovative drugs and devices. Future tragedies caused by a synthetic organism or an autonomous robotic device will increase calls for federal oversight of those industries.

Accidents undermine the public's willingness to buy into the assumptions supporting a broad policy or large government project. Once in a while those assumptions get questioned without waiting until something goes wrong. That happens very seldom, but should happen more often. The ability to act without being forced to act by a crisis indicates a rare form of intelligence.

Many of the basic assumptions that must be inspected are so deeply ingrained that they seldom get questioned. A few examples from the military-industrial-research complex, and health care-academic-research complex will suffice to illustrate this point.

## IF WE DON'T THEY WILL

On July 12, and then again on August 2, of 1939, the Hungarian physicist Leó Szilárd (1898-1964) travelled from his home in New Jersey to visit his friend and colleague Albert Einstein, who was spending the summer in a cottage on the North Fork of Long Island. Szilárd described an experiment in which he and Enrico Fermi (1901-1954) started a nuclear chain reaction using uranium. The experiment demonstrated the feasibility of creating a nuclear bomb. Einstein expressed in German that he had not thought of this possibility. Szilárd went on to convey his and other colleagues' belief that physicists working in Nazi Germany, such as their colleague Werner Heisenberg, had already begun a program to build a bomb.

During his second visit, Szilárd presented Einstein with drafts of two different letters informing President Roosevelt of the possibility of "extremely powerful bombs" being created by a nuclear chain reaction in uranium, and of their suspicion that the Germans had already started on such a weapon. Einstein's signature on a letter would carry particular weight, not only because he was the world's most famous scientist, but also because of his well-known pacifist views. President Roosevelt read one of the letters Einstein signed. This led directly to Roosevelt authorizing the massive Manhattan Project, which employed 130,000 people at its height and built the first atomic bombs at a cost of $2 billion (more than $27 billion in 2015 dollars).

The German forces had already surrendered to the Allies two months before the first bomb was tested at 5:30 AM on July 16, 1945. Documents later revealed the Germans decided as early as 1942 that research on nuclear fission would not help their war effort. Ironically, the Nazis abandoned the bomb at roughly the same time as the Manhattan Project began to ramp up. With the German surrender, the original justification for building a bomb disappeared. Nevertheless, the U.S. dropped an atomic bomb on Nagasaki, and then a second one on Hiroshima, in August 1945. Einstein later regretted having signed the letter.

Szilárd was neither the first, nor would he be the last, to believe an advanced weapon system must be built before the enemy acquires it. In rooms around the world, where plans for future wars and national security are formulated, a military strategist will utter thoughts similar to those of Szilárd in support of the development of a proposed weapon. If a potential enemy has already initiated a program, then our side must follow suit to keep pace. Our adversary's planners, of course, make the same argument in pointing to our weapons development program. The circular nature of these arguments has not gone unnoticed.

Few countries can compete with the U.S. in the funding of basic and exploratory research to develop and demonstrate the feasibility of innovative weapons. In other words, the U.S. is the world's leading driver of an accelerating and ever-escalating arms race. By its grants, DARPA drives military research, much of which is dual use, meaning it will also be adapted for non-military applications. In the 1940s, the U.S. proved that a nuclear chain reaction could be controlled and a bomb of unprecedented power created. With that proof of concept, published research, and espionage, the U.S.S.R. would duplicate the feat at a fraction of the cost. The deployment and rapid proliferation of drones offers a more recent example.

The logic of staying ahead or keeping pace with potential adversaries rests on the dangers of a weak defense and contentions that weaker foes will be intimidated by one's power, adversaries with similar strength will be restrained by the fear of mutually assured destruction, and technological supremacy will prevail. These three positions are all flawed. Al-Qaeda and ISIS militants were not cowed by the U.S.'s overwhelming military superiority. Mutually assured destruction did not stop the former U.S.S.R. from keeping pace in the nuclear arms race. Technological supremacy is not a substitute for good strategic planning.

New weapons systems can both increase security and destabilize delicate geopolitical balances and the prospect of putting additional arms control agreements in place. The advent of the first Ohio-class submarines armed with Trident missiles in 1988 functioned as a tipping point that destabilized the Cold War. The Trident significantly shortened the time between launch and strike, which was advantageous if the U.S. intended to strike the Soviet Union first, but a serious disadvantage in that it undermined the time for the Soviets to determine whether a strike was actually occurring or a false alarm. There would be little or no opportunity for Soviet leaders to be brought into the decision-making process. The likelihood rose of a nuclear war being started because of a reactive response to false data. By the 1990s, the future of humanity became dependent on the state of Soviet computer technology, sensors, and warning systems. In retrospect, it was revealed that these systems were seriously deficient.

The intricacies of military planning are far beyond the scope of this book. Two takeaway points, however. First, technological solutions to military challenges are quickly replacing well-developed strategies for dealing with geopolitical tensions. Dennis Gormley, an expert in international security studies, put it well while pointing out the strategic flaws and negative consequences for arms control in the U.S. buildup of a quick strike intercontinental missile force. He writes, "the notion of prompt use of a highly precise intercontinental-range missile within an hour's decision time powerfully conveys the long-standing American preference for dabbling with technological solutions to the exclusion of clearheaded strategic thinking."

Second, the logic of staying ahead in the arms race does not have an endgame other than eventual destruction. The pace of change gets continually ramped up with no significant checks on unanticipated consequences. What will happen when a military planner proposes building the equivalent of Skynet? In the *Terminator* movie franchise, Skynet was the military artificial intelligence system that became self-aware and acted through cyborgs and war machines to exterminate the human race. Skynet is fiction. Nevertheless, how can a policy directed at maintaining military superiority turn away from the continual and escalating development of weapons systems that are just as threatening to humanity's survival as the adversaries they are meant to restrain? Without a universally recognized inflection point, can cautious minds prevail and put in place effective checks on the expansion of high-tech weaponry?

The new frontline is cyber warfare, less destructive of life than conventional warfare, yet capable of disabling critical infrastructure and destabilizing core institutions. The Stuxnet virus used to disable computers in Iran is often referred to as the first cyber bomb. But, in fact, ongoing cyber disruption and espionage by states including Russia, China, and the U.S. has become common. Anything dependent upon information technology comes with security risks. For example, forty million credit card numbers stolen from the Target computer system in 2013, and the hacking of eighty-three million accounts containing customer data from JP Morgan computers in 2014, highlight the simple fact that our assets are not safe. Sometimes computer vulnerabilities are exploited by criminals and sometimes by governments for purposes ranging from the establishment of a strategic advantage to stealing industrial secrets. In May 2014, the U.S. Justice Department publicly accused China of stealing sensitive materials from the computer systems of five companies and a trade union.

In a world dependent on information technology, there is no security when one nuclear bomb exploded high in the atmosphere will generate an electromagnetic pulse (EMP) capable of destroying the use of electronic systems over a vast area. The threat of a counterattack serves as a deterrent to the use of an EMP. But non-state actors or countries backed against the wall may not feel similar restraint. Unfortunately, there is limited security and no safe harbor in a world dependent upon technological wizardry.

The logic supporting the development of high-tech weaponry is weak, but so too is the will to question its core assumptions. Within the Pentagon and NATO, challenges to the expansion of our weapons systems remain subservient to the reigning mind-set, which is wholly about formulating effective strategies for these institutions to fulfill their missions. Only on occasion will questions about the worthiness of the mission get a full analysis. Do the ends truly justify the consequences of adopting these technological means?

Powerful vested interests will fight anyone with the courage to challenge whether the prevailing strategies for defense and security make sense. In 2013, the professed whistleblower Edward Snowden began leaking a massive trove of classified data he copied while working as a computer professional at the National Security Agency. In response, the administration of President Obama did everything in its power to stop him from finding safe haven. The Snowden case is far from simple, but in making the American people aware of how much personal data about each of us our government was collecting, he performed a service. Regardless whether one trusts a particular administration, the tools information technologies afford could be used to destroy the rights of citizens by a future government.

There have been occasions when cooler minds prevailed. In 1982, a confluence of incidents set in motion a strong worldwide nuclear freeze movement. *The Fate of the Earth*, a book written by Jonathan Schell and serialized in *The New Yorker* magazine, had an unusually profound influence on the formation of the movement. In his book, Schell clearly demonstrated the absurdity of the logic behind the nuclear arms race. One can hope that a similarly benign inflection point will arise to help us in questioning the logic behind building future military technologies.

The public is concerned with three core issues in the development of weaponry: 1) Will it make us more or less secure? 2) Can we live with what our military will do to safeguard our privileges? 3) How will our society change as military technologies are adapted for domestic purposes? These three concerns each offer opportunities to question assumptions driving the reliance on technological solutions to provide security.

## HEALTH CARE AD ABSURDUM

The military is not the only realm in which assumptions about the need for ever more innovative technologies are problematic. Health care is another. Expenditures on new tools and treatments is a major contributor to growing health care budgets. The yearly rate of growth fluctuates, but continues to increase by an average of 6–7 percent yearly, and will do so for decades to come. Health care economists concur that about 50 percent of the annual rise in the cost of medical services can be attributed to new technologies or the intensified use of older technologies. In the U.S., health care reached 17.9 percent of GDP in 2012, and by the end of this decade it will exceed 20 percent of GDP. This absurdly imbalanced state of affairs cannot continue. Zeke Emanuel, formerly in the Obama administration working on health care reform and presently at the University of Pennsylvania, points out that U.S. healthcare by itself is the fifth largest economy in the world – larger than the GDP of France and just a half trillion dollars less than the GDP of the economic powerhouse Germany.

Discussing health care in terms of money might be dismissed as minimizing the importance of providing the ill with the best care we can afford.

After all, is not health care a priority that should be placed above all others? But money spent on medical technology is money that is unavailable for other goals including education, welfare, and security. Furthermore, the U.S. is apparently getting fewer bangs for its health care buck than other countries. Resources, including time and intelligence, are expended on maintaining a medical-research complex that is bloated, chaotic, and fails in many respects to provide proper care to all citizens.

Heartwarming anecdotes of individual lives saved by new medical miracles tell one side of the story. The other side is told through heartless numbers that pass on both bad and good news. For example, deaths from colon and rectal cancers have slowly but progressively decreased over the past thirty years, and yet the cost for treating these diseases continually rises. In the U.S., the cost for treating colorectal cancer will rise an estimated 89 percent from 2000–2020.

The availability of high-tech equipment and the threat of lawsuits have lead U.S. doctors to order a growing number of expensive diagnostic tests. Most doctors know that all the tests they request are not necessary for the care of their patients. But what else can they do to protect themselves and their institutions in a litigious environment? Insurance premiums protecting doctors and hospitals are high and rising. In many other countries, socialized medicine or caps on malpractice settlements serve as a way to keep health care costs manageable. While such policies are represented as being unfair or harmful to some people, there is little or no evidence that, for example, the quality of health care in Canada (10.9 percent of GDP in 2012) is worse than that in the U.S. Some would argue that it is better, at least in covering the basic medical needs of all Canadian citizens.

More research promises better technology, better health care, and more resources directed toward determining which treatments are most effective. In turn, the growth of research contributes to a spiraling rise in costs. Developing new treatments is expensive, and companies must be given an opportunity to recoup and be rewarded for their investment. Thus, new pills will be costly even when the medicine can be produced inexpensively. The availability of many new medications adds to the overall expenditures by insurance providers, while the sheer number of new expensive diagnostic, surgical, or health monitoring tools adds to a hospital's budget. One da Vinci robot that assists in performing surgery can cost $1.5–$2.2 million. Furthermore, patients ask for the use of the latest gadgetry even when, for example, some surgeries can be performed as well, and less expensively, without the assistance of a surgical robot.

On the one hand, funding every promising piece of medical research is far too expensive. On the other hand, there are countless scientific questions that could and should be researched but are not because of a lack of funds. For example, hospitals and insurance companies all wish to provide treatments whose efficacy has been demonstrated through properly performed investigations. But much of the research necessary to provide evidence-based medical care has not yet been performed. Herein lies the health care dilemma. To underscore the financial crisis caused by the rising cost of health care could jeopardize investment in the medical research directed at providing new treatments and for analyzing the efficacy of existing practices.

Decades of predictions as to how better technology will lower the cost of health care turned out to be wrong. In its latest incarnation, this prediction gets entangled in the belief that medical science sits on the verge of solving all human ailments. Techno-optimists propose that the accelerating pace of scientific discovery means that a much longer and healthier life will be within reach soon. Furthermore, they contend that the exponential decline in the cost of gene sequencing, fMRI scans, and other diagnostic tools means the overall cost of health care will decline.

The contention that the sum total of new technologies under development will lower the cost of health care should be treated with disbelief and labeled as foolish. There is no end in sight for the escalating cost of health care. Each dream realized spawns new, often expensive, applications. Consider the respirocyte, just one of the hypothetical nanomachines proposed by Robert Freitas, the author of the multi-volume *Nanomedicines*. Respirocytes, if they can be developed, will function as artificial blood cells, roughly the same size as a blood corpuscle only much more efficient. Each respirocyte would carry more then two hundred times the oxygen of a blood cell. In theory, they might make it possible to stay underwater for an hour without a tank, drive to the hospital during a heart attack, and run a marathon without muscle fatigue. The cost of developing respirocytes can be justified as basic research with special applications. But once they are developed, how might they be distributed? Will everyone have a right to her own vial of respirocytes just in case of a heart attack, and if so, at what cost? Each new medical technology poses issues of cost and fair distribution. While some medical treatments get less expensive over time, many do not. Even when the price goes down, the cumulative cost goes up as the technology gets distributed more broadly.

Apportioning the latest medical technologies becomes a particularly touchy issue when dealing with end-of-life care. Medical services required in the last two years of life are dramatically greater than at any other period. Europe and Canada have been better than the U.S. in forging methods to ration health care in light of limited budgets, but in all countries there is a natural desire to give loved ones the best care possible. This holds true even when buying only days or hours of extended life at exorbitant expense.

Common sense indicates the need to find methods to rein in health care costs through various schemes for rationing services. Simply doing the research to determine which procedures work or are necessary, and then supporting doctors and insurance companies in refusing to do or fund other procedures would be a good start. A more rational health care system will, however, require additional oversight by governments, and as the debate over the Affordable Care Act (Obamacare) has illustrated in the U.S., that is a politically loaded topic.

The war between head and heart finds no respite when the health of a family member, friend, or oneself is at stake. Emotional factors frustrate and political factors obstruct any rational reform of health care services. Any reorientation of the health care system requires a questioning of the values that maintain the status quo. For example, when end-of-life care is predicated on measuring the quality of a life by the days lived, or the quality of one's love by not allowing a terminally ill loved one to die, then death with dignity is the first victim. In other words, the values that underpin the present medical system entail trade-offs that compromise alternative values, which, upon inspection, could be viewed as worthy of being given greater importance. Moderating the rate of adopting new technologies need not be seen as reducing care for our loved ones. It simply expresses a different attitude as to what actually constitutes care.

## BUYING TIME

Many different strategies must be employed to effectively manage the impact of emerging technologies. But it all begins with questioning the assumptions and values that drive the techstorm forward at an increasing pace. Simply recognizing the underlying assumptions offers a good first step. That recognition puts one in touch with the forces coalesed around those assumptions, and to a small degree begins the process of defusing their might. The next steps in disarming those forces become apparent through analyzing the trade-offs. What losses accrue when we presume that there is no alternative other than buying into the need to build every weapon system that can be perceived, or to fund each promising piece of medical research?

Is, for example, extending the life of an elderly aunt for a brief period of time more important than improving a niece's young life through better education? Where in one's list of priorities should rebuilding failing infrastructures, such as highway bridges, be placed?

Differing institutional structures prioritize values differently. Alternative responses to a challenge are not necessarily right or wrong. They often embody somewhat different sets of values, either of which might be judged acceptable. Passions run high, so discussing trade-offs in health care and national security is difficult. The example of defending against an asteroid strike will illustrate that making a hard choice does not necessarily mean selecting between a right and a wrong course of action.

There was no advance warning of the meteor blast in the atmosphere above Chelyabinsk, Russia on February 15, 2013. Roughly fifteen hundred people were injured, primarily by the flying shards from windows that the sound wave shattered. A direct hit by the meteorite on Chelyabinsk would have wiped out its population of more than one million souls. This meteor was not among those tracked by the NASA Spaceguard program, which scours the asteroid belt to identify any that might be on a course to strike the earth. NASA believes it will soon have identified 90 percent of the 120 meter wide (393.7 feet) or larger asteroids whose orbits come anywhere near the earth. The Chelyabinsk meteorite was too small to be tracked, a mere 17 meters (58 feet) wide, and yet it weighed ten thousand tons.

The Spaceguard program plans to build missiles that will destroy or redirect asteroids up to 1 kilometer (.6 miles) in diameter. Chicago or Rome would easily be leveled by a meteor that size. Asteroids large enough to destroy a city strike the earth about once every three hundred years. Once every hundred million years, give or take tens of millions, a meteorite 10 km (6 miles) in diameter or larger strikes our planet, and such a strike could extinguish all human life. An asteroid that struck the earth off the Yucatan Peninsula sixty-five million years ago is believed to have led to the extinction of the dinosaurs. Given the low probability that such a disastrous event will happen soon, would it be prudent to forgo investing the great sums of money necessary to ward off such an event? Or given that it might actually happen next year and lead to humanity's extinction, should we escalate investing in technologies that could destroy or deflect the path of a large meteor or a comet? Should, for example, accelerating the development of technology to combat the existential risk posed by a comet strike take precedence over research directed at curing cancer? How should we prepare for a low probability event, which, if it occurs, will have a very high impact?

There is no one correct answer for each of these questions. There is only the hard choice of selecting which value or system of values, among many, will inform our actions.

Space authorities have not ignored the prospect of an extinction-level event, but this does not answer the hard question of how much energy and funding should be directed to counter the danger. Nor does it solve the problem of convincing policy planners to commit funds to projects meant to avert such low-probability events. Legislators are notoriously shortsighted, unless they can be assured that funds will be spent in the district they represent. Furthermore, legislators can find plenty of justifications, in addition to the low probability of such an event occurring, to delay funding. One such excuse might be the possibility that some future technological fix, such as an anti-gravity device, will be invented that would enable redirecting the course of large meteorites without the use of nuclear-armed missiles.

There is nothing like a near-miss by an asteroid, such as the one that exploded over Chelyabinsk, to focus policy makers' minds. Before that event, various asteroid defense programs had been proposed, but now, more of them are likely to be funded. An international program to coordinate asteroid and comet defense was endorsed by a working group of the United Nations General Assembly in October of 2013. That program includes a UN International Asteroid Warning group as well as an advisory group to plan space missions to deflect the course of an asteroid. Identifying the larger asteroids and their orbits is the first order of business. To help on that front, in 2018 a privately funded space telescope that will contribute toward identifying and tracking more than a million asteroids will be launched.

In one intriguing strategy, the course of a one-kilometer-wide or smaller meteorite might be redirected to miss the earth by the gravitational pull of a spacecraft positioned nearby. To date, this theory has not been tested, although advocates call for a demonstration international asteroid deflection mission to be launched by 2023. Each strategy explored will entail trade-offs in the form of the funds available to address other problems.

Once a dangerous asteroid has been identified, its course might be altered years before it is expected to strike the earth. Even then, there will be political as well as technological challenges. In slowly changing the course of an asteroid, the actual spot it would strike the earth will migrate and cross national boundaries. Will, for example, Russia agree to allow an asteroid headed toward Europe to be redirected for a time so that it would strike Russian soil, if continued efforts to change its path fail?

The path of an asteroid can be identified years in advance and, with this time, might only require a relatively small nudge to alter its course and miss the earth. Deflecting a potential comet strike will be a much more difficult challenge. Once a comet enters the solar system, the time available to identify whether its course intersects with that of the earth, and to deploy an effective strategy if it does, is greatly reduced. Less time means much more force will be required to alter a comet's course. In other words, course corrections are easier when the challenge gets identified and addressed significantly in advance.

Course adjustments for asteroids and comets work well as a metaphor for course adjustments in the trajectory of any technological development. Inflection points exist even in outer space. Early action requires less force to reorient the unfolding pathway for any technological development. Acting early buys time and lowers cost. However, even when the inflection point where the course of a potentially harmful technology can be altered is identified, the difficulties in obtaining political support for the necessary adjustment should never be underestimated.

Something truly undesirable, like a world economic collapse, could dramatically alter the drivers of the storm. Short of such a horrific event, it would be naïve to believe that the forces propelling forward research on defense and health care can be dramatically slowed. Nonetheless, some of the push that accelerates the pace of technological drivers is the result of our buying into dubious predictions. The questioning of core assumptions defuses the artificial momentum these forces acquire through our support. But the first stage in this process is difficult—recognition of the assumptions and the courage to take off the rose-colored glasses.

# { 12 }

# Terminating the Terminator

THE IDEA TO PROPOSE A PRESIDENTIAL ORDER LIMITING THE development of lethal autonomous robots (killer robots) popped into my mind as if someone had placed it there. The date was February 18, 2012, and I was waiting for a connecting flight home in the U.S. Airways departure terminal at Reagan Airport. My gaze traveled along the panoramic view across the tarmac where the Capitol Dome and the Washington Monument rose high above the tree line along the Potomac River. And suddenly, there it was, an idea I felt compelled to act upon. In the following days, I wrote up and began circulating a proposal for a presidential order, declaring that the U.S. considers autonomous weapons capable of initiating lethal force to be in violation of the Laws of Armed Conflict (LOAC).

For decades, Hollywood has supplied us with plenty of reasons to be frightened about the roboticization of warfare. But now that drones and autonomous antimissile defense systems have been deployed, and many other forms of robotic weaponry are under development, the inflection point where it must be decided whether to go down this road has arrived.

For many military planners, the answer is straightforward. Unmanned drones were particularly successful for the U.S. in killing leaders of al-Qaeda hidden in remote locations of Afghanistan and Pakistan. Some analysts believe unmanned air vehicles (UAVs) were the only game in town, the only tool the U.S. and its allies had to successfully combat guerrilla fighters. Furthermore, drones killed a good number of al-Qaeda leaders without jeopardizing the lives of soldiers. Another key advantage: reducing the loss of civilian lives through the greater precision that can be achieved with UAVs in comparison to more traditional missile attacks. The successful use of drones in warfare has been accompanied by the refrain that we must build more advanced robot weapons before "they" do.

Roboticizing aspects of warfare gained steam in the U.S. during the administrations of George W. Bush and Barack Obama. As country after country follows the U.S. military's lead and builds their own force of UAVs, it is clear that robot fighters are here to stay. This represents a shift in the way future wars will be fought, comparable to the introduction of the crossbow, the Gatling gun, aircraft, and nuclear weapons.

What remains undecided is whether robotic war machines will become fully autonomous. Will they pick their own targets and pull the trigger without the approval of a human? Will there be an arms race in robotic weaponry or will limits be set on the kinds of weapons that can be deployed? If limits are not set, eventually the natural escalation of robotic killing machines could easily progress to a point where robust human control over the conduct of warfare is lost.

The executive order I envisaged and proposed could be a first step in establishing an international humanitarian principle that "machines should not be making decisions about killing humans." A vision of the president signing the executive order accompanied the idea, as if from its inception this initiative was a done deal. Like many dreamers I was wooed to action by the fantasy that this project would be realized quickly and effortlessly. Life is seldom that easy. To this day, I have no idea whether my proposal or any other campaign would lead to a ban of lethal autonomous weaponry. Nevertheless, from that first moment it was clear to me that an opportunity to pursue a ban presently exists but will disappear within a few years. The debate focuses upon whether the development of autonomous killing machines for warfare should be considered acceptable. But in a larger sense, we have begun a process of deciding whether people will retain responsibility for the actions taken by machines.

The first formal call for a ban of lethal autonomous robots had been formulated fifteen months earlier (October 2010) at a workshop in Berlin, Germany organized by the International Committee for Robot Arms Control (ICRAC). At that time, ICRAC was nothing more than a committee of four scholars (Jürgen Altmann, Peter Asaro, Noel Sharkey, and Rob Sparrow). The gathering they organized included arms control specialists, experts in international humanitarian law (IHL), roboticists, leaders of successful campaigns to stop the development of nasty weapons such as cluster bombs, and most of the small community of scholars that had written about lethal autonomous robots. Colin Allen and I were invited to par-

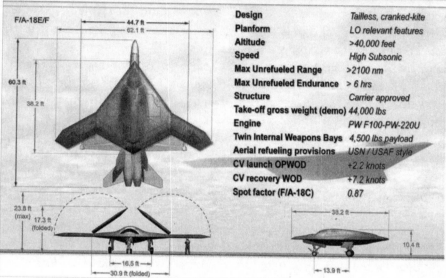

| Design | Tailless, cranked-kite |
|---|---|
| Planform | LO relevant features |
| Altitude | >40,000 feet |
| Speed | High Subsonic |
| Max Unrefueled Range | >2100 nm |
| Max Unrefueled Endurance | > 6 hrs |
| Structure | Carrier approved |
| Take-off gross weight (demo) | 44,000 lbs |
| Engine | PW F100-PW-220U |
| Twin Internal Weapons Bays | 4,500 lbs payload |
| Aerial refueling provisions | USN / USAF style |
| CV launch OPWOD | +2.2 knots |
| CV recovery WOD | +7.2 knots |
| Spot factor (F/A-18C) | 0.87 |

**Northrop Grumman X-47B**

ticipate for the work we had done in co-authoring the book *Moral Machines: Teaching Robots Right From Wrong.*

The word autonomous in reference to robots denotes systems capable of initiating actions with little or no ongoing human involvement. The drones that were used successfully by the U.S. in Afghanistan and Pakistan are un-manned but remotely controlled by personnel often thousands of miles away. They are not what others and I are worried about. Officers at the re-mote location made decisions as to when a legitimate target is in sight, and then authorize missiles to be dispatched. A human is "in the loop" making decisions about who to kill or what to destroy. Increasingly, however, more and more functions are being turned over to computerized systems. For example, in 2013, the Northrop Grumman X-47B, a prototype sub-sonic aircraft with two bomb compartments and a 62-foot wingspan, autonomously took off from and landed on an aircraft carrier. The proposed ban on autonomous lethal robots is focused upon ensuring that in the future, selecting a target and pulling the "trigger" is always a decision made by a human and never delegated to a machine. There must always be a human in the loop.

Today's computers do not have the smarts to make discriminating decisions such as who to kill or when to fire a shot or a missile. Thus, a ban is directed at future systems that have not yet been deployed, and in nearly all cases, have not yet been built. There is still time to make a course correction. Nevertheless, there already exist dumb autonomous or semi-autonomous weapons that can kill. For example, a landmine is autonomous and will go off without a human making a decision to harm the specific individual who trips the device. Unfortunately, all too often it is children who trip landmines. In addition, defensive weaponry including antiballistic missile systems such as the U.S. Navy's Phalanx or Israel's Iron Dome can autonomously intercept incoming missiles well before military personnel would have time to make a decision. A ban would probably not include defensive weaponry, although often the difference between declaring a weapon system defensive or offensive is merely a matter of what direction the weapon points. Nor would a ban affect autonomous weapons viewed as the last line of defense in an area of known hostility. The Samsung Techwin, a stationary robot capable of gunning down anything that moves in the Demilitarized Zone (DMZ) separating North and South Korea, has been deployed since 2010. These stationary robots are unlikely to have even modest success should North Korea decide to send a million troops across the DMZ to invade the South.

The proposal for a presidential order that I wrote up and circulated received only a modest degree of attention and support. However, I'm certainly not alone in calling for a ban. Soon after the November 2012 presidential election, Human Rights Watch (HRW) and the Harvard Law School International Rights Clinic entered the campaign with a high-profile report that called for a ban of lethal autonomous robots (LARs). Three months later, HRW and a coalition of other nongovernmental organizations (NGOs) launched the international campaign to ban killer robots. That campaign is directed at activating worldwide support for an arms control agreement covering robotic weaponry. In addition, a growing community of international experts advocates the need for robot arms control. In a 2013 report, Christof Heyns, U.N. Special Rapporteur on extrajudicial, summary or arbitrary executions, called for a moratorium on the development of LARs as a first step toward considering an international ban.

These efforts have had significant success in catalyzing governments around the world to give a ban serious attention. In May 2014, the U.N. Convention on Certain Conventional Weapons (CCW) convened a meeting in Geneva to discuss the dangers autonomous weapons pose. One hundred seventeen nations are party to the CCW, which restricts the use of specific weapons deemed to cause unjustifiable harm to combatants or civilians. On November 14, 2014, the CCW voted to continue deliberations over LARs, an important first step in acknowledging the importance of the issue.

Opponents of robotic weaponry contend that their use could lower the barriers to starting new wars. The potential loss of one's own troops has been one of the few major deterrents to starting wars. Autonomous weaponry contribute to the illusion that wars can be started and quickly won with minimal costs. Once a war begins, however, not only the lives of soldiers, but also those of civilians will be lost. Furthermore, a decision about who or when to kill by an autonomous weapon could accidentally initiate hostilities. They could also be dangerous from an operational perspective if robotic weapons, for example, escalated an ongoing conflict or used force indiscriminately or disproportionally. For a military commander, the possibility of autonomous systems acting in a manner that escalates hostilities represents a loss of robust command and control.

In addition to saving the lives of one's soldiers, two other ostensibly strong arguments get put forward as objections to banning LARs. The first considers LARs to be a less lethal option than alternative weapons systems. Presuming LARs are more accurate than other available weapons systems, they will cause less loss of civilian lives (less collateral damage). Shortsighted contentions such as this do not fully factor in the future dangers once many countries have robotic armies. The long-term consequences of roboticizing aspects of warfare could far outweigh the short-term benefits.

The second argument proposes that future machines will have the capacity for discrimination and will be more moral in their choices and actions than human soldiers. Ronald Arkin, a roboticist at Georgia Tech, takes this position. Arkin has been working toward the development of means for programming a robot soldier to obey the internationally established Laws of Armed Conflict (LOAC). He contends that robot soldiers will be better at following the LOAC because "the bar is so low." In this Arkin refers to research that shows, for example, that human soldiers will not squeal on their buddies even when atrocities have been committed. Nevertheless, the prospect is low for developing robot soldiers any time soon with the ability of making an appropriate judgment in a complex situation. For example, a robot would not be good at distinguishing a combatant from a non-combatant, a task that humans also find difficult. Humans, however, bring powers of discrimination to bear for meeting the challenge; capabilities that will be difficult, if not impossible, for robots to emulate. If and when robots become ethical actors that can be held responsible for their actions, we can then begin debating whether they are no longer machines and are deserving of some form of personhood. But warfare is not the place to test speculative possibilities.

Widespread opposition to killer robots already exists according to a 2013 poll of a random sampling of a thousand Americans by Charli Carpenter, a Professor of Political Science at the University of Massachusetts. Overall, 55 percent of respondents oppose the development of autonomous weapons, with 39 percent strongly opposed. In other words, there exists substantial national and international public support for a ban. But without a continual campaign actively championed by a large segment of the public, policy makers will dismiss support for a ban as the product of fears generated by science fiction. Interestingly, 70 percent of respondents in the active military rejected fully autonomous weapons. Nevertheless, military planners have retained the option to build autonomous killing machines.

Traditional forms of arms control that depend upon inspection protocols provide a poor model for limiting the use of robotic weapons. The difference between an autonomous system and one requiring the actions of a human to proceed may be little more than a switch or a few lines of software code. Such minor modifications could be easily hidden from an inspector, or added in the days following the inspector's departure.

Furthermore, arms control agreements can take forever to negotiate, and there's no reason to think that negotiations to formulate procedures for the verification and inspection of robotic weaponry will be any different. Continual innovations in lethal autonomous weaponry would also require a need to periodically revise the arms agreements that had been negotiated.

Ban proponents are also up against powerful forces in the military-industrial complex who want sophisticated robotic technology funded by defense budgets. Weapons production is a lucrative business, and research funded by defense budgets can often be sold to our allies or spun off to develop non-military technologies. During the years before a ban is enacted, there will be an alignment of countries and corporations that have a vested interest in continuing the development of robotic weaponry and in defeating any attempts to limit their use. That is why an inflection point exists now, before autonomous weapons become a core weapons system around which major powers formulate their defense strategy. In the U.S. plans to reduce active personnel and increase the deployment of military technology were announced in 2014. Half of the bombers on aircraft carriers are slated to become a version of the unmanned X-47. Future models will fly in swarms and even make autonomous decisions about bombing targets. The inflection point for limiting lethal autonomous weapons that presently exists could easily disappear within a matter of a few years. The length of time the window stays open is dependent upon whether or not the campaign to ban killer robots gains enough clout to impact the willingness of governments to invest in developing the required technologies.

An international ban on LARs is needed. But given the difficulty in putting such an agreement in place, the emphasis should initially be upon placing a moral onus on the use of machines that make life-and-death decisions. A long-standing concept in just war theory and international humanitarian law designates certain activities as evil in themselves —what Roman philosophers called *mala en se*. Rape and the use of biological weapons are examples of activities considered *mala en se*. Machines making life-and-death decisions should also be classified as *mala en se*.

Machines lack discrimination, empathy, and the capacity to make the proportional judgments necessary for weighing civilian casualties against achieving military objectives. Killer robots are *mala en se*, not merely because they are machines, but also because their actions are unpredictable, cannot be fully controlled, and attribution of responsibility for the actions of autonomous machines is difficult if not impossible to make. Delegating life-and-death decisions to machines is immoral because machines cannot be held responsible for their actions.

## AUTONOMOUS CARS AND LIABILITY

Killer robots offer but one example of the move to place decisions for which humans were once responsible in the hands of machines. Increasingly, autonomous computers and robots threaten to undermine the foundational principle that a human agent (either individual or corporate) is responsible, and potentially accountable and liable for the harms caused by the deployment of any technology. Turning to non-military applications for robots will make this point clearer. Systems developed for most domestic applications should not be banned. In many instances their benefits will far outweigh the risks. Nonetheless, they can set deeply troubling precedents when they dilute the core principle of a responsible human or corporate agent.

As of May 2014, a fleet of self-driving Toyota Priuses modified with hardware and software designed by Google had logged over seven hundred thousand miles on California highways and city streets. None of these autonomous cars caused an accident. The continuing media coverage of Google cars creates the impression that within the next five to ten years consumers will be able to purchase fully autonomous vehicles. Visions of a car capable of serving as a designated driver that delivers one safely home after a wild party is a delicious fantasy, but Google's cars are not that. They require an operator in the driver's seat ready to take over when the vehicle's computerized Chauffeur, the name given Google's automated driver, encounters a situation it cannot handle. With as little as ten seconds' notice, the human driver needs to be fully engaged, and will in all likelihood be held responsible if an accident occurs. This rules out napping and texting while Chauffeur drives. The sheer boredom and inattention that will naturally result from having little to do will also likely hamper one's ability to respond to a potentially dangerous situation on short notice.

The U.S. National Highway Traffic Safety Authority estimates that human error is the "certain" cause of fully 81 percent of all crashes. So advocates for self-driving cars commonly propose that these vehicles will dramatically lower fatalities. The accuracy of that claim should be questioned. No doubt many accidents caused by human drivers are a result of daydreaming, distraction, or simply falling asleep. And that may well be more accidents than self-driving cars have from either computer error or a driver unable to fully engage and avert an accident once the automated Chauffeur warns him to take over. According to Google's estimates, the current version of Chauffeur only makes a mistake severe enough that it requires the intervention of a human driver to avert an accident on an average of once every 36,000 miles. This may be somewhat reassuring in that the average driver requires two or three years to cover that distance. However, human drivers only get into an accident once every 500,000 miles, injuries occur once every 1.3 million miles driven, and fatalities every 90 million miles. Therefore, it is very difficult to know whether self-driving cars dependent on the readiness of a backup human driver will truly lower fatalities. Yet the simple fact that an autonomous vehicle can apply brakes much more quickly than a human could respond to a dangerous incident, suggests that autonomous vehicles will be significantly safer than cars with human drivers.

Now consider the possibility that there may indeed be fewer deaths from accidents involving vehicles with self-driving capabilities, but the deaths in some of these incidents are of different people than those who would die if a human were driving. The trade-offs in who dies will pose new issues for both the law and for insurance providers. Let's say, for instance, that in order to avoid hitting a bus or a bystander, the autonomous car drives off a bridge and kills the driver. Not a pleasant scenario, but one in which a few altruistic drivers might do the same. In a 2013 article, the ethicist Patrick Lin proposed that under the moral math of the "greatest good for the greatest number," perhaps self-driving cars should be programmed to kill the driver rather than two pedestrians. Mathematical determinations of right and wrong, however, fail to be reassuring. Certainly such situations will be extremely unusual. But do we want robotic cars making decisions about whether we live or die?

Who will be held responsible when an accident occurs? Car manufacturers will be averse to marketing autonomous vehicles before the liability issues are resolved. Presently, however, the law will hold the manufacturer liable for most accidents caused by an autonomous vehicle. Drivers will be uncomfortable if they can be held liable for an accident that takes place because of their inability to take the wheel on short notice. In addition to solving various technological challenges, finding means to manage the liability and legal consequences if something goes wrong has slowed manufacturers from introducing fully autonomous vehicles. For many situations the liability issues cannot and should not be decided by manufacturers or engineers. Legislatures, courts, and insurance providers must decide them.

None of these issues will necessarily stop enterprising companies from finding means to sell self-driving cars. They will petition state governments and lobby for new laws. In the meantime, they are likely to redefine what it means to call a vehicle autonomous, using vague language such as "driver assistance." Already, safety features such as autonomous parking and emergency braking are being introduced into high-end vehicles.

Certainly manufacturers will make sure that their liability is limited. Even without the liability issues fully resolved, self-driving cars will be an attractive option for special purposes. Older drivers who, for example, no longer trust their own ability to stay constantly engaged may be thrilled by the prospect of having a vehicle that provides driving assistance, even if it means they must make their best effort to keep their eyes on the road.

A major accident, however, in which a self-driving vehicle is implicated, could set back the industry for decades. Accidents happen. Engineers are certainly working diligently to limit the number of accidents caused by autonomous vehicles. They know that the acceptance of self-driving cars will rest upon the public's willingness to accept a horrific accident if it has been clearly demonstrated that autonomous autos will be significantly safer than those driven by people.

No-fault insurance for vehicles that fall short of the holy grail of full autonomy offers one possible way of working around these societal considerations. If autonomous vehicles truly provide a net advantage, then legislatures could enact policies that will ease the development of the industry. To be sure, the whole robotic industry could be spurred along by legislation that lowered the liability for system failures and accidents by autonomous systems in situations that had not been anticipated by designers and engineers.

How important is it to travel quickly down the proverbial highway toward increasingly autonomous robots? Lowering liability to speed up the development of robotics comes with trade-offs. It opens the door for manufacturers to introduce products that have either not been fully tested or have known defects. Companies will emphasize the difficulty in establishing who is responsible for complex intelligent systems as they petition to dilute liability for system failures. When do the rewards justify the risks?

Practical ethicists and social theorists raise many serious concerns regarding the dangers inherent in diluting corporate and human responsibility, accountability, and liability for the actions of increasingly autonomous systems. In 2010, as an expression of their concerns, five rules were proposed as a means of reinforcing the principle that humans cannot be excused from moral responsibility for the design, development, or deployment of computing artifacts (objects created by humans that execute a computer program). More than fifty scholars in fields related to technology and ethics undersigned the five rules. For our purpose, stating the first of the five rules will be sufficient:

Rule 1: The people who design, develop, or deploy a computing artifact are morally responsible for that artifact, and for the foreseeable effects of that artifact. This responsibility is shared with other people who design, develop, deploy, or knowingly use the artifact as part of a sociotechnical system.

If this first rule was codified into law, the manufacturer of a gun would be accountable for its use in a murder, and the manufacturer of cigarettes accountable for all the lung cancers of smokers. Expressed differently, the five rules are too broad in scope, and somewhat impractical. Their application would not be easy and would significantly slow the development of the robotics industry. And yet they highlight the concern of a cross-section of legal theorists and ethicists about diluting responsibility for the actions of robotic devices, or introducing machines whose actions no one can fully predict. Whether rules, such as those proposed for computing artifacts, can or should be translated into liability law remains an open question. A difficult policy debate lies ahead. Should accountability and liability for computers and robots be lowered in order to stimulate the development of a potentially transformational industry? Or should existing protections be maintained even if this arrests the willingness of companies to introduce products that offer significant benefits with low or uncertain risks?

When Google's robotic car project first came to light in 2009, it was attacked in a TV commercial for the Dodge Charger. In the advertisement, a deep, ominous voice declares, "Hands-free driving, cars that park themselves, an unmanned car driven by a search-engine company. We've seen that movie. It ends with robots harvesting our bodies for energy." Cars will not harvest our bodies for fuel, but the ad captures the recurring image that any step toward increasingly autonomous robots is another step down a slippery slope that eventually leads to conflict between humans and the artificial entities they create. How much attention does this speculative fear warrant? Certainly fully autonomous AI systems are well beyond today's technology. And even if one believes they will eventually be realized, the challenges of ensuring that superior robotic systems will be friendly to humans might be solved between here and there. But simply put, much of the public, while intrigued by robots, remains uncomfortable about where robotic technology may lead.

## SERVICE ROBOTS

A Czech playwright, Karel Čapek, coined the term "robot" for his 1920 play *R.U.R.*, in which a rebellion by initially happy slave androids leads to the extinction of humans. Čapek's robots, like the golems of Jewish folklore, were inanimate matter that somehow became anthropomorphic beings. With the development of computers in the 1940s and '50s, scientists perceived pathways to actually imbue machines with intelligence, and the term "robot" took on its more contemporary meaning.

Robot slaves without the onus of slavery remain an extremely attractive dream. And yet, a recurring plot for science fiction portrays the vision that slave robots will eventually acquire enough intelligence to want their freedom.

The service robots available today are limited purpose machines that function as companions, therapeutic trainers, guides in museums, and task-specific assistants in hospitals. A robot designed to lift patients in and out of a hospital bed can ease the workload of a nurse, but may also take away jobs from hospital orderlies. With an aging population, Japan is particularly interested in developing robots capable of taking care of the homebound and elderly.

As designers and engineers develop more sophisticated service robots, they will increasingly be confronted with questions about what the robots should do when confronted with situations that have ethical or legal significance. Service robots will eventually have a degree of autonomy and therefore will pose issues similar to those arising from the deployment of autonomous weapons and automobiles.

Social theorists such as M.I.T.'s Sherry Turkle raise questions as to which tasks are appropriate for robots. Does the use of robotic caregivers for the homebound and elderly reflect badly upon modern society, and is it heartless or abusive? Certainly robotic care is better than no care at all. Some people find the use of robots as sex toys offensive. Others lament the sensibilities and lessons lost in substituting "robopets" and "robocompanions" for animals or people. The increasing use of "robonannies" in some Asian countries to tend infants and children raises serious concerns. Computing scholars Noel and Amanda Sharkey propose that the extended use of robots as nannies may actually stunt an infant's emotional and intellectual development, because the robot will fail to provide the kind of rich verbal and nonverbal interaction human caregivers offer.

Robot toys and companions cannot interact with their charges in the same way a human companion would, which forces manufacturers to consider how to design the systems to respond appropriately in ethically significant situations. A few years back, the manufacturer of a speaking robot doll considered what the doll should say if treated abusively by a child. The engineers knew how to build sensors into the doll that would alert the system to abusive treatment. After analyzing the issues and consulting with lawyers, they decided that the doll would say and do nothing. What forms of interaction are appropriate for a robot? How should a robot nanny respond to a child that relates to the nanny in a way that would be inappropriate or even physically violent if the nanny were human? Would you want the robot to say, "Stop that! You are hurting me," presuming (as is probable) that the robot has no capacity to feel pain? While well-intentioned, such a statement by a robot can sound absurd and lead to unintended consequences.

Consider, too, whether a robot that serves as a companion to a child should intervene when the youngster puts herself in harm's way. Are there circumstances where inappropriate intervention by the robot might do more harm than good? Would programming a robot to tell a child to stop endangering herself be a good or a bad idea? What if the child ignores the directive? Should the robot discipline the child? A robot that instructs but cannot follow up with discipline may well be teaching the wrong lessons. But the child may not trust a robot that disciplines.

Many similar scenarios have been introduced. What should a robot caregiver do if an elderly patient refuses medicine? Or what if the robot enters a room and discovers that the person under its supervision is hysterical? How would the robot know whether the fear on the face of its charge was caused by the robot itself or by some other factor? Some of these situations suggest the need for capabilities that present-day robots do not have. But all of the ethically significant scenarios raise similar questions. What tasks are acceptable for autonomous robots to perform? Who is responsible or liable if something goes wrong? Do we truly want to introduce increasingly autonomous robots into the commerce of daily life? And who is or should be making decisions about the introduction of such technologies in the first place?

The appropriateness and ability of robots to serve as caregivers is commonly misunderstood by the public or misrepresented by those marketing the systems. The limited abilities of present-day robotic devices can be obscured by the human tendency to anthropomorphize a robot whose looks or behavior faintly mimic that of a human. A professional association or regulatory agency that evaluates the capabilities of systems and certifies their use for specific activities would be helpful. However, ongoing evaluation of systems would also be very expensive, as the development of each robotic platform is a moving target—existing capabilities undergo refinement and new capabilities are constantly being added to systems.

Looming over this discussion is the prospect of autonomous robots having to make their own judgments. Engineers and designers will not be able to predict in advance all the situations increasingly autonomous systems will encounter. Evaluating a new context and determining the appropriate course of action is a capability that will be extremely difficult to program into a computerized system. Whose values or what values should be programmed into the ethical decision-making systems of robots? Sophisticated, sensitive, and verbally communicative robots such as *Star Trek*'s Data may or may not be possible to create. Between here and there will be generation upon generation of robots with more limited decision-making abilities. Will we be able to trust that their actions are safe and appropriate?

## LIFE AND DEATH BY ALGORITHM

In a two-day 2011 competition, Watson, the IBM computer named after the company' s founder, handily dispatched the two most successful human competitors, Brad Rutter and Ken Jennings, on the TV game show *Jeopardy*. Watson had been designed by IBM specifically to win this competition. Its massive four-terabyte database of facts was impressive, but more striking was the artificial intelligence system's ability to parse the initial question. The game show functions on a kind of reverse logic, where contestants are given an answer and asked to determine the question asked. Essentially, however, it is still a question-and-answer show where determining what answer is being sought demands an analysis of the initial statement, a task that even human players get wrong.

The early founders of artificial intelligence believed communicating with natural language would be an easy task for computers to learn. More than a half-century of research demonstrated that mastering language sufficiently so as to communicate in the same way as humans would be among the harder tasks for AI. The programmers at IBM had finally demonstrated that, in a manner of speaking, a computer could "understand" questions.

After performing a search on its database, Watson responded with an answer that its software evaluated as having the highest probability of being correct. Watson's unique advantage lay in its ability to buzz in (indicate it had an answer to the question) within 5–10 milliseconds (thousandths of a second). Even twitchy-fingered human stars like Rutter and Jennings generally need about 190 milliseconds (0.19 sec) to send a signal from their brain to their finger to press the buzzer. In *Jeopardy* a contestant cannot push the buzzer until a light signals the question is complete. However, when Rutter or Jennings have anticipated the answer to a question while it is still being read, they have pressed a buzzer within 10–20 milliseconds after the question is completed.

Intriguingly, Watson made one serious error during the competition. In the Final Jeopardy category titled "U.S. Cities," Watson answered the statement, "Its largest airport was named for a World War II hero; its second largest, for a World War II battle," with the question "What is Toronto?" While answering the question with a Canadian city was dumb, Watson was smart enough to recognize its weakness in this category, and only bet $947, which was subtracted from the computer's daily total. In this context, the mistake was insignificant, however, a similar mistake by Watson in the future could have a more serious outcome. For an encore, IBM is retooling Watson to function as a medical advisor.

Watson's first medical project is to provide assistance in decisions about the management of lung cancer treatments at Memorial-Sloan Kettering Cancer Center. Watson is not the first computer system to assist in the diagnosis, prognosis, and treatment plan for patients suffering from serious diseases. The Apache Medical System, a computerized program that calculates the severity of a seriously ill patient's disease, has been adopted by more than forty hospitals. The first version of the Apache system dates back to 1981. A partnership between Microsoft and General Electric is also developing systems in the burgeoning market for medical diagnostic systems. Computerized medical diagnostic systems afford the advantage of quickly comparing an individual's medical history and recent test data to a massive database of past cases. On the surface, the challenges posed by Apache or a medical diagnostic system such as Watson can sound similar to the Doc-In-A-Box discussed in Chapter 3. However, there is one big difference here. Diagnoses and prognoses by computers that demonstrate a high degree of accuracy threaten to replace the judgment of even gifted physicians. While these systems are marketed as adjuncts to the judgments of doctors, in some situations they can dilute the authority of physicians who have traditionally been considered responsible for quality medical care.

Even though a disease can often be treated appropriately through a computerized diagnosis, we should never forget that it is a person's fate that is being decided by the algorithmic analysis of numbers. Within the field of medicine, a vigorous debate has ensued over the proper role for computerized diagnoses and treatment plans. Over the long run, statistics may indeed tell the story. If computers are better than doctors in diagnosing most diseases and recommending successful courses of treatment, then so be it. Improving quality of life for patients and the efficiency of health care will be served. But we don't have that certainty yet. And in the meantime, there remains an undying belief that those uniquely human skills—experience, intuition, a sixth sense, and knowledge of factors the computer might not have considered—will prevail over raw numbers. Even advocates for computerized diagnoses acknowledge that the systems are not always correct. The challenge for hospital administrators lies in maintaining room for human judgment to prevail over algorithmic analysis. Unfortunately, in a litigious age, doctors may feel uncomfortable going against the recommendations of semi-intelligent systems, particularly when an audit trail noting the advice produced by the software will be available to enterprising lawyers.

The life-support systems for terminally ill patients are not connected to diagnostic systems. But, with its growing influence, a computer doc might eventually have the *de facto* ability to pull the proverbial plug. In the best of both worlds, trained human specialists work in a team with diagnostic data and computerized advisors. In higher-quality hospitals this already occurs. A fear exists, however, that increasingly there will be pressure on the human decision-makers to defer to the machine's advice, particularly when it can be demonstrated that the computerized prognosis is statistically more accurate than that of the average physician. Human decision-makers need to be empowered and protected when they have the courage to override the actions or suggestions of robotic systems. Such empowerment will only make sense when health care workers recognize their own limitations and acknowledge their need for assistance.

Doctors deferring to the advice of diagnostic computers could become the norm, and perhaps it should. But if something goes wrong, who is responsible? Who should be held legally culpable and financially liable if the doctors, hospitals, company marketing the computer, and the experts and engineers who designed the system are sued? Insurance providers and courts are already beginning to deal with those questions. Precedents will be set by court decisions, and in this manner lines will be drawn that clarify the legal responsibility of the various parties.

Driverless cars, service robots, and medical diagnostic programs all illustrate different ways in which smart or increasingly autonomous computer systems can dilute human responsibility and, ultimately, human control. The stakes and social ramifications of going down this road are high and overshadow the desirability or benefits that might be accrued from each of the individual technologies. In none of these three instances, however, are the risks comparable to the introduction of lethal autonomous robots (LARs). And yet, together with LARs, all of these technologies share the prospect that human responsibility and authority can be undermined if we fail to be careful about which forms of artificial intelligence we adopt and how we implement them. In this respect, the threat killer robots pose is an extreme example of a threat we must confront in many different spheres. More importantly for our discussion, the inflection point for LARs is now.

## A RED LINE

*The Terminator* is clearly science fiction, but it speaks to a powerful intuition that the robotization of warfare is a slippery slope—the end point of which can neither be predicted nor fully controlled. Machines capable of initiating lethal force and the advent of a technological singularity are future-oriented themes, but function as placeholders for more immediate concerns. The computer industry has begun to develop systems with a degree of autonomy and capable of machine learning. Engineers do not know whether they can fully control the choices and actions of autonomous robots.

The long-term consequences of roboticizing warfare must neither be underestimated nor treated cavalierly. But given that arms manufacturers and a few countries are focused on the valuable short-term advantages of developing killer robots, the momentum behind the robot arms race is likely to quickly spin beyond control. Setting parameters now on the activities acceptable for a lethal robot to perform offers one of the few tools available to restrain a robot arms race that mirrors that of the Cold War. One difference, however, is that, unlike nuclear weapons, robotic delivery systems will be relatively easy to build.

In the same November 2012 week in which Human Rights Watch called for a ban on killer robots, the U.S. Department of Defense (DoD) published a directive titled, "Autonomy in Weapons Systems." The timing of the two documents' release could have been coincidental. Nevertheless, the directive read as an effort to quell any public concern about the dangers posed by semiautonomous and autonomous weapons systems. Unlike most directives from the DoD, this one was not followed up by detailed protocols and guidelines as to how the testing and safety regimes noted in the document should be carried out. The DoD wants to expand the use of self-directed weapons, and in the directive it explicitly asks the public not to worry about autonomous robotic weaponry, stating that the DoD will put in place adequate oversight on its own. Military planners do not want their options limited by civilians who are fearful of speculative possibilities.

Neither military leaders nor anyone else wants warfare to expand beyond the bounds of human control. The directive repeats eight times that the DoD is concerned with minimizing "failures that could lead to unintended engagements or loss of control of the system." Unfortunately, this promise overlooks two core problems. First, the DoD has a specific mission and cannot be expected to handle oversight of robot autonomy on its own. During the war against al-Qaeda in Afghanistan and Pakistan, military and political necessities caused Secretary of Defense William Gates to authorize the deployment of second-generation unmanned aircraft before they had been fully tested. In fulfilling its primary mission, the DoD is prepared to compromise other considerations. Second, even if the American public and U.S. allies trust that the DoD will establish robust command and control in deploying autonomous weaponry, there is absolutely no reason to assume that other countries and non-state actors will do the same. Other countries including North Korea, Iraq, or whoever happens to be the rogue government *du jour,* will almost certainly adopt robotic weapons and use them in ways that are entirely beyond our control.

No means exist to ensure that other countries, friend or foe, will institute quality engineering standards and testing procedures before they use autonomous weapons. Some country is likely to deploy crude autonomous drones or ground-based robots capable of initiating lethal force, and that will justify efforts by the U.S., NATO, and other powers to establish superiority in this class of weaponry—yet another escalation that will quicken the pace of developments leading toward sophisticated autonomous weapons.

The only viable route to slow and hopefully arrest a seemingly inexorable march toward future wars that pit one country's autonomous weapons against another's is a principle or international treaty that puts the onus on any party that deploys such weapons. Instead of placing faith in the decisions made by a few military planners about the feasibility of autonomous weapons, we need an open debate within the international community as to whether prohibitions on autonomous offensive weapons are implicit under existing international humanitarian law. A prohibition on machines making life-and-death decisions must either be made explicit and/or established and codified in a new international treaty.

Short of an international ban, a higher-order principle establishing that machines should not be making decisions that are harmful to humans might suffice. Such a principle would set parameters on what is and what is not acceptable. Once that red line is drawn, diplomats and military planners can go on to the more exacting discussion as to the situations in which robotic weapons are indeed an extension of human will and intentions, and those instances when their actions are beyond direct human control. A higher-order principle is something less than an absolute ban on killer robots, but it will set limits on what can be deployed.

There are no guarantees that such a principle will always be respected. The recent use of chemical and biological weapons by the Syrian President Bashar al-Assad on his own people to quell a popular revolution provides ample demonstration that any moral prohibition can be ignored. Furthermore, robot delivery systems will be available to non-state actors who may control just one or two nuclear weapons, and feel they have little to lose. Each generation will need to work diligently to keep in place the humanitarian restraints on the way in which innovative weapons are used and future wars are fought.

My proposal for an executive order from the U.S. president was largely ignored. And yet it may come to the fore again, particularly if activity on banning lethal autonomous weapons makes headway in international forums. U.S. legislators have been poor in their willingness to endorse international treaties. A domestic initiative could still be necessary for the U.S. to join a worldwide ban. However, before any action is taken, a ban will need to become a major issue upon which presidents or presidential candidates are forced to take a stand.

Under those circumstances, President Barack Obama or his successor could sign an executive order declaring that a deliberate attack with lethal and nonlethal force by fully autonomous weaponry violates the Laws of Armed Conflict. This executive order would establish that the United States holds such a principle and is already implicit in existing international law. A responsible human actor must always be in the loop for any offensive strike that harms a human. An executive order establishing limits on autonomous weapons will reinforce the contention that the United States places humanitarian concerns as a priority in fulfilling its defense responsibilities. NATO would soon follow suit, heightening the prospect of an international agreement that all nations will consider computers and robots to be machines that should never make life-and-death decisions.

Drawing a red line limiting the kinds of decisions that computers can make during combat reinforces the principle that humans are responsible for going to war and for harming each other. Responsibility for the death of humans should never be shirked off or dismissed as the result of error or machines making algorithmic decisions about life and death. The broader meaning of restraint on life-or-death decisions being made by military computers lies in a commitment that responsibility for the action of all machines, intelligent or dumb, resides with us and with our representatives.

## COMPUTERS COMPUTERS EVERYWHERE

For thousands of years, machines have been extensions of human will and intention. Bad design and flawed programming have been the primary dangers posed by most computerized tools deployed to date, but this is rapidly changing as computer systems with some degree of artificial intelligence become increasingly autonomous and complex.

The investment industry was the first to be conquered by computer decision-makers. Arguably, the financial markets have become hostage to the computerized technology they deploy. Winners on Wall Street depend upon algorithmic trading that outpaces the decision-making capabilities of any human. When the computers fail or act erratically, investors lose money and firms collapse. Blaming the failure on the computer or even the programmers does not save anyone's job, recreate lost wealth, or quell lawsuits.

Computer decision-makers are even encroaching on the world of sports, a less risky prospect, to be sure, but one that is still scary for purists and players alike. Always obsessed with stats, the fascination of baseball franchises with the new analytics was not totally surprising. The decision by the general manager of the Oakland Athletics baseball team, Billy Beane, to use statistics to pick players rather than the subjective wisdom of scouts and managers, would certainly have been Beane's responsibility if his 2002 team had a lousy record. Given the success of his strategy, dramatized by the movie *Moneyball,* in which Brad Pitt plays Beane, many other ball clubs have followed suit. Algorithmic analysis of players' past performance now encroaches upon decision-making in basketball, soccer, and football. At least in sports, the numbers help pick among human athletes who must then go on to play the game.

Task by task and industry by industry, computers are taking over decision-making and replacing human workers. Big data drives marketing. Informatics facilitates medical research and helps lower health care costs. Robots that roam through warehouses picking items off shelves fulfill Internet orders. And Amazon tests autonomous drones that use GPS coordinates to deliver a package up to ten miles from a distribution center. Eventually, deliveries may occur within thirty minutes of an order. In theory, no people need be directly involved in fulfilling your Internet purchase of a new blouse, book, ball, or tablet computer.

A reliance on increasingly autonomous systems offers one trajectory in the development of computerized technology. Over time, that course leads to a dilution of responsibility by humans for the unforeseeable actions of complex and occasionally unpredictable systems. An alternative course, a much better course, explicitly retains responsibility for the benefits and harms caused by technologies in the hands of people. That alternative course will require adjustments in the way technological artifacts get engineered.

In flagging the dangers posed by advances in artificial intelligence, four eminent scientists (Stephen Hawking, Max Tegmark, Stuart Russell, and Frank Wilczek) stated succinctly in a February 2014 op-ed: "So, facing possible futures of incalculable benefits and risks, the experts are surely doing everything possible to ensure the best outcome, right? Wrong." We are sleepwalking—surrendering control of the future to smart computers and other emerging technologies. Those who recognize the problems either fail to see them as their challenge, or simply lack the power to do anything.

The case studies highlighted so far in *A Dangerous Master* illustrate the risks of relying upon complex systems that are unpredictable and therefore beyond our full control. The sheer pace of development and the plethora of new technologies mean that the deployment of beneficial and risky tools and techniques far exceeds putting in place effective oversight mechanisms. The simple fact that every technology under development can be both beneficial and risky requires that added attention be directed at weighing the trade-offs and finding methods to limit the harms.

These last two chapters signaled a shift in our narrative. Examining and challenging the false assumptions that propel us forward can defuse even the inexorable forces driving the accelerating pace of development. This chapter provided an example of a presently available inflection point in the march toward dangerously autonomous killing machines. That march can be dramatically slowed and perhaps fully arrested by putting in place a ban on lethal autonomous robots. It is not necessary to build every conceivable weapons system. If a tenth as much energy were devoted to arms control agreements as is dedicated to building risky weapons systems, we would all live in a much safer world.

Slowing the pace of technological development by revising budgets and holding fast to core values expands opportunities to recognize and evaluate inflection points. There are plenty of astute critics stuck in academic cubbyholes, working for nongovernmental organizations (NGOs), or writing blogs who struggle to get our attention. But when the pace of change is rapid and the media flits from crisis to crisis, there is no possibility of evaluating which dire warnings warrant serious consideration.

Defusing false assumptions and acting on inflection points is not enough. Something more proactive is needed. The following three chapters outline key elements to ensure technology does not slip beyond our control: putting in place an engineering culture that designs for responsibility, forging oversight mechanisms which are effective and flexible, and building a cadre of informed citizens whose voices can counterbalance or augment that of the experts. In a democratic society, everyone, not just the experts or those desiring the benefits, deserves a voice in how humanity navigates a dangerous future.

# { 13 }

# Engineering for Responsibility

THE MAGNITUDE 7.0 EARTHQUAKE THAT STRUCK HAITI ON January 12, 2010, killed at least 230,000 people and leveled 70 percent of the buildings and other structures in the capital city, Port-au-Prince. Roughly 1.4 million people were displaced from their homes. A month later, a magnitude 8.8 quake—the sixth largest ever recorded—struck off the coast of Chile and released five hundred times more energy than the Haitian quake. A whole Chilean city moved 10 feet westward. Despite that, the Chilean quake only took a total of five hundred lives. Adjusting for population density, the damage of the extremely powerful Chilean earthquake was 1 percent of the Haitian event. The primary difference between the impact of the two calamities—building codes. In a TED talk, the building activist Peter Haas declared that, "[Haiti] was not a natural disaster. It was a disaster of engineering." Poor engineering leads to disasters. Quality engineering averts them.

In comparison, the potential harms caused by emerging technologies will result from the introduction of advanced engineering innovations that fail in ways that cannot be fully anticipated. Waiting until after a failure and then addressing the revealed problem exemplifies the time-honored approach to uncertainty. Anticipating and addressing potential problems offers a more responsible approach. Uncertainty, however, comes in many guises. While some failures could never be anticipated, others fall within categories of failures that are known. The introduction of new species into ecosystems changes those environments, on occasion, for the worse. Some nanomaterials will be toxic. In other words, a good number of the dangers emerging technologies pose are generally discernible, even if the specific occurrence of a problem cannot be determined in advance.

Any intimation of a potential technological failure suggests additional safety mechanisms that engineers can build into advanced systems. Building in those mechanisms entails costs and slows the development, marketing, and implementation of tools and techniques, something companies are rarely eager to accept. The trade-off between expediency and safety typically defaults to expediency, unless tempered by the recognition of a truly serious high probability danger. But no responsible company wants to be held liable for harm to many individuals or for major damages.

Any failure of a mechanical or computational system indicates a failure in design, whether because of a technical flaw or human error. Engineers take the lead in resolving technical failures. Negligent actions by humans fall upon the shoulders of management. As mentioned repeatedly in earlier chapters, an innovative technology is best understood as comprising one element in a sociotechnical system. Sociotechnical systems include people, relationships between people, other technologies, the physical context or surroundings, values, assumptions, and established procedures. Reworking any of these elements can solve a problem. Engineering and management solutions are most effective when understood as aspects of reengineering the overall system. Similarly, many problems can be prevented through a better understanding of the weaknesses in the sociotechnical system.

Before considering broader approaches to reengineering the non-technical elements within a complex sociotechnical system, we will examine ways to remediate problems through the design of the computational, biological, or mechanical components. Knowledge accumulated by engineers can turn a large disaster into a small one and a tragedy into an inconvenience. But developing methods to limit the harm from unknown or vaguely speculative possibilities requires engineers to think about safety in totally new ways.

Creative approaches to engineering innovative tools and techniques offer ways to address many of the dangers discussed throughout *A Dangerous Master*. Those creative approaches fall within two broad categories: mechanisms related to the physical design of systems, and those related to the values that become criteria used to determine the quality of a technology. The former category includes features designed specifically for when the system fails or is about to fail. The latter category can go beyond instrumental criteria (safety, reliability, efficiency, and ease of use) to specify methods for introducing societal values (responsibility, protecting the environment, and protecting human subjects) into the design process. A few examples will help illuminate the different ways in which innovative engineering practices can diminish future harms.

## EUNUCHS AND KILL SWITCHES

Monsanto had a worthy idea when it considered marketing genetically modified crops with suicide genes. Plants that cannot reproduce solve the serious worry that they might contaminate the genome of similar plants from ancestral stock. As discussed in Chapter 6, however, Monsanto's strategy was perceived as a means to profit by forcing farmers to purchase new seed each planting season.

Arresting a genetically modified plant's ability to reproduce provides an example of interference in the operation of a device in situations where its use is dangerous. Guns come with safety switches, fuses blow, and sensors can automatically shut down delicate electronic equipment.

Good engineers take pride in designing safe products. Until recently, safety meant the creation of reliable and predictable components and systems with features to accommodate known, potentially harmful events. Air bags deploy during a crash. Breathing masks drop from overhead when oxygen levels in the cabin of a plane fall. As technologies get more complex and autonomous in their activity, the meaning of safety also changes. Safety goes beyond the proper functioning of the technical components to encompass the proper functioning of the sociotechnical system and of the societal impact of adopting the technology. The education of an engineer increasingly incorporates training directed at heightening her sensitivity to the potential impact of the tools she builds. In the past, city planners or other specialists in social engineering might have considered societal concerns. In the new paradigm, engineers at all stages of developing a technology need to think through how it will be used. Considering the needs of people with frailties has lead engineers to design telephones with large buttons or chairlifts that transport those with weak legs up staircases. The design of a synthetic organism must take into account the environmental impact if it migrates and finds a niche in a different ecosystem.

Science fiction has turned autonomous intelligent computers and harmful nanomachines into the contemporary version of runaway trains. Bill Joy, co-founder and chief scientist at Sun Microsystems, famously stirred up fear of self-reproducing future technologies with an April 2000 article published in *WIRED* magazine titled "Why the future doesn't need us." In the article, Joy took the grey goo scenario, where self-reproducing nanobots consume all the organic matter on the planet, from nerd fantasy to a broadly disseminated vision that engendered fear. Additional visions of designer pathogens reproducing and intelligent robots manufacturing their own army aroused public debate on questions of whether the dangers justified relinquishing further research in genomics, artificial intelligence, or nanotechnology. Those such as Joy, who championed relinquishing potentially dangerous research, lost the debate.

How could super-intelligent robots be stopped from cloning themselves into a million-soldier army? Would interfering with their supply chain, their access to basic materials and components, be sufficient? Will future smart systems have the ability to override built-in safeguards? The need to invent methods to ensure that future robots smarter than humans would be friendly to us inspired a small network of computer scientists and mathematicians to create the Singularity Institute for Artificial Intelligence, later renamed the Machine Intelligence Research Institute. So far they have made more progress in elucidating the problems then in developing specific strategies to mitigate the dangers.

Biologists and nanotechnologists also began to reflect upon means to assure that their future creations would not run amok. The spread of a pathogen or a synthetic organism that causes an illness already presents challenges. The introduction of a synthetic gene into a flower's DNA in the laboratory makes for an interesting controllable experiment. The release of the same plant outside of the laboratory could have unanticipated consequences such as filling a more important plant's ecological niche and thereby leading to that plant's extinction. Nevertheless, synthetic bacteria, single-celled organisms, and multicelled organisms will be disseminated because they serve valuable purposes, such as breaking down pollutants. How might synthetic organisms be safely introduced into environments outside the laboratory?

The Harvard biologist George Church has put forward a number of novel proposals to arrest or reverse the potentially harmful effects of synthetic organisms. As one of synthetic biology's leading scientists and visionaries, Church demonstrates sensitivity to dangers posed by this emerging field of research. In one proposal, he suggests that biologists develop a new form of synthetic DNA that operates through novel metabolic (chemical) processes. Furthermore, the synthetic biological device or organism created by this new form of DNA should also function through altered metabolic processes. In this manner, the genes from the synthetic chromosome would not function if transposed to native or hybrid forms of DNA. And the resulting synthetic organism could not crossbreed with existing organisms to create a new mutation. Through this approach geneticists might use the benefit derived from a synthetic organism without making it possible for that organism to directly alter existing organic material. But an organism created from synthetic DNA could still take over an ecological niche inhabited by a more important plant.

Building in the high-tech equivalent of an OFF switch can fortunately halt the activity of any device. For example, introducing an essential component made of a material that dissolves through a known chemical reaction would disable a nanomachine. The dissolvent would have to be harmless in general use and not readily available in the environment where the nanodevice operates.

Research on kill switches for technologies such as nanomachines will require serious funding. But introducing such mechanisms during the early stages of development offers a means to make a device or organism safer, even when we still don't know all the applications for which it will eventually be used or the dangers it poses. These mechanisms will not, however, be integral to the design of technologies built for intentionally destructive purposes. A terrorist will leave the kill switch out of the biological or computational device. Fortunately, terrorists represent the exception, not the norm. Most of the dangers arise from accidents and unanticipated consequences.

## DESIGNING FOR RESPONSIBILITY

Even when kill switches and other new designs decrease the likelihood that technologies will run amok, there is still a chance that they will cause harm. The last chapter asked how should responsibility be apportioned if a computer that can make choices and initiate actions causes physical harm or property damage. Usually a court decides this question. The harms caused by present-day computers clearly fall within the existing body of tort law and are the responsibility of the manufacturers who build and market the systems or the user who adapts a system for specific applications. Unusual cases establish legal precedents.

With increasingly autonomous systems a new legal issue will come to the fore. When the manufacturers or the people working with an intelligent machine cannot know how it will act in unfamiliar situations and something harmful happens, who is responsible? The manufacturers will continue to be responsible for most situations, but perhaps not all accidents. In the distant future, it might be possible to build conscious software agents capable of making sophisticated and sensitive choices and knowing what they are doing. In the meantime, holding the machine responsible for its actions is nonsense. Insurance companies could elect to sell no-fault coverage to shield either manufacturers or users from situations where a robot's actions surprise everyone, presuming that such incidents are truly outliers. But, as discussed, allowing people and corporations to abrogate their responsibility is a bad idea that opens the door to the creation of potentially dangerous computing devices. Any commonsense approach to safety requires that a person or corporation be held responsible for all harms caused by a technology.

Rather than waiting until after products are marketed and accidents occur, it would be helpful if the question of responsibility for computing artifacts gets addressed during the machine's design. Engineers could treat responsibility as an extension of their mandate to make the products they build safe. Jeroen Van den Hoven, a professor at Delft University in The Netherlands, proposes that engineers design for responsibility. By this he means that the question of who or what will be responsible for the tools they create serves as one of the design specifications. Just as using components that will not overheat would presently be a design specification, so also should ensuring that someone will be accountable if the machine causes an accident.

Consider an engineer designing a robot that would be a helpmate to a homebound elderly person. Taking responsibility for the robot's actions into account at the start of the design process could determine features of the system, or even the computational platform upon which to build the robot. A design team might reject platforms that were heavily reliant on machine autonomy and favor platforms in which decisions reside with the human members of a support team. When, for example, the robot encounters an unfamiliar situation, it calls a medical support operator able to view the environment remotely through a camera mounted on the robot. The human member of the support team, not the robot, decides what to do in the unfamiliar situation. And the human operator, her supervisor, and the company they work for bears responsibility for a bad decision.

Engineering for responsibility illustrates just one example of how values can be introduced into the design process. For years there have been calls for engineers to make an ethicist or a social theorist part of their design team. Not as a naysayer, but as a participating designer, sensitive to the societal and ethical factors that enter into how innovative tools will be used. Values have always entered into the design process, often as the norms or implicit values of the engineers or the companies for whom they work. Value-sensitive engineering makes values an explicit and conscious element of a technology's design. For example, in designing a coffee pot to be used by the elderly, engineers have taken into account the user's frailty. They placed the pot on a pivot wheel that enables the user to easily tilt the pot and thereby pour a cup of coffee. Similar concerns arise when designing robots to perform tasks for the homebound.

The social theorist Aimee van Wynsberghe notes that introducing a robot that helps pick up patients from their beds in a hospital setting changes the role of a number of different actors, and the way in which a variety of secondary functions get performed. It changes the care practice. If, for example, nurses are no longer involved in moving patients or changing their linen, the nurse loses an opportunity to directly observe the patient's physical and mental state. On the other hand, robots that lift patients can reduce a nurse's workload, and perhaps reduce the number of orderlies hired. Having a member of the design team sensitive to all the dimensions of providing health care could lead to implementing important features and capabilities into the robot. And, by extension, the introduction of the robot into the hospital can be accompanied by team training that sensitizes all parties to how essential elements of the care practice will be altered.

Demand from consumers already directs engineers to embed values such as ease of use, aesthetic qualities, and protection from sensitive material into new products. Additional social values such as fairness, privacy, security, and the autonomy or dignity of the individual, can also be treated as explicit elements in the design of a tool or technique. Engineers excel at problem solving. Once they know the problem that demands their attention, good engineers will find a solution. Each new value might well direct engineers toward very different solutions to the problem they wish to solve, and therefore very different designs. In turn, the value would be used as one more item on the checklist to assess whether the device had been designed successfully.

## MORAL MACHINES

In *Moral Machines: Teaching Robots Right From Wrong,* Colin Allen and I mapped a new field of inquiry known as machine ethics or machine morality. This field explores prospects for developing computers and robots capable of factoring moral considerations into their choices and actions. Machine ethics provides one more example of how values can be engineered into technologies in order to ensure that their actions will be safe and appropriate.

Engineers already program values and moral behavior into computers and robots (hereafter collectively referred to as "robots"). The engineers determine in advance the various situations robots will encounter within tightly constrained contexts. They then program the system to act appropriately in each situation. Colin Allen and I referred to this kind of moral behavior as *operational morality*. However, designers will not be able to perceive in advance all the new options encountered by increasingly autonomous robots. At unanticipated junctures, the robots will need their own ethical subroutines to evaluate various alternative courses of action. For example, a service robot may need to decide whether to recharge its battery before or after performing a series of tasks such as delivering medication or the newspaper to the individual it assists. Perhaps the tasks are prioritized and those priorities change according to the time of day, the health of the individual, or other factors. Robots capable of making ethical determinations are *functionally moral* even if they have little or no actual understanding of the tasks they perform. Eventually, robots may learn to align their behavior and values with that of people, and progress to apply rules and principles to a broad array of challenges.

A few basic question shape research within machine ethics. Do we need robots that make moral decisions? When? For what? Do we want robots making moral decisions? Then there is the age-old question of whose morality or what morality should be implemented. In theory, any set of moral principles could be programmed into a robot and different value systems might be implemented for different contexts. And finally, is moral aptitude something that can actually be built into computational devices? From the perspective of applied ethics and engineering, a great deal of attention is directed at this latter question. Can robots be designed so they follow rules, such as the Ten Commandments or the Laws of Armed Conflict? Or can robots follow procedures, such as calculating which of various alternative courses of action represents the greatest good for the greatest number?

The challenge of building moral machines leads readers of science fiction to immediately think of Isaac Asimov's (1920–1992) Three Laws of Robots. In a collection of science fiction novels and stories, Asimov proposed that all robots be programmed with these laws:

1. A robot may not injure a human being or, through inaction, allow a human being to come to harm.

2. A robot must obey the orders given to it by human beings, except where such orders would conflict with the First Law.

3. A robot must protect its own existence as long as such protection does not conflict with the First or Second Law.

Later, he added a Zeroth Law: A robot may not harm humanity, or allow humanity to come to harm.

The Three Laws are straightforward and arranged hierarchically. Nevertheless, in story after story, Asimov illustrates how difficult it would be to design robots that followed these simple ethical rules. For example, a robot programmed with just the three laws will not know what to do when confronted with conflicting orders from two different people. Through his stories, Asimov illustrated that a simple, rule-based morality would be far from sufficient for ensuring that smart machines act in morally appropriate ways.

Similar hard problems arise when exploring ways to implement any norms, principles, or procedures for making moral judgments. In addition, a second set of very hard problems confronts engineers intent on developing moral machines. They entail the robots' understanding of the ethical dilemma. For example, how does the robot recognize that it is in an ethically significant situation? How will it discern which pieces of information it possesses are essential for making its judgment, which are unimportant, and whether it has sufficient information to even make a judgment? What capabilities will it require to obtain the additional information it needs? Might it need the ability to empathize with a human to act appropriately in certain circumstances? Is empathy a capability that can be built into a machine?

Small, preliminary grants laid the foundations for researchers to study approaches to implementing moral decision-making in robots. Then, in May of 2014, the U.S. Office of Naval Research awarded the first large grant of $7.5 million to university researchers at Tufts, RPI, and Brown to explore methods for building robots capable of discerning right actions from wrong ones. New markets will be opened up if robots can be designed to obey laws and act morally. On the other hand, if companies and academic engineers fail to design robots that follow laws or are sensitive to human values, consumers will not trust their actions, and will demand that governments impose limits on how robots can be used.

## RESILIENCE

The management of complex systems that adapt to changing circumstances, which we discussed in Chapter 3, epitomizes the most daunting of engineering challenges. Various fields such as resilience engineering and complexity management have emerged to study new approaches for tackling the uncertain behavior adaptive systems periodically exhibit. The known unknowns include the extent to which adaptive systems can even be successfully managed.

Eventually, all mechanical systems fail. People working with complex systems make mistakes. Persistent human error is often a symptom of a badly designed system. Biological systems die. One growing trend in engineering has been toward anticipating failure and planning for its occurrence, particularly for tasks that are central to keep critical operations from failing.

Maintaining electricity is critical for many businesses as is keeping an Internet site up and running for a company whose services are supplied through the Web. Redundant components in a nuclear power plant carry the load when one fails. Backup arrays copy all computer data in a manner that ensures no information gets lost when one or more storage drive fails. In the 1990s, IBM introduced high-end computer systems that monitor the performance of key components, and even call in repair technicians to replace parts well before they malfunction.

Mechanical and computation systems display brittleness when the demands placed upon them exceed their performance boundaries. In an automobile, this could manifest in a vibration that increases until some component malfunctions. In a computer, brittleness causes a lockup or crash. Brittleness in the form of constant lockups warns users that their computer will need maintenance. Feedback to manufacturers from users that a device is brittle indicates that they must design an upgrade with improved resilience.

Resilient systems have the ability to recover from a failure. Years ago, when a computer crashed you lost all the data that had not been copied to a hard drive. Today, very little gets lost. Designing systems that recover from a failure quickly, effortlessly, and with minimal degradation in functionality, has evolved into a totally new approach to safety and risk assessment. The more resilient systems have the ability to adjust to disturbances and continue operating even when the situation was unanticipated.

Engineers believe a trade-off exists between dimensions in which a system is robust and those where it exhibits fragility. An eggshell illustrates a container designed by nature that can handle remarkable internal and external pressures. But when tapped strategically, the eggshell reveals its fragility and shatters. Systems are designed to serve a particular purpose, and when successful, the finished product displays the robust capacity to fulfill that goal. In designing the Internet, attention was given to ease the addition of individual nodes and their ability to freely communicate with each other. No one node or group of nodes plays an essential role in maintaining the overall system. The Internet will continue functioning throughout the world even when power failures in one region cut off local businesses, homes, and mobile devices from service. Unfortunately, the very design that made the Internet so robust and accessible also enabled vulnerabilities that expose users to viruses and breaches of security.

The trade-offs between robust features and fragile dimensions cannot be eliminated, and yet creative design strategies carry the promise of minimizing degradation when a system fails. Commonly, a failure occurs in a secondary dimension whose frailty was the result of designers' focus upon fulfilling primary goals in a robust manner. A better understanding of the overall operations of a system, with particular attention to its frailties will suggest compensating measures to minimize the effects of something going wrong. For example, lighter laptop computer systems are easy to transport and can be aesthetically pleasing, but the device may not be adequately protected from being dropped on a corner. Therefore we add protective cases that, in turn, take away from the lightness and the aesthetics. Usually these secondary concerns are where the problems lie.

The field of resilience engineering offers a very important new approach to the study of complex adaptive systems. Indeed, it goes beyond the study of the individual complex mechanical, biological, or computational sub-systems to study the larger system of systems, especially those with both human and mechanical participants. In a sociotechnical system, the adaptive ability of the human participants (managers, operators, and technicians) plays a central role. The people in the overall system compensate for the inherent brittleness of mechanical components, and therefore become the responsible agents if a failure occurs. The more resilient sociotechnical systems are those best able to deal with and respond to both the expected and unexpected demands placed upon them. In concert with technological aids, it is the human actors who assess a disruptive situation, recognize the specific challenge at hand, decide what to do, and when.

Chapter 3 opened with the story of an unmanned air vehicle (UAV) that ran off a runway after a failure in coordination between the software-initiated actions and those of the human operators. Increasing the autonomy of the unmanned aircraft is a commonly suggested solution to such accidents. But as pointed out, this would only increase the responsibility of the human operators. They would then need to understand the operations of an even more complex computerized partner to fully coordinate all operations. Outmaneuvering complexity is not easy. Many proposed solutions merely make the overall system more complex, and therefore, more unpredictable.

Certainly some of the human load can be transferred to the more computational sub-systems. But a better approach entails considering the managers, operators, and technology as a team, and turning the computerized component into a good team player. Counter to intuition, the reduction or elimination of disasters builds upon what can go right, for example, improving teamwork, rather than outmaneuvering what can go wrong. Forging a working relation between human actors and our machines heightens the capacity for complex sociotechnical systems to adapt when unanticipated events occur.

Our discussion of resilience has focused upon the management of man-made systems. But similar problems arise when dealing with humans, non-human animals, and environments. Indeed, much of our understanding of complex adaptive systems comes from the study of the adaptive skills and resilience of organisms and environments. Reptiles, for example, regrow severed limbs. A cold-blooded organism adapts its internal temperature to the temperature in its surroundings. I periodically visit Mount St. Helens, site of a powerful July 2008 volcanic eruption. On the first visit, I was overwhelmed by the vast destruction. But now I marvel at how the plant and animal life demonstrate the remarkable resilience to slowly recover from such a devastating blow.

Through the study of resilience, we might hope to understand how human activity and man-made systems alter and tax the environment. Figuring out how we and our technological creations can fit in and join in partnership with natural systems offers a truly worthy goal for resilience engineering.

Considerable research must be directed at a better understanding of complex adaptive systems and means to maintain or improve their reliability. The research will be expensive and its success uncertain. Early funding of research directed at technological solutions to the very problems that innovative technologies will create is a responsible approach to the dangers of innovative research. The presumption that for every technological problem there will be a technological fix is not sufficient. Let us put money up front into researching possible technological fixes. This can also reveal which problems are not fixable. Either we make the investment, or we should begin robust disaster planning.

Engineering for responsibility is not a panacea for all challenges. It merely provides a way forward, given that we cannot fully know the impact of an innovative line of research early in its development. It provides bottom-up solutions. Creative engineering approaches can transform some of the possible catastrophes discussed throughout *A Dangerous Master* into relatively insignificant events.

# Tools for Governance

BEFORE THE ARRIVAL OF TRAINS, THE CLOCK TOWER IN ONE village might register a time that differed significantly from that in a neighboring town. Railroad schedules imposed a compelling reason to coordinate timekeeping. The standardization of time illustrates one of many secondary and tertiary societal impacts accompanying the introduction of a technology directed at reshaping transportation. Cities connected by railroads grew rapidly and new towns sprouted up along the tracks. Railways hauled in the raw materials needed to feed the continual appetite of growing industries and, in turn, delivered finished goods to ever-expanding markets. Large farms able to convey produce quickly to population centers reshaped the ecology of the Midwest. In reviewing the transformative impact of railroads, Arizona State University's Braden Allenby notes that, "Any technology of enough significance to be interesting will inevitably destabilize existing institutions, power relationships, social structures, reigning economic and technological systems, and cultural assumptions."

Nonetheless, Allenby argues that uncertainty in both how a new technology will develop and in its societal impact make efforts to regulate innovative applications foolhardy. Indeed, he perceives premature regulations as not only encumbering development, but also often doing so in an unnecessary and harmful manner. In principle, he rejects early precautionary regulations based on unproven hype or fear. Dystopian projections often get invoked merely to "protect existing economic interests and worldviews." But he emailed me: "It is okay to regulate human cloning right now, because the risks of something going wrong are so high." On the other hand, Allenby argues that drones in domestic airspace should not be banned without clear evidence that they pose challenges which existing privacy law fails to manage in a satisfactory manner.

In contrast, Victoria Sutton confronted a need for early (upstream) regulatory guidelines while serving in the administration of President George W. Bush. In his January 2003 State of the Union address, Bush called for a governmental initiative to develop cars powered by pollution-free hydrogen. The Bush administration learned from earlier biotech initiatives that fear of unknown regulatory decisions would slow the participation of angel investors. So to jumpstart what became known as the hydrogen economy, they needed to reassure investors that regulations would neither impede progress nor unduly tax future earnings. Hydrogen-powered cars pose many unique problems that they were compelled to address up front. The volatility of pure hydrogen gas when mixed with air means leaks easily lead to explosions. Safe refueling hydrogen generators would be needed in each home and in parking garages.

At the time, Sutton held the position of Chief Counsel of the Research and Innovative Technology Administration at the U.S. Department of Transportation. She confronted the fact that many different government agencies and regulatory authorities held overlapping jurisdictional responsibility for aspects of hydrogen energy policy and the use of hydrogen vehicles for transportation. So to spur the development of the hydrogen economy, she convened an ad hoc committee with both scientists and lawyers representing each of the governmental jurisdictions involved. She requested that the participating agencies and departmental representatives be "named" in order to assure accountability. Sutton's ad hoc committee fashioned a regulatory framework, and all the agencies and departments with oversight responsibility "signed off" on a statement published in the Federal Register.

The early-stage regulatory framework paved the way for industry to invest in research. But the advent of hydrogen-powered vehicles in the U.S. still awaits technological breakthroughs. Even then, a refueling infrastructure must be put in place to make the vehicles practical for long journeys. It is difficult to imagine that a hydrogen-refueling infrastructure can be built without significant funding and regulatory support from the federal government. The Obama administration directs more attention to stimulating the growth of sun and wind power as alternative energy sources. Future administrations may look to other sources to meet the country's energy needs. Nevertheless, research continues on hydrogen vehicles and hydrogen-based fuel cells. Which future sources of clean energy win the day as successors to biofuels largely depends on which technology proves to be safe and cost-effective.

Dr. Sutton presently teaches at Texas Tech University School of Law. Her initiative while in the George W. Bush administration remains a unique example of how input from many discrete yet overlapping federal agencies can be coordinated to stimulate the upstream development of an emerging technology. She believes that similar coordination could spur the progress of other emerging technologies. However, in the development of technologies whose social impact will be far-reaching and disruptive, coordination must go beyond the federal government to encompass the concerns of all stakeholders. Harmonization with the policies of other countries will also be required. Not all parties, however, will perceive coordination as in their interest.

To regulate or not to regulate? That question dominates discussions about the governance of emerging technologies. Ideological debate swirls around the value of bureaucratic headaches created by the promulgation of new laws and regulations. And yet, the example above from the George W. Bush administration illustrates that even anti-big government politicians can find regulations valuable. Nevertheless, minimization of government bureaucracy rules the contemporary political landscape, and serves as one major reason why oversight of innovative fields has fallen so far behind technological development.

The U.S. government, for example, failed to provide effective oversight for nanotechnology, even as it provided billions of dollars for research. Richard Denison, the lead scientist at the Environmental Defense Fund, put it well when he stated, "The real danger is continuation of a government oversight system that is too antiquated, resource-starved, legally shackled and industry-stymied to provide to the public and the marketplace any assurance of the safety of these new materials as they flood into commerce."

Concerned stakeholders have been forced to rely upon nongovernmental means to set standards for responsible nanotechnology. Anxious to proceed with research and frustrated by the government's inability to formulate guidelines, the DuPont Chemicals Company decided to create its own recommendations for managing risks. In this effort, it elected to work together with the Environmental Defense Fund (EDF), a highly regarded and influential advocacy group. The resulting six-step process outlines ways to identify, assess, and manage risks. Other parties quickly adopted their Nano Risk Framework, which soon became the *de facto* standard. To further its acceptance, EDF and DuPont translated the framework into Mandarin, French, and Spanish.

The Nano Risk Framework illustrates an increasing reliance on soft governance mechanisms devised by groups outside of government. Soft law or governance includes industry standards, codes of conduct, certification programs, laboratory practices and procedure, statements of principles, and social norms. Standards set by insurance companies, for example, force individuals and institutions to put in place safe practices in order to get coverage. Unlike government regulations, soft law can be adopted rapidly and revised or discarded just as quickly. However, only a few of these soft governance approaches carry means to punish bad behavior. Furthermore, the group that creates a soft governance tool is seldom representative of or credible to all parties. Thus, reliance on soft governance alone is insufficient to ensure safety.

In contrast, governments have many means to enforce their will. Departments and agencies initiate suits for unlawful activity, shut down dangerous facilities, and can withdraw government financing of research. But as we've seen, governments are slow to pass laws and approve standards for new technologies because industries and legislators are generally fearful that regulations will slow research. Furthermore, once enacted, governmental regulations that appear warranted in earlier stages of development can quickly become outmoded. And yet, by that time, the regulatory regime has become entrenched and ossified into a bureaucratic burden.

The effective oversight of emerging technologies requires some combination of both hard regulations enforced by government agencies and expanded soft governance mechanisms. Indeed, actively combining the two approaches could solve many of the problems arising from overregulation and the ossification of outdated laws that become "frozen in place." That view is championed by Gary Marchant, a professor at the Sandra Day O'Connor School of Law at Arizona State University, and a proponent of the role soft governance can play in managing technological innovation. Yet, even with an increasing reliance on soft governance mechanisms, a core problem remains: coordinating the separate initiatives by the various government agencies, advocacy groups, and representatives of industry. Gary Marchant and I have offered a proposal to meet that challenge. We recommend the creation of issue managers for the comprehensive oversight of each field of research. We call these issue managers Governance Coordinating Committees. These committees, led by accomplished elders who have already achieved wide respect, are meant to work together with all the interested stakeholders to monitor technological development and formulate solutions to perceived problems. Rather than overlap with or function as a regulatory body, the committee would work together with existing institutions.

The notion of bringing coordination to the many disparate parties involved in the development of innovative technologies and the shaping of technology policy can sound naïve, particularly given the rapid rate of discovery and improvement in products and processes. Governments traditionally move at a snail's pace, while soft governance mechanisms can be enlisted to respond more quickly. Governance coordination will only work if it provides an effective means to address the ever-increasing gap between the appearance of a technology and the management of its societal impact.

## THE PACING PROBLEM

The pacing problem refers to the time gap between the introduction of a new technology and the establishment of laws, regulations, and oversight mechanisms for shaping its safe development. But the lag in government oversight also affects the removal of obsolete regulations. In 1958, the U.S. Congress enacted laws charging the FDA to regulate the use in food of substances determined to cause cancer in laboratory animals. The intention of Congress was good, and yet the determination that an agent caused cancer had been founded upon scientific theories later proven inaccurate. By the early 1990s, it was already apparent that the laws, known as the Delaney Clause, were out-of-date, and yet they have not been withdrawn. Throughout government, regulatory agencies are charged with upholding obsolete laws at great expense. With limited staff and resources, the attention given to outmoded laws detracts from the oversight of often more pressing issues.

There has always been a pacing problem. The accelerating pace of development alone widens the gap. However, legislative gridlock, the ossification of regulatory agencies, and the glacial pace at which courts clarify law compound the problem. As technological development accelerates, regulatory agencies slow down and become rigid, burdened by red tape and unfunded mandates. New regulatory needs go unaddressed. For example, genetic testing proliferates with limited protection for consumers. Online privacy, while hotly debated, lacks effective regulation.

In addition to the rate of development, emerging technologies cover a vast array of applications and present pervasive uncertainty as to their risks,

benefits, and future directions. Uncertain risks make it hard to set priorities or design solutions. The sheer breadth of the different kinds of applications for enabling technologies, such as nanotech and infotech, overwhelm traditional approaches to governance. Existing regulatory agencies are poorly equipped to manage the issues, which go beyond traditional health and environmental concerns to encompass broad societal concerns including privacy, fairness, and human enhancement. Adaptive tools for oversight are needed to reform outdated legislation and quickly respond to the unfolding possibilities and dangers.

Regulations, often thirty to forty years old, were designed to cope with yesterday's challenges. Major environmental legislation passed in the 1970s has not been updated to cover new concerns such as those posed by nanomaterials. For example, the 1976 Toxic Substances Control Act (TSCA) regulates the introduction of new chemicals. The creation of TSCA took place during a period when the detrimental impact of chemicals on the environment was considered to be the result of their use in large quantities. The Environmental Protection Agency attempts to apply TSCA to the oversight of nanomaterials, but the existing regulations are ill suited to substances with potency in miniscule amounts. To date, all efforts to reform TSCA to provide better oversight for nanomaterials have failed to pass.

Dedicated leaders of underfunded regulatory agencies rightfully fear being blamed for new tragedies. And yet they are seldom given the regulatory authority, staff, or resources to expand their duties to provide sufficient oversight for emerging challenges. Indeed, their proscribed mandate often restricts their ability to respond to the societal impact of new technologies. As mentioned in Chapter 6, when the FDA concluded in 2008 that milk and meat from cloned animals present no unique health or safety issues, tens of thousands of citizens submitted objections on societal ethical grounds. The FDA's single-line response that "the agency is not charged with addressing non-science based concerns such as the moral, religious, or ethical issues associated with animal cloning for agricultural purposes, the economic impact of products being released in commerce, or other social issues unrelated to FDA's public health mission," failed to satisfy anyone. On the one hand, the First Amendment prohibits mixing church and state. Parochial values or the "yuck" factor should not influence governance. On the other hand, excluding broad ethical considerations from the process of formulating new regulatory guidelines can mean the government never deals with serious long-term considerations—many of which are worth discussing precisely because of the associated ethical issues.

Legislators, focused upon improving productivity, shortsightedly ignore gathering storms. A belief that the acquisition of benefits occurs quickly and risks exist more in the future, supports their negligence. As noted in Chapter 12, the roboticization of warfare will be pursued because of its short-term benefits with little regard for the longer-term impact of pursuing this form of weaponry. All too often Congress waits for the next crisis to occur before it will take concerted action. But disasters lead to legislation that targets forestalling their repetition, and only periodically at putting in place thoughtful means to prepare for a landscape changing at hyper-speed. In the case of roboticizing warfare, a backward-looking policy will come too late to bend a trajectory already determined within the military-industrial complex. Good politicians understand that they will be judged by how well they respond to crises. Unfortunately, being seen as active during an actual crisis carries more weight with the chattering classes than having staved off a crisis that never develops.

In the White House, the Office of Science and Technology Policy (OSTP) oversees emerging fields of research. In spite of legislative gridlock, OSTP endeavors to shape the progress of technological innovation. Much of its energy targets the establishment of new programs. Given their limited time in an administration, it is understandable why staff at OSTP get excited about opportunities such as starting President Obama's BRAIN initiative to advance research in neuroscience. Such programs warrant high-profile announcements and publicity. However, the day-to-day work of the staff at OSTP receives little attention. It entails the coordination of communications across the multiple agencies responsible for overseeing technological development. OSTP also acts on the recommendations of the President's Council of Advisors on Science and Technology and fosters various public-private partnerships.

The potentially disruptive impact of emerging technologies is taken seriously by the staff of OSTP, but given the legislative stalemate, their scope for action remains limited. However, through the clarification of regulatory guidelines and the application of codes to new forms of research, the various agencies within the executive branch of government continually shape technological development. Tom Kalil, Deputy Director for Technology and Innovation at OSTP, put it so well when he titled an unpublished internal brief, "Saving the world—one document at a time."

A similarly entangled web of "soft law" complements the matrix of existing and often overlapping government agencies and statutes. The various parties proceed piecemeal with their own initiatives and pay little attention to how the actions of other bodies affect the same technology's development. This patchwork system leads to both overlaps and serious gaps.

Differences between domestic oversight and the regulatory authorities, or lack thereof, in other countries adds an additional layer of complications and gaps. Issues such as climate change and management of the Internet are international in scope. Concerns over human cloning or enhancements in sports can only be managed through international approaches. Multinational corporations avoid new regulations by moving controversial activities overseas. The overlap and gaps created by a hodgepodge of institutions, laws, and guidelines (domestic and international) involved in governing the emerging technologies cries out for some means to coordinate the various actors, initiatives, and approaches.

## COORDINATING COMMITTEES

The issue managers proposed by Gary Marchant and myself would serve as "orchestra conductors." A committee that comprehensively coordinates the development of an individual field would oversee the industries that field creates. These Governance Coordinating Committees (GCC) would neither replicate nor usurp the tasks fulfilled by other institutions such as regulatory agencies. They would, however, search for gaps, overlaps, conflicts, inconsistencies, and synergistic opportunities among the various public and private entities already involved in shaping the course of an emerging field. The

GCC's primary task: to forge a robust set of mechanisms that comprehensively address challenges arising from new innovations, while remaining flexible enough to adapt those mechanisms as an industry develops and changes. Members will be charged with keeping the committee lean and responsive and, wherever possible, utilizing existing resources and institutions. A robotics coordinating committee, for example, would work to integrate the activities of governmental agencies and soft law mechanisms to promote safe practices, methods for testing and certifying products, and means to managing risks.

The GCC is meant to be a good-faith broker. Its recommendations will only carry weight if the public, industry, legislatures, and the executive branch perceive it as a credible authority. Each GCC can expand its effectiveness and influence by helping the various parties appreciate that cooperation in building a robust set of mechanisms lie in their interest.

Avoiding hard regulations as a solution to gaps will be among the mandates for the committee's staff. But when necessary, a GCC might use the carrot-and-stick approach to persuade an industry that it should commit to a voluntary system of self-regulation to avoid the burden of government regulations. However, if they fail, legislators and administrators within the executive branch of government, knowledgeable that the GCC looked first for other means to address gaps in oversight, should take the committee's recommendations seriously. Either that, or be held responsible if anything goes wrong.

The carrot of recommending that lawmakers lower or limit liability risks might be held out to industry as an incentive to commit to strong self-imposed standards. The robotic industry, for example, might be amenable to the establishment of rigorous testing and certification procedures for devices, if this, in turn, led to no-fault insurance policies that covered their risk when something unanticipated occurs.

Government agencies, we believe, will perceive cooperation with GCCs that do not usurp their authority as being in their interest. Overwhelmed and under-resourced agencies could be buffered by the GCC from additional unfunded mandates. When, for example, a GCC persuades industry to put in place robust self-regulation, the government agency will be relieved of some additional responsibilities.

Other stakeholders might also look toward the GCC as a focal point for their concerns. Scholars, social critics, and nongovernmental organizations all complain that issues they raise go unheard and unaddressed. They will not necessarily appreciate decisions made by the GCC, but they will appreciate knowing that some oversight body factored their concerns into the process of forging robust oversight of an industry.

The GCC could also serve as a library or repository for all of the publicly available research on new fields of discovery and the societal challenges they pose. It is never easy for either the public or media to distinguish hype and unwarranted fear from credible information. The GCC would monitor scientific progress and report when scientific thresholds are about to be crossed that make more speculative threats plausible.

The press could also turn to the GCC for an honest appraisal of new research. Responsible reporters work hard under deadline to understand complicated issues. Irresponsible media sensationalize stories by playing upon rumors, distortions, fears, and biases. Even when they try to be fair, they do so by presenting both sides of an issue, contributing to the illusion that opposing positions are of equal worth, which is seldom the case. A credible and well-functioning GCC could ease access to a truly balanced understanding of the issues.

The network of individual GCCs might be collectively viewed as a modular structure. Committees responsible for individual fields could compare notes and identify successful or failed approaches. Committees might work together to coordinate the management of technologies that are the product of converging fields of research. Furthermore, they could work with similar bodies in other countries to harmonize methods for international governance of specific technologies.

## IMPLEMENTATION CHALLENGES

GCCs offer a way forward. The path toward their implementation is less clear. Many questions remain unanswered. How will a GCC acquire influence (power), authority, or legitimacy? Who will select or appoint the staff? What steps must be taken in order to establish a committee's credibility? To whom will a committee be accountable? How will it get funding? The funding source will, for example, influence how the members and staff are chosen. A GCC might be established as a governmental institution or as a private entity. Each approach comes with advantages and disadvantages. A private entity could be more easily insulated from political pressure, but from within the government a GCC is likely to wield more power.

Representation from the relevant stakeholders—government, business, nongovernmental organizations, scientists, the public, and policy experts—would certainly help build a committee's credibility. Credibility and respect might also be established by recruiting administrators from among retired individuals. If someone who had acquired widespread admiration through her work in business, academia, nonprofits, or the military took on a committee as her personal mission, that would go a long way toward establishing credibility. However, as a committee gains power and respect, it will be difficult to insulate it from political influence or at least the suspicion of bias. Influence and authority are extremely difficult to maintain in the modern world, particularly when those with differing opinions are quite willing to take dishonest potshots at credible leaders.

These implementation challenges might lead some to believe that Governance Coordinating Committees are too complicated or naïve, given the present political environment. Perhaps the proposal *is* both complicated and naïve. Nevertheless, we cannot do without some kind of solution. Something like GCCs will be required to adaptively manage the challenges new technologies create. Gary Marchant and I believe the implementation challenges can be worked out. Therefore, we recommended a pilot project to explore the viability of a GCC. The pilot project would be for robotics or for synthetic biology. Both of these fields of research are young and not yet hampered by a preponderance of regulations.

There is a clear need for a new governance model. A pilot project affords the opportunity to work through the implementation challenges and study whether a GCC represents an effective approach to governing the governance of the emerging technologies.

# { 15 }

# Guild Navigators

FRANK HEBERT'S CLASSIC SCIENCE FICTION NOVEL *DUNE* AND its five sequels describe the "spice" melange as the ultimate enhancer. It provides longer life, heightens awareness, and increases vitality. More important to the plot of *Dune*, melange offers limited prescience. Prescience holds particular importance for the Spacing Guild that has acquired a monopoly on interstellar travel. Melange enables the Guild's navigators, who pilot enormous ships at speeds faster than light, to perceive safe pathways through the vast reaches of space-time as they accurately traverse the galaxy from one solar system to another. At each juncture in the journey, the prescience-enhanced navigators can discern which future will result from every course of action.

In a manner of speaking, each of us is a Guild navigator, making decisions in the present that will give form to the future. If we were blessed with prescience, navigating the future would be easier. We could see the consequences of each course of action—which choices maximize benefits and which minimize harms, whether inequalities actually serve humanity as a whole or exacerbate pain and suffering, and how best to nurture the character, freedom, and flourishing of each individual. Perhaps with full foreknowledge there would be no need for ethical debate, as the best course of action would reveal itself in stark relief.

Of course, we lack prescience. There are many unknowns. Some choices will lead to unintended consequences. Nor do any of us have sole control over the vessel we hope to direct. Indeed, we control very little, but through the network of relationships that knits humanity together, we influence everything.

All good navigators understand that many routes can be charted to an intended destination. In science fiction, skilled navigators scan the horizon for both known and unanticipated obstacles—warfare in one sector of the galaxy, space pirates in another, and the sudden appearance of a wormhole. Only foolhardy navigators routinely select the fastest or most direct route. On our planet, the challenges confronted are perhaps somewhat more mundane, and yet, of profound importance to humanity's future. Engineers, diplomats, and wise citizens share in the understanding that a circuitous pathway and an unconventional method often offer the best solution to a problem.

One question has lain at the heart of *A Dangerous Master*. Do we, humanity as a whole, have the intelligence to navigate the promise and perils of technological innovation? This book has been more about the journey than about getting to any particular destination. Indeed, the destination is often less important than the road traveled. In finding "two roads diverged in a yellow wood," Robert Frost "took the one less traveled." In finding her way through Wonderland, Alice stopped to ask directions from a large Cheshire cat perched in a tree.

> Alice: "Would you tell me, please, which way I ought to go from here?"
> Cheshire Cat: "That depends a good deal on where you want to get to."
> Alice: "I don't much care where——
> Cheshire Cat: "Then it doesn't matter which way you go."
> Alice: "——so long as I get *somewhere*."
> Cheshire Cat: "Oh, you're sure to do that."

A small but growing community of transhumanists desires a future with freely available enhancements or the technological means to immortality. But like Alice, most of us have no idea what humanity's future should look like. Nevertheless, we each hold strong opinions about what we do not want. Those opinions, however, are often unrealistically dismissive of change and overlook creative courses of action. In *Children of Dune*, the third novel in the Dune series, Hebert cites an excerpt from "The Spacing Guild Handbook." He writes:

> Any path that narrows future possibilities may become a lethal trap. Humans are not threading their way through a maze; they scan a vast horizon filled with unique opportunities. The narrowing viewpoint of the maze should appeal only to creatures with their noses buried in the sand.

Cognizant of the limits on our vision and our control over the future, *A Dangerous Master* downplays ends while focusing more upon means. Two different approaches or sets of tools have been outlined for managing technological innovation. One set integrates values and ethics into the way technologies are designed and engineered, and the other approach addresses policy considerations.

By embedding shared values in the very design of new tools and techniques, engineers improve the prospect of a positive outcome. The upstream embedding of shared values during the design process can ease the need for major course adjustments when it's often too late.

Governance Coordinating Committees provide a flexible, adaptive, and comprehensive mechanism for the oversight and management of each major new field of research. A large toolkit of individual policy mechanisms, from government funding and regulations to industry standards and customized insurance policies, are already available for tackling the oversight of emerging technologies. As any good craftsman knows, having and picking the right tool for the task eases effort and produces quality results. Governance Coordinating Committees, or some similar policy mechanism, can ease selecting the correct tool to address each challenge. They can help us navigate the oversight of innovation in a manner that will avoid the haphazard construction of an unwieldy system of laws, regulations, and bureaucracies.

Comprehensive monitoring of technological innovation facilitates the recognition of key inflection points. These inflection points offer opportunities to alter a technology's development in consequential ways. A subtle early adjustment in trajectory eventually leads to an alternate destination.

The character of the future we will actually create rests upon the values embodied in present-day actions, not just in speculative notions of technological possibilities. Speculative visions of the future certainly inspire individual action. But admirable goals alone do not ensure a safe passage.

## A HUMANLY MANAGEABLE PACE

The necessity of maintaining the accelerating adoption of technology within a humanly manageable pace has been championed throughout this book. Humanly manageable refers to a pace that allows for informed decision-making by individuals, institutions, governments, and humanity as a whole. A more deliberative and careful process of technological development permits opportunities to put in place solutions to many (but not all) of the dangers raised in these pages. In the case of government oversight, informed deliberation should actually speed up the implementation of adaptive mechanisms for guiding innovative fields of research. And yet, the need for reflection will in all probability moderate, and perhaps slow, the accelerating implementation of new tools and techniques. As stated earlier, the cavalier adoption of technologies whose impact will be far-reaching and uncertain is the sign of a society that has lost its way. Moderating the adoption of technology should not be done for ideological reasons. Rather, it provides a means to fortify the safety of people affected by unpredictable disruptions. A moderate pace allows us to effectively monitor risks and recognize inflection points before they disappear.

Time and the march of science cannot be stopped. But we do have tools to modulate the rate of development. Ethics, law, and public policy are imperfect instruments, and yet they provide means to fortify our commitment to foundational values.

Theorists note the feedback loop between high-speed technologies, the accelerated rate of social transformation, and the ever-increasing pace of daily life. They perceive speeding up as a characteristic of modernity and of capitalism. Capitalism's demand for increased productivity and efficiency makes it the predominant driver of acceleration. The sped-up pace of work and daily life increases productivity, but in the contemporary economic climate, it has not led to quality of life improvements such as wage growth for the average citizen. Furthermore, speed actually undermines productivity when, for example, an accelerating pace accompanied by poor oversight leads to a disruptive crisis, such as the BP oil spill in the Gulf of New Mexico. The cleanup, the shutdown of all offshore drilling in the Gulf, the environmental impact, and the setback to the fishing and tourist industries over a large region were all extremely costly. The losses in capital alone from a disruptive crisis can quickly wipe out the economic benefits generated by limiting the regulation of transformative technologies. And this economic calculus fails to include the pain and suffering to people harmed by each disruption. The human toll can easily outweigh the losses of capital. As usual, that burden falls most heavily on the poor. And yet, the illusion that accelerated development serves humanity as a whole will endure as long as powerful interests feel this state of affairs serves their personal and corporate goals.

Too much speed not only places a burden on economic systems; it also undermines democratic institutions. Increasing numbers of problems go unsolved. A heady pace of change makes everything a crisis. A "state of emergency" becomes the normal state of affairs. When a society is in a state of permanent crisis, there will always be justifications and rationalizations for expediency in the form of careless policies. Executive prerogatives move from foreign policy to domestic issues. Note, for example, how President Obama issued an executive order in 2014 to resolve the serious issue of immigration reform, when a stalemated Congress was unable to act. This all gets compounded when multiple disruptions, from many quarters, converge over a short time period. Eventually, there will be a "legitimization crisis," a concept proposed by the German social theorist Jürgen Habermas, to represent a time when the government is perceived as being unable to solve problems.

Finally, the accelerating pace undermines the quality of each of our lives. Some individuals find pressure to keep pace exhilarating. Increasingly, however, the crescendo in the chorus of complaints about stress from having more on one's platter than can be managed effectively, has risen to a deafening pitch. Time-saving gadgets don't help when one must learn how to use hundreds of individual features. Stress leads to sleep disorders, disease, and a sense that one's life is out of control.

An analysis such as this can be dismissed as unwarranted doom and gloom, or as a call for creative vigilance in the oversight of technological development. The latter motivated me to write this book.

Moderating, if not actually slowing down, the adoption of innovative technologies bespeaks a society for whom care takes precedence over making a buck. It is about diligence in attending to details and vigilance in analyzing a technology's potential impact. Care, however, is not risk aversion. There will always be risks. Indeed, the misguided attempt to eliminate all risks is one factor that feeds the accelerated development of technological solutions that are only vaguely feasible. Care is one of those universal values that must be reinforced, and yet, it cannot always be honored because care requires time and energy. But carelessness represents one of the forms that evil takes. It is an abrogation of responsibility that surrenders control over the future to economic imperatives and the blind activity of emerging technologies.

An improving economy offers a short-term reprieve. Without it, we will continue to operate in crisis mode and fail to take stock of new technologies before they pose serious risks. Any normalization in our economy should be used as an opportunity to strengthen safety and testing procedures for potentially dangerous systems. This alone will moderate the pace at which risky areas of research progress.

The mechanisms we institute to ensure deliberative and careful processes of technological development do not need to be those I recommend. But if we fail to institute effective approaches, all bets are off. I can certainly appreciate skepticism that the relatively modest measures I have put forward are adequate. Many readers may continue to be skeptical that even modest mechanisms could be implemented in the present political climate or in light of the powerful drivers of technological development.

After making a prediction of an increase in disasters and crises, I question whether we and our leaders have the will and intention to make the hard choices necessary to limit those crises and to minimize their harm. If we are unable or unwilling to put even modest measures in place, we have an answer to that question.

Without our active engagement, the unfolding of technological development takes an inevitable course. When we assume moderate measures are useless, we have already signed away our future. Failure to maintain a humanly manageable pace implicitly surrenders the planet to high-speed robots and enhanced *techno sapiens*. They will be the only species capable of prevailing in an adrenaline-soaked world of accelerating change.

## ENGAGEMENT

*A Dangerous Master* has been an invitation for you to engage the challenges arising from our adoption of new technologies—to be among the representatives for us all in navigating the future. A good number of political leaders and self-appointed experts believe they should decide for us all. Clearly, many of their voices offer one-sided guidance. Government commissions and countless scholars call for more public engagement in decisions about which technologies should be embraced, which should be regulated, and which must be rejected. Sincere requests for public input from political leaders predominate over disingenuous appeals. Responsible leaders appreciate that many of the decisions at hand are too big to be made by elected officials without input from an informed and engaged electorate.

But how to get a larger share of the public involved, or for that matter, what kinds of public engagement are helpful? Generating a critical mass of informed citizens has never been easy. There are too many subject matters and too little time for most of us to do more than scratch the proverbial surface. Generally, we rely upon elected officials and the experts they consult to do the heavy lifting. Nevertheless, we should reject contentions that we must depend solely upon the experts because the issues posed by the emerging technologies are too complex for mere mortals to fathom. There are certainly experts with specialized knowledge who have studied the intricacies of individual problems. From them we can learn a lot. But when it

comes to thinking comprehensively about the brave new world we are entering, there are no experts. There are only people with somewhat more information than you, who are being guided by their own intuitions, beliefs,
values, and desires. We can defer to the experts' opinions, or as an informed
public, we can take a lead in helping to set priorities.

In Europe and the U.S., a new model for determining the priorities of an
informed electorate is under development. The consensus conference or citizen panel allows a representative cross-section of citizens to tap into the
knowledge of experts, but then independently formulate their own opinions. In the 1980s, Denmark was the first country to experiment with forums that enable representative citizens to study and report on issues of
importance, such as nanotechnology. More recent experiments in other
countries have adapted this approach for studying what does and what does
not concern the public in the adoption of new technologies. The representative citizens call upon and question experts in their pursuit of becoming
knowledgeable about the topic under inspection. The experts provide input,
but the citizen panel makes the actual decisions about priorities. The citizens' concerns are then passed on to lawmakers.

The broader electorate may or may not be willing to defer to the judgment of a citizen panel on serious issues, such as the acceptability of releasing synthetic organisms into the environment or the adoption of new
designs for nuclear power plants. With some public education, however, the
citizen panel might be perceived as a viable alternative to relying solely upon
the views of politicians, experts, and vested interests.

Turning this model into a robust means to gauge informed attitudes in
a large country such as the U.S. will not be easy, but it can be done in a very
cost-effective manner. Regional forums, television programming, and Web-
based surveys could be enlisted to broaden the representative sampling. Analytical tools already exist for determining the depth of each respondent's
understanding and whether, collectively, they truly represent a cross-section
of the populace. The media would report strongly held concerns. Whether
elected officials act on those concerns is another matter. At the least, the
public gets an opportunity to make its priorities known to its representatives. Thankfully, politicians do generally act on clear expressions of their
constituents' will.

We have entered a pivotal inflection point in history, where our judgments about advances in science and technology will determine the place of
our species in the future, or whether we will have any place at all. The focus
in this book has been upon whether we enter this transition consciously

attentive to the ramifications of our choices. Or whether we will distractedly surrender the future to forces already set in motion.

*A Dangerous Master* has been a cautionary tale. It has emphasized what can go wrong as we adopt new technologies for a reason: in order to instill a degree of discomfort and caution. On the highway, a caution sign alerts the driver to the need for care and attention. For navigating the future, cautionary tales serve a similar purpose.

In the *Dune* saga, the Guild navigators lived within a tank suffused with a cloud of melange. Over time, their bodies mutated from that of humanoids to bloated, insect-like creatures. The warning is timely. Our humanity could be lost. Not because our bodies might be altered, but because we fail to actively affirm the more admirable qualities and values that have distinguished our species. There are clearly dangers in relating to the development of transformative technologies passively. We need not be at the mercy of the mechanical unfolding of technological possibilities.

Conscious engagement entails attention, energy, and intelligence as well as the will to reject forces that undermine the human spirit. The most dangerous of those forces woo us to sleepwalk into the technological wonderland. However, with proper attention, opportunities for creative action emerge and the time and space to ground the future in universal principles expands. Navigating the future successfully will require our full conscious engagement.

# ACKNOWLEDGMENTS

In the evolution of ideas, countless individuals contribute their insights, intuitions, and scholarship. It is impossible for me to even know, let alone acknowledge, all the people who influenced the creation of *A Dangerous Master*, especially given the many topics touched upon in the text. Nevertheless, I have tried to acknowledge individuals and important sources that shaped my thinking in the notes section and bibliography, and am certain I missed many. Here, I would like to express my appreciation to the people who played a direct role ensuring the clarity, accuracy, and final production of *A Dangerous Master*.

Paul Starobin read an early version of the opening chapters, and told me I "buried the lead on page 14." This critique was seminal in focusing the outline for the book. Paul, a gifted journalist and writer, also introduced me to Andrew Stuart, his outstanding agent, who became my agent.

I'm grateful to Joe Calamia, an editor at Yale University Press, whose encouragement and help pushed me to finish the proposal for what was initially named *Navigating the Future*.

Special appreciation goes to TJ Kelleher, my editor at Basic Books, whose close reading of the manuscript and countless suggestions tightened the chapters and overall trajectory of the book. Quynh Do, another editor at Basic, helped at later stages. Together they have championed *A Dangerous Master*.

Throughout the writing, my friend Rodney Parrott copyedited the various drafts, made helpful editorial suggestions that gave individual chapters form, and most importantly did so with constant encouragement. Chris Bosso, Rosalind Dickenson, and Hillary Bowman read various chapters and each caught factual errors and made sure the writing was balanced. My sister Amei, a gifted and seasoned writer who now produces films about major artists, was immensely helpful for assisting me in finding my own approach to the creative process.

Three graduate students provided yeomen's service fact checking, editing, and making sure the bibliography conformed to APA standards. Evie Kendal,

Derek So, and Juan Carmona are all first-class scholars-in-the-making. Carol Pollard introduced me to these three assistants and Hillary Bowman, when they were interns in the fabulous Yale summer program for bioethics interns, which Carol runs.

Basic Books has a truly talented production team. Marco Pavia oversaw production, interacted with me constantly, responded to all my requests for additions and alterations, and is a joy to work with. I have never met Brent Wilcox (page design), Matt Auerbach (copy editor), Lauren Grober (proof reader), or Robert Swanson (indexer), but am most appreciative for what they have contributed to the quality of the final publication. Clay Farr (Vice President), Cassie Dendurant Nelson (Director of Publicity), and Nicole Jarvis (Marketing) are all working with TJ and Quynh to help interested readers find their way to *A Dangerous Master*.

Spouses commonly get mentioned at the end of these lists, but in my case Nancy's help went far beyond support for me in this project. Not only did I bend her ear over the more than three years it took to write the book, but she even performed a meticulous final edit on the book galleys, catching many errors that had been missed by previous editors. Her expertise in research ethics is reflected in the quality of the writing on that subject. I knew that her final editing and appreciation of the completed book meant that I could now relax and pass it on to readers like you.

# NOTES

These notes provide enrichment for themes discussed in the chapters, as well as an opportunity to acknowledge additional individuals that have contributed to my thinking. A more complete list of sources can be found in the bibliography. Many sources are easy to determine from the text, and therefore, have not been repeated here in the notes, on the assumption that interested readers will find the pertinent item in the bibliography.

## CHAPTER 1: NAVIGATING THE FUTURE

Four months before his January 31st keynote address at the Transmediale conference in Berlin, Otto Rössler outlined his concerns in a paper titled "Abraham-Solution to Schwarzschild Metric Implies that CERN Miniblack Holes Pose a Planetary Risk." Available at: http://www.wissensnavigator.com/documents/ottoroesslerminiblackhole.pdf

Alex Knapp, a *Forbes* staff writer, determined that by July 2012, "the total cost of finding the Higgs boson ran about $13.25 billion."

More recent controversies around the claim that Hawking radiation has been measured are beyond the scope of this book.

Global catastrophic risks have received considerable attention in recent years. Martin Rees, Nick Bostrom, and Richard Posner, among others, contend that we underestimate the likelihood of such events occurring, and are therefore ill-prepared for the advent of a serious disaster.

My agent Andrew Stuart suggested the term "techstorm" as a title for an earlier draft of this book.

I'm indebted to Dr. Sydney Speisel for his assistance in clarifying the role that the two influenza outbreaks had in prompting world health authorities to upgrade pandemic preparedness.

The number of genetic sequences in the human genome that had been patented, or should have been patentable, was already a source of debate before the Supreme Court's Myriad decision. The figure of 41 percent was taken from Rosenfeld and Mason (2013).

While the Supreme Court ruled against patenting naturally occurring genes, it upheld patents for man-made variants of those genes. Furthermore, it left open the possibility that new applications for naturally occurring genes might be patented.

Gary Marchant brought to my attention that the continuing patenting of human genes in Europe and elsewhere will complicate the application of the Supreme

Court's ruling, and also complicate decisions by companies whether to invest in research to discover patentable human genes.

Bonnie Kaplan and Nick Bostrom founded the Technology and Ethics study group, at the Yale Interdisciplinary Center for Bioethics, in the fall of 2002.

## CHAPTER 2: A PREDICTION

Two hundred seventy-five dollars is merely an estimate for what a good quality 3-D home printer might cost within a few years. This estimate could be low depending on how good a printer is needed to fabricate a functional gun.

The actual percentage of market activity represented by high frequency trading varies, as do estimates regarding its size.

Alexis Madrigal draws upon research from Nanex when stating that:

It appears that an algorithm, presumably Knight's, began "buying at the offer price and selling at the bid, which means losing the difference in price." On one stock, Exelon, the algorithm resulted in the loss of about 15 cents a trade, repeating the same mistake 2,400 times a minute. This is all part of a very quick and twisted robot logic of buy high, sell low. It is not something any human being would consciously do, and yet, here we are. It happened.

According to a 2000 article by David Brown, the agriculture editor for *The Telegraph*, "BSE has since killed about 179,000 cattle in the UK. Another 4.4 million have been destroyed as a precaution. It has cost the taxpayer more than £5 billion in consumer safeguards, compensation payments and aid to the beef industry."

Two Japanese reactors were restarted in 2014 but subsequently shut down after requests from local governments. At the time of this writing, it is unknown whether any reactors will actually be restarted in 2015.

Garry Kasparov played chess against two different versions of Deep Blue, the first of which he beat in a 1996 competition. The updated version, which beat Kasparov in 1997, is sometimes referred to as Deep Blue II.

## CHAPTER 3: THE C WORDS

David D. Woods brought to my attention his and Erik Hollnagel's research on and commentary about the Global Hawk failure; particularly, how this incident illustrates difficulties in managing what are referred to as *joint cognitive systems*. My discussion of this incident also draws on Michael Peck's explanation of the incident.

According to James Reason, operators at the Chernobyl plant "wrongly violated plant procedures and switched off successive safety systems, thus creating the immediate trigger for the catastrophic explosion in the core."

In her *New York Times* article titled, "In BP's Record, a History of Boldness and Costly Blunders," Sarah Lyall accuses BP executives of cutting corners and not learning from prior mistakes.

There have been disputes in the death toll of the 1984 explosion at the Union Carbide chemical plant in Bhopal. All parties agree that the immediate death toll exceeded 2,200. The government of Madhya Pradesh claims a total of 3,787 deaths due to the release of methyl isocynate and other chemicals.

The quote by Charles Perrow is taken from an abstract he submitted for a September 2009 presentation to Yale's Technology and Ethics study group.

Various clips of Erin Burnett and Jim Cramer on CNBC during the "flash crash," can be found on the Internet. YouTube offers a 9:36-minute segment under the title, "FLASH CRASH! Dow Jones drops 560 points in 4 Minutes! May 6 2010."

David Bolinsky's March 2007 TED Talk titled, "Visualizing the Wonder of a Living Cell," discusses and demonstrates segments of the video. Available at: https://www.ted.com/talks/david_bolinsky_animates_a_cell

## CHAPTER 4: THE PACE OF CHANGE

Anthony Atala, director of the Wake Forest Institute for Regenerative Medicine, and his team, printed a bladder—the first organ to be tranplanted in a human being. During a TED Talk he demonstrates the early stages of printing functioning liver tissue. Available at https://www.ted.com/talks/anthony_atala_printing_a_human_kidney

The announcement of the groundbreaking 3D printing of a bioresorbable splint that saved the life of an infant, can be accessed at http://www.uofmhealth.org/news/archive/201305/baby's-life-saved-groundbreaking-3d-printed-device

*What Technology Wants,* authored by Kevin Kelly, develops his claim that technology grows like an organism. Kelly refers to the collection of technologies as a *technium,* which has a life of its own.

Marc Andreessen notes in *The Wall Street Journal* that "practically every financial transaction…is done in software," and contends that health care and education are the next industries that will experience a "fundamental software-based transformation."

Ray Kurzweil's predictions regarding the arrival of the technological singularity in the near future have varied over the years; these predictions also depend upon the differing definitions as to what constitutes a super-intelligent computer system. More recently, Kurzweil has focused on 2028 or 2029 as the year when a computer system will first exceed the intelligence of a human.

In February 2009, Eric Horvitz, then President of the Association for the Advancement of Artificial Intelligence (AAAI), and Bart Selman, a professor of computer science at Cornell University, convened a AAAI Presidential Panel on Long-Term AI Futures. The meeting explicitly considered speculative concerns. The August 2009 report summarizing the gathering states, "There was overall skepticism about the prospect of an intelligence explosion as well as of a "coming singularity," and also about the large-scale loss of control of intelligent systems." It was at a similar Puerto Rican January 2015 gathering, organized by the Future of Life Institute, that Bart Selman stated, "a majority of AI researchers now express some concern about super-intelligence due to recent breakthroughs in computer perception and learning." He later emailed me:

It's amazing what a difference five years have made. Most of us who have been in the field for 20 or 30 years or so treated various forms of perception, such as computer vision and speech recognition, as essentially unsolved (and possibly unsolvable) problems. (Even my computer vision colleagues would jokingly say "our stuff doesn't work.") However, with the recent work on deep neural nets, trained on massive computer systems, using massive amounts of data, we see computers are starting to hear and see. You now have demos where you point a camera at a room, and all objects and people in the image are now nicely labeled. Completely infeasible, up to about five years ago! So, now researchers are thinking of combining work from other areas such as reasoning and decision making with the new perception capabilities. This will lead to very powerful systems. So, AI researchers are feeling a qualitative shift in the field.

Research at DeepMind, a UK company acquired by Google in 2014, played a leading role in expanding AI possibilities. Utilizing a relatively new approach known as "deep learning," the team at DeepMind designed algorithms in which a computer system *observes* the data flow from other computers playing old video games such as *Breakout*, and develops strategies to play and creatively win the various games. The excitement around deep learning is similar to excitement created by earlier approaches to AI. While most AI researchers expect significant progress utilizing this approach to computer learning, major technological thresholds have yet to be crossed.

## CHAPTER 5: TRADE-OFFS

Due to conflicting data, various dates between 1923-1933 are cited for the patenting of aerosol cans. Additionally, the date is dependent on whether Nowegian or U.S. application grants are used.

The role of *Silent Spring* in the controversy surrounding the banning of DDT should not be read as diminishing Rachel Carson's truly significant contribution to environmentalism.

The NASA Website contains graphs of the global mean surface temperature as recorded by the Goddard Institute for Space Studies. It indicates an increase between 1.1 and 1.4°F (0.6 to 0.9°C) between 1906 and 2005, and a temperature rate increase that has nearly doubled in the past fifty years. Available at http://earthobservatory.nasa.gov/Features/GlobalWarming/page2.php

Harvard's David Keith is a leading advocate for testing methods to mitigate the effects of global climate change by seeding the stratosphere with sulphates or nanoparticles. The models measuring the effect of this strategy have largely been developed by Keith and his colleagues.

Raymond T. Pierrehumbert refers to geoengineering as "a rather desperate and alarming prospect as a solution to global warming." He has also called geoengineering "barking mad."

Justin Doom notes that other human activities, particularly the domestication of cats, causes more bird deaths than wind turbines.

Many scholars, including Andrew Maynard, director of the University of Michigan Risk Science Center, consider the distinction between nanotechnology and chemistry to be spurious.

While the classification of a molecule as a nanoparticle is often restricted to a maximum of 100 nanometers, many particles exceeding this limit are, correctly or not, designated as nanomaterials.

The $500 million figure for getting approval of a drug is often repeated; however, the actual cost of acquiring FDA approval will vary from application to application. There is no empirical proof of the $500 million cost that this author could find.

I'm indebted to Charles Perrow for bringing to my attention the challenge of removing fuel rods from the damaged Unit 4 at Fukushima.

Comparisons between the present and future cost of differing forms of energy are difficult to make and often vary from region to region. For example, producing power from wind is less costly in the north-to-south wind corridor of midwestern U.S. states than it is in other regions. However, with the completion of a nationwide energy grid, transporting electricity from regions where it is produced cheaply, to areas in need, will be much more efficient and costs will be substantially lower. See *Smart Grid (R)evolution: Electric Power Struggles* for a fuller discussion of these matters.

The two examples of new forms of nuclear power generation are summarized from Diamandes and Kolter (2012), and Naam (2013). The Myhrvold quote was taken from Diamandes and Kotler.

## CHAPTER 6: ENGINEERING ORGANISMS

Biologists differ on the feasibility of a synthetic organism finding an ecological niche without being outcompeted by naturally-occurring organisms.

Refer to "Hacking the President's DNA" for the included quotes of Edward You. Available at http://www.theatlantic.com/magazine/archive/2012/11/hacking-the-presidents-dna/309147/

The Christian presumption that the earth was only a few thousand years old dates back to a notion popularized by Bishop James Ussher (1581-1656). He calculated the creation occurring 4,004 years before the birth of Jesus, as measured by the Julian calendar. Ussher made this determination by counting the generations in the Bible backwards. Given such a relatively short history, it is impossible to imagine that species had enough time to evolve.

The assignment of flat percentages to genetic similarities between species is problematic, but, nevertheless, serves to illustrate how much evolutionary ancestry is shared.

It is important to note that long before there were techniques for directly altering genetic material, new species were developed through breeding techniques that produced offspring from similar species of plants or animals. Indeed, techniques for breeding hybrid plants predate written human history. The hybrid species would not only combine differing characteristics of the parents, but their genetic material would be altered as well.

In the U.S., the rejection of GMOs is often characterized as irrational. However, the evidence as to whether GMO plants pose health risks is mixed. See Krimsky and Gruber (2014) for a fuller discussion of this matter.

In defending the Presidential Commission report, Dr. Amy Gutmann (2011), the chair of the commission, writes:

> The commission calls this strategy 'prudent vigilance.' Some commentators mistook these conclusions as a pass on any restraint of this emerging science. Rather, the commission called not only for more coordinated agency oversight and monitoring of risks and benefits, but also for experts and policy-makers to actively and openly engage in public dialogue as the science evolves, so that all concerned citizens can understand and offer their own perspectives on what lies ahead. The commission worked to model such public outreach in its deliberations, and in its conclusions underlined the responsibility of experts, policy-makers, and federal agencies to carry forward this critical work of public feedback, education, and outreach.

The discovery of one hundred fifty human animal hybrids genetically engineered in the U.K. was reported by Martin and Caldwell (2011).

"The official [2005] USDA survey of animal use [in research] indicated 1.2 million animals, but this does not include mice, rats, birds, fish, reptiles or amphibians. Our estimate . . . suggests the actual number may be closer to 17.3 million in the USA." (Taylor et al. 2008)

## CHAPTER 7: MASTERY OF THE HUMAN GENOME

Dr. Alexandre Mauron, Professor of Bioethics at the University of Geneva, asked, "Is genome a secular equivalent of the soul?" He notes that "genes are popularly associated with stable unchanging defining characteristics of an individual."

Christian fundamentalism in modern times is usually associated with sects that give a strict literal interpretation to biblical texts. In the nineteenth century, the broad Christian establishment read biblical texts literally.

According to Sofair and Kaldjian (2000), of the three hundred fifty thousand people sterilized by the Nazis, "37% were voluntary, 39% were involuntary (done against the person's will), and 24% were nonvoluntary (consent was granted by a guardian for persons who could not choose or refuse sterilization)." They also point out the continuation of sterilizations in the U.S. between 1943 and 1963.

Sickle-cell anemia is the first known example of a mutation that affects a protein. This process was identified by Linus Pauling (1901–1994), an American biochemist, who won both a Nobel Prize in Chemistry and the Nobel Peace Prize.

Embryonic stem cells were first identified in 1981. However, their isolation, accomplished by researchers in 1998, was an important step in turning regenerative medicine into a viable possibility.

The quoted letters between Isadora Duncan and George Bernard Shaw come from a version of this tale first presented by the architect Oswald C. Hering at a 1925 Interfraternity Conference in New York. For further discussion of this and similar tales, visit http://quoteinvestigator.com/2013/04/19/brains-beauty/

## CHAPTER 8: TRANSCENDING LIMITS

The quote from Martine Rothblatt concerning Remodulin's loss of effectiveness was made during a talk at Central Connecticut State University on April 17, 2013. See Skoro-Sajer et al. (2008) for the longevity of patients surviving on intravenous trepostinil (the scientific name for Remodulin).

The research of Susan Lederer, Professor of Medical History and Bioethics at the Univerity of Wisconsin, has been particularly helpful to me in understanding the misuse of the "yuck factor" for justifying moral repugnance based on culturally endorsed prejudices.

The quote from Robin Hanson about the societal impact of *ems* was extracted from an email he sent me on June 19, 2014.

The topic of technological unemployment, particularly once robots have cognitive capabilities functionally equivalent to those required from humans to perform a vast array of jobs, has been bandied about within technological circles for decades.

An early draft of the manuscript for Jerry Kaplan's *Humans Need Not Apply: The Real Implications of Artificial Intelligence* was shared with me. The book will be published in July 2015 by Yale University Press.

## CHAPTER 9: CYBORGS AND TECHNO SAPIENS

Braden Allenby, Professor of Civil and Environmental Engineering at Arizona State University, is the first person I heard use the rich concept that the human body is being turned into a "design space."

Andy Clark's vision of natural-born cyborgs was inspired by the theory of embodied cognition, developed by psychologists and roboticists. In this view, the human mind is shaped by the structure of the human body. For roboticists, such as Rodney Brooks, this means true artificial intelligence can only be realized in systems connected and responding to the outer world through a body. The theory of embodied cognition explicitly critiques theories of intelligence that perceive the mind as constructing an inner world dissociated from the body.

Kevin Warwick's representation of himself as a cyborg sometimes overshadows his very serious research, which studies ways to interface the human nervous system with computing technologies. The fact that it took him eight weeks of practice to get a consistent signal from an implanted chip emerged from a conversation we had in 2013 at a conference in Pisa, Italy.

Before enlisting subjects for an experiment, researchers must first acquire informed consent from each individual. When giving informed consent, the prospective subject reads and signs a document that describes the research and possible dangers of participating; this may include potential side effects of a drug, or loss of privacy if personal information is inadvertently disclosed. In addition, before signing the document, participants have the opportunity to ask questions concerning any part of the research project. In practice, however, it can be difficult for subjects to understand everything about a research study. First, they do not necessarily read over the entire document before signing. Second, even when elicited, many participants fail to voice their questions and concerns.

I have heard Chalmers Clark outline his views on advocates for research subjects during a number of workshops, and they are discussed in an unpublished paper outlining his collaborative work with Robert (Brad) Duckrow MD, an Associate Professor of Neurology and of Neurosurgery at the Yale School of Medicine. They refer to to the inclusion of an advocate as a "triangular approach" to improve the priority of patients in research contexts. An advocate would ensure that subjects fully understand their rights as research participants, and guards them from blithely ignoring serious warnings regarding potential dangers. While the advocates represent the subjects and not the researchers, they help to ensure the integrity of the research. However, advocates add an additional level of expense to research and may discourage potential subjects from ultimately participating.

## CHAPTER 10: PATHOLOGIZING HUMAN NATURE

The World Transhumanist Association (WTA) was beset by political infighting and by concerns that the term "transhumanism" was acquiring negative connotations. In 2008, the WTA was rebranded as Humanity+.

Overseen by J. Hughes (Executive Director), The Institute for Ethics and Emerging Technologies (IEET, http://www.ieet.org) is not explicitly a transhumanist forum, even though the vast majority of its contributors and readers are tech enthusiasts who embrace technologies that would enhance human capability. IEET has evolved into a vital forum for debating societal challenges posed by both real and speculative tools and techniques. It includes fellows such as Patrick Lin, Director of the Center for Ethics & Emerging Sciences Group at CalPoly, and myself, both of whom identify ourselves as ethicists (or philosophers), and not as transhumanists.

The symptoms of PTSD can disappear, but often continue throughout life. Therapeutic methods for treating the disease have shown only moderate success. In addition to the suffering, the economic cost of treating former soldiers suffering from PTSD adds to the overall expense of going to war.

Propranolol inhibits the activity of noradrenaline (norepinephrine). Noradrenaline is a hormone that acts as a neurotransmitter and contributes to the consolidation of memory. The effectiveness of Propranolol is being studied by the military as both a prophylactic, and as a tool to help those suffering from PTSD reconsolidate their memories in a way that improves their ability to function.

The series of twenty-four Robert Sapolsky lectures sold by The Great Courses (http://www.thegreatcourses.com), and titled "Biology and Human Behavior: The

Neurological Origins of Individuality, 2nd Edition," are an excellent way for readers to get an overview of his research and perspective.

Scholars have been unable to trace the quote about the "death of millions a statistic" to Stalin. According to the *Christian Science Monitor*, if he indeed said this, "he would likely have been quoting a 1932 essay on French humor by the German journalist, satirist, and pacifist Kurt Tucholsky . . . Tucholsky quotes a fictional diplomat" who says something quite similar while speaking on the horrors of war. See http://www.csmonitor.com/USA/Politics/2011/0603/ Political-misquotes-The-10-most-famous-things-never-actually-said/The-death -of-one-man-is-a-tragedy.-The-death-of-millions-is-a-statistic.-Josef-Stalin

The William Irwin Thompson quote from *The American Replacement of Nature* originally referred to the twentieth century, but can certainly be applied to the twenty-first.

Michaelson made his inaccurate prediction during a talk at the University of Chicago; however, this did not stop him from receiving a rightfully deserved 1907 Nobel Prize in Physics.

"Dark energy" and "dark matter" are names physicist have given to explain observations that challenge prevailing cosmological models. The Universe's expansion can only be understood by positing that roughly 95% of the energy and matter it contains cannot be observed. The matter we observe and which makes up the earth, stars, and other planets composes only 5% of the Universe. Dark energy and dark matter must exist if Einstein's theory of gravity is accurate, and if they do not exist, than some new synthesis is needed to replace relativity theory.

## CHAPTER 11: DRIVERS OF THE STORM

Various terms, such as the Laws of Armed Conflict (LOAC), international humanitarian law (IHL), and the Geneva Conventions, are often used interchangeably to designate internationally agreed upon methods for conducting war in a manner that is just, and minimally harmful to non-combatants. Norms for making war more ethical and clarifications regarding acceptable grounds for going to war (just war theory) have developed over thousands of years. These principles have been codified in modern times through international agreements known as the Geneva Conventions (1864, 1906, 1929, and 1949), the Hague Conventions (1899 and 1907), and additional protocol amendments (1977 and 2005).

While *jus ad bellum* concerns the justification for going to war, IHL focuses on laws governing the conduct of warfare (*jus in bello*) and the treatment of prisoners of war.

The NeXTech war game sketched out military challenges arising within different scenarios, and then proposed various technologies that could be used in response. For example, in response to a rogue nation with a biological weapon of mass destruction, one option was the preemptive use of a synthetic organism to neutralize the destructive biological agent—a kind of "counter-virus." We divided into three groups to discuss the options. Those with executive and strategic planning experience adjourned to a room to evaluate the options from the perspective of public policy. Another room housed experts in military law and international humanitarian law (IHL). I was a member of a third group that explored the ethics of all options—which of these were right to pursue, and which were immoral? In my mind, the use of a counter-virus was a form of medicine. But many others thought that a preemptive strike with a biological agent, whether for good or evil, violated the Biological Weapons Convention. After about an hour, the three groups came together to continue the discussion and share their findings. For reasons of public policy, international law, or morality, a substantial number of the new weapon systems being discussed were not endorsed as viable options. Indeed, dropping an old-fashioned bomb, that would incinerate everything, on a plant that is producing a biological weapon of mass destruction, was deemed by many in attendance as more effective, and moral, than the use of technological wizardry that could save a few civilian lives.

Cal Poly's Patrick Lin summarizes the third NeXTech gathering with an article published in *The Atlantic*. Lin was particularly struck by the way in which the exercise accentuated "that ethics, policy, and law may come to radically different conclusions. When they do converge on a solution, they often focus on different issues . . . In wargaming, we saw substantial disagreement not just at the intersection of ethics, policy, and law but also within each community, adding to the complexity of the exercise."

"In September 2014, health spending increased to a seasonally adjusted annual rate (SAAR) of $3.06 trillion from its value of $3.05 trillion in August. September's health spending accounted for 17.4% of GDP, which is unchanged since December 2013." From a November 2014 summary report produced by the Altarum Institute. Available at: http://altarum.org/sites/default/files/uploaded-related-files/CSHS_SpendingBrief_November2014_04.pdf

In 2008, I noticed widely disparate claims regarding the percentage of Medicare spent on futile means to keep individuals alive for a few days or weeks at the end of life. In the eyes of some bioethicists, these expenditures are a waste of money. I asked Dan Callahan, often referred to as the father of bioethics, and a leading authority on the challenges of health care, whether he had a more authoritative estimate. He emailed back:

The essence (and confusion) of the issue is this. First, for almost three decades now about 20% of Medicare costs go to those who die, some 5%. It was once widely thought that this money was wasted. However, that is not necessarily so. The figures are based on retrospective data, that is, data taken from records of death. Those records do not show whether those who died were expected to die; some probably were and others not. Hence, it is not possible to say that the cost of those who died was wasteful or useless. In fact, another study some fifteen years ago (not a Medicare study) showed that the most expensive patients were those who were not expected to die but developed unexpected complications and were thus treated aggressively and expensively. Second, the most expensive category of patients, old and young, are those in the last year of life, and they incur the largest portion of US health care. Again, however, this does not show waste, only that those in their last year of life are the most expensive, as might be expected.

The subject of asteroid defense, and especially issues in defending against comet strikes, was impressed upon me by my colleague Joel Marks.

## CHAPTER 12: TERMINATING THE TERMINATOR

The Harvard Law School International Human Rights Clinic and Human Rights Watch 2012 report is titled *Losing Humanity: The Case Against Killer Robots*.

Dramatic phrases from science fiction such as "killer robots" turn off the policy-making establishment, for they appear to trivialize the very serious business of national security. The organizations supporting the ban on killer robots campaign recognize the seriousness of the issues at stake, but have made the calculated assessment that this campaign will only move forward with broad public support. Linking the campaign to concerns raised by science fiction draws attention. However, any opposition to autonomous weapons will be counterproductive when it is merely based upon naïve assumptions that these weapons will inevitably lead to *Terminator* or *Matrix*-like scenarios. Only proposals that are supported by sound and practical arguments will engage policy-makers—especially when there are plenty of counterarguments and vested interests directed at thwarting a ban.

The assessment claiming that battlefield ethics of U.S. soldiers is low comes from a report prepared by the Office of the Surgeon Multinational Force-Iraq and the Office of The Surgeon General of the U.S. Army Medical Command. The Mental Health Advisory Team (MHAT) IV recommended that soldiers and marines receive battlefield ethics training before deployment after finding that 10 percent of personnel reported mistreating noncombatants. Ronald Arkin refers to findings in MHAT IV to support his contention that battlefield robots, programmed to follow the Laws of Armed Conflict, could surpass the ethics of soldiers.

The possibility of robotic vehicles killing people that would not have been harmed by human drivers presents a particularly thorny issue. Ryan Calo (2014), an Assistant Professor at the University of Washington School of Law, writes:

Driverless cars are likely to create new kinds of accidents, even as they dramatically reduce accidents overall . . . What happens when a robot confronts a shopping cart and a stroller at the same time? You or I would plow right into a shopping cart—or even a wall—to avoid hitting a stroller. A driverless car might not. The first headline, meanwhile, to read "Robot Car Kills Baby To Avoid Groceries" could end autonomous driving in the United States—and, ironically, drive fatalities back up.

The five rules outlining moral responsibility for computing artifacts were developed by the Ad Hoc Committee for Responsible Computing. The Committee's reflections were lead by Keith Miller, presently the Orthwein Endowed Professor for Lifelong Learning in the Sciences at the University of Missouri, St. Louis.

The 2011 Dodge Charger "Slippery Slope" commercial refers to the Charger as the "leader of the human resistance"; moreover, it claims that technology is "never neutral." Available at: http://www.youtube.com/watch?v=QWy6A6bLSW0

Service robots will be able to discern certain ethical challenges—presuming that the designers and engineers anticipate the challenge, build in the necessary sensors, and program the software for appropriate responses to ethical situations. But one response may not suit everyone. For example, some parents might want a robot to tell a youngster to "stop" if she is relating to the robot in ways that would hurt a human. Other parents might reject having a robot reprimand a son or daughter. Engineers could design software that offers parents a choice regarding the manner in which a robot caregiver responds to a child. During setup, the parents would be introduced to a variety of ethically charged situations. They would be informed about the ramifications of different alternatives, and the responsibility of placing a child in situations where a robot may need to take such actions. A setup procedure would provide an excellent opportunity for manufacturers of companion robots to educate parents on what they can and cannot expect from such devices. Parents become educated on the proper use of a robonanny, and the manufacturer protects itself from certain forms of liability.

The service robots introduced to date are limited in purpose and have not captured a significant share of the home market. JIBO, a personal robot developed by a team lead by Cynthia Breazeal, will be shipped in 2015 and is anticipated to have considerable impact. Breazeal has been a central figure in the development of social robots, noted for her work with the robots Kismet and Leonardo. While JIBO is a stationary device with limited movement, it displays various features that will elicit users to anthropomorphize the device. More importantly, it will help perform a wide variety of tasks. Similar to a smart phone, it will serve as a platform upon which developers can build additional functionality through apps. During a two month crowdfunding campaign on INDIEGOGO in 2014, 4800 JIBOs were pre-ordered and $2,289,606 was raised. The original target of the campaign was $100,000, a goal reached within four hours.

The "Autonomy in Weapons Systems" directive was signed by former Undersecretary of Defense Ashton Carter. On December 5, 2014, President Obama nominated Carter to be Secretary of Defense, which might be interpreted as a sign of increasing reliance on robotic weaponry.

## CHAPTER 13: ENGINEERING FOR RESPONSIBILITY

Engineering ethics has become an important addition to the curriculum of engineering schools; some even require a course in the field to fulfill degree requirements. Generally, these courses emphasize the obligations that engineers owe to their clients and society, while avoiding potential conflicts of interest. However, more ambitious programs consider ways to address values and societal concerns in the design of products.

Batya Friedman has taken a lead role in advocating for value-sensitive design. Together with David Hendry she co-directs the Value Sensitive Design Research Lab at the University of Washington.

The Resilience Engineering Association states that:

The term Resilience Engineering is used to represent a new way of thinking about safety. Whereas conventional risk management approaches are based on hindsight and emphasize error tabulation and calculation of failure probabilities, Resilience Engineering looks for ways to enhance the ability at all levels of organisations to create processes that are robust yet flexible, to monitor and revise risk models, and to use resources proactively in the face of disruptions or ongoing production and economic pressures. http://www.resilience-engineering-association.org

## CHAPTER 14: TOOLS FOR GOVERNANCE

Prior to the enactment of the National Nanotechnology Initiative (NNI), there were concerns expressed by Langdon Winner, a Professor of Humanities and Social Sciences at Rensselaer Polytechnic Institution, and Rochelle Hollander, the Director of the Center for Engineering, Ethics, and Society at the National Academy of Engineering, among others, about the societal impact of the field. These concerns lead to the commitment of a small portion of NNI funding to research the ethical, legal, and societal impact of nanotechnology. According to David Guston (2014), only .3 percent of NNI funding went to science and technology studies. "At least some in the U.S. Congress sought much higher levels of societal research spending, as a report from the U.S. House of Representatives (2006) specified that 3 percent of expenditures should be so dedicated. . ." Nevertheless, that .3 percent stimulated significant research that went beyond nanotechnology to include ethical and governance issues that cut across other emerging technologies.

The Nano Risk Framework, jointly developed by the Environmental Defense Fund (EDF) and Dupont, can be downloaded from the EDF website at http://business.edf.org/projects/featured/past-projects/dupont-safer-nanotech/

I'm indebted to Phillip Rubin, a longtime colleague, who as of this writing serves within the Science Division of OSTP as Assistant Director of Social, Behavior, and Economic Sciences. Over our twelve-year friendship Rubin and I have shared countless discussions on the challenges of emerging technologies. More recently, he helped me understand the development of science policy within the White House. The example of Tom Kalil's internal brief was passed on to me by Rubin.

In 2014, Ryan Calo authored "The Case for a Federal Robotics Commission." In the paper, he focuses on robotics issues that can be resolved at the federal level. In many respects, Calo's proposal mirrors concerns voiced by Marchant and myself in outlining the need for Governance Coordinating Committees. However, this is not a situation where various proposals are in competition with one another. Hopefully, in bringing together a variety of concerns, and methods to address these concerns, a new kind of oversight mechanism will be envisioned and realized.

## CHAPTER 15: GUILD NAVIGATORS

In 1997, Richard Sclove, of the Loka Institute, ran the first citizens' technology forum in the U.S. There has been signicant progress in adapting the Danish model in the U.S. thanks to Patrick Hamlett and Michael Cobb, both scholars at North Carolina State University. Along with ASU's David Guston, one of their projects "selected from a broad pool of applicants a diverse and roughly representative group of seventy-four citizens to participate at six distinct sites across the country." All six groups deliberated on issues posed by nanotechnologies and human enhancements.

ASU's David Guston and Jennifer Kuzma, co-director of the Genetic Engineering & Society Program at North Carolina State University, are among the U.S. researchers who continue to explore effective methods for adapting citizen panels to the needs of large countries with diverse populations. Guston believes that citizen forums significantly contribute to the anticipatory governance of emerging technologies. In the fall of 2014, ASU and the Boston Museum of Science developed a participatory technology assessment for NASA's Asteroid Initiative. Kuzma has experimented with low-cost means to inform large numbers of citizens about the issues, in order to elicit their attitudes and determine their priorities. She concludes that effective tools for acquiring public participation in helping direct science policy can be implemented with limited resources.

The World Wide Views process, led by the Danish Board on Technology, enlisted sites around the world to acquire citizens' input on global warming and biodiversity. See http://www.wwviews.org

Readers will notice that scholars at Arizona State University (ASU) figure prominently in the development of approaches to address the ethical and governing challenges of emerging technologies. Michael Crow, President of ASU, has made science policy a key focus for the university, and is very supportive of scholars working on emerging technologies and their societal impact. I am particularly appreciative of the support from colleagues at ASU including, but certainly not limited to, Braden Allenby, Gaymon Bennett, Joel Garreau, David Guston, Joseph Herkert, Ben Hurlburt, Gary Marchant, Jason Robert, and Daniel Sarewitz.

# BIBLIOGRAPHY

Adams, M. B. (1990). *The wellborn science: Eugenics in Germany, France, Brazil, and Russia*. New York, NY: Oxford University Press.

Ad Hoc Committee for Responsible Computing. (2010). *Moral responsibility for computing artifacts: The rules* (version 27). Retrieved from https://edocs.uis.edu/kmill2/www/TheRules

Alderson, D. L., & Doyle, J. C. (2010). Contrasting views of complexity and their implications for network-centric infrastructures. *IEEE Transactions on Systems, Man and Cybernetics – Part A: Systems and Humans, 40*(4), 839–852.

Al Jazeera. (2013, December 2). Amazon plans to deliver packages using drones. *Al Jazeera America.* Retrieved from http://america.aljazeera.com/articles/2013/12/1/amazon-will-deliverpackagesbydroneinfourorfiveyears.html

Allen, C., Smit, I., & Wallach, W. (2005). Artificial morality: Top-down, bottom-up and hybrid approaches. *Ethics and New Information Technology, 7,* 149–155.

Allen, C., & Wallach, W. (2011). Moral machines: Contradiction in terms, or abdication of human responsibility? In P. Lin, G. Bekey, & K. Abney (Eds.), *Robot ethics: The ethical and social implications of robots* (pp. 55–68). Cambridge, MA: MIT Press.

Allenby, B. (1996). *Information, complexity and efficiency: The automobile model.* Livermore, CA: Lawrence Livermore National Laboratory.

Allenby, B. (2009). The industrial ecology of emerging technologies. *Journal of Industrial Ecology, 13*(2), 168–183.

Allenby, B. (2013). The dynamics of emerging technology systems. In G. E. Marchant, K. W. Abbott, & B. Allenby (Eds.), *Innovative governance models for emerging technologies* (pp. 19–43). Cheltenham, UK: Edward Elgar Publishing.

Allenby, B., & Sarewitz, D. R. (2011). *The techno-human condition.* Cambridge, MA: MIT Press.

Allhoff, F. (2009). Risk, precaution, and emerging technologies. *Studies in Ethics, Law, and Technology, 3*(2), 1–27.

Allhoff, F., Evans, N., & Henschke, A. (Eds.). (2013). *Routledge handbook of ethics and war.* Oxford, UK: Routledge.

Allhoff, F., Lin, P., Moor, J., & Weckert, J. (Eds.). (2007). *Nanoethics: The ethical and social implications of nanotechnology.* Hoboken, NJ: John Wiley & Sons.

Allianz and OECD. (2005). *Small sizes that matter: Opportunities and risks of nanotechnologies.* Retrieved from http://www.oecd.org/science/nanosafety/37770473.pdf

Alonso-Zaldivar, R. (2011, July 28). Health care costs to account for one-fifth of U.S. economy by 2020: Report. *Huffington Post.* Retrieved from http://www.huffingtonpost.com/2011/07/28/health-care-costs-economy-us_n_911917.html

Anderson, K., & Waxman, M. C. (2013). Law and ethics for autonomous weapons systems: Why a ban won't work and how the laws of war can. *Jean Perkins Task Force on National Security and Law Essay Series.* Stanford University: The Hoover Institution.

Anderson, M., & Anderson, S. L. (2008). *EthEl: Toward a principled ethical eldercare robot.* Retrieved from http://citeseerx.ist.psu.edu/viewdoc/download?doi=10.1.1.177.5971&rep=rep1&type=pdf

Anderson, M., & Anderson, S. L. (Eds.). (2011). *Machine ethics.* New York, NY: Cambridge University Press.

Andreessen, M. (2011, August 20). Why software is eating the world. *The Wall Street Journal.* Retrieved from http://online.wsj.com/news/articles/SB10001424053111190348090457651 2250915629460

Andrianantoandro, E., Basu, S., Karig, D. K., & Weiss, R. (2006). Synthetic biology: New engineering rules for an emerging discipline. *Molecular Systems Biology, 2*(1), 1–146.

Angelica, A. D. (2011, February 18). The buzzer factor: Did Watson have an unfair advantage? [Blog]. *Kurzweil: Accelerating Intelligence.* Retrieved from http://www.kurzweilai.net/the-buzzer -factor-did-watson-have-an-unfair-advantage

Ariely, D. (2009). *Predictably irrational: The hidden forces that shape our decisions.* New York, NY: HarperCollins.

Aristotle. (1908). *Nicomachean ethics.* (W. D. Ross, Trans.). Oxford, UK: Clarendon Press. (Original work published 350 BC)

Aristotle. (1958). *Aristotle's metaphysics.* (W. D. Ross, Trans.). Oxford, UK: Clarendon Press. (Original work published 1907)

Arizona State University. (n.d.). Projects doc-in-a-box: Making medicine more predictive. *ASU Biodesign Institute.* Retrieved from http://biodesign.asu.edu/research/projects/doc-in-a-box

Arkin, R. C. (2007). Robot ethics: From the battlefield to the bedroom, robots of the future raise ethical concerns. *Research Horizons,* Winter/Spring, 14–15.

Arkin, R. C. (2009). *Governing lethal behavior in autonomous robots.* Boca Raton, FL: CRC Press.

Asaro, P. M. (2006). What should we want from a robot ethic? *International Review of Information Ethics, 6*(12), 9–16.

Asaro, P. M. (2012). On banning autonomous weapon systems: Human rights automation, and the dehumanization of lethal decision-making. *International Review of the Red Cross, 94*(886), 687–709.

Asimov, I. (1942). Runaround. *Astounding Science Fiction, 29*(1), 94–103.

Asimov, I. (1950). *I, robot.* New York, NY: Gnome Press.

Associated Press. (2013, September 28). Chicago tylenol murders remain unsolved after more than 30 years. *Fox News.* Retrieved from http://www.foxnews.com/us/2013/09/28/chicago-tylenol-murders-remain-unsolved-after-more-than-30-years

Association for Molecular Pathology v. Myriad Genetics, Inc., 569 U.S. (2013). Retrieved from http://supreme.justia.com/cases/federal/us/569/12–398

Atala, A., Kasper, F. K., & Mikos, A. G. (2012). Tissue engineering: Engineering complex tissues. *Science Translational Medicine, 4*(160), 1–10.

Auletta, K. (2009). *Googled: The end of the world as we know it.* New York, NY: Penguin Press.

Australian Government Department of Health. (2013). *Literature review on the safety of titanium dioxide and zinc oxide nanoparticles in sunscreens.* Retrieved from http://www.tga.gov.au/pdf/sunscreens-nanoparticles-review–2013.pdf

Axelrod, R., & Hamilton, W. D. (1981, March 27). The evolution of cooperation. *Science, 211*(4489), 1390–1396.

Azevedo, F. A., Carvalho, L. R., Grinberg, L. T., Farfel, J. M., Ferretti, R. E., Leite, R. E., ... Herculano-Houzel, S. (2009). Equal numbers of neuronal and nonneuronal cells make the human brain an isometrically scaled-up primate brain. *Journal of Comparative Neurology, 513*(5), 532–541.

Baars, B. (1997). *In the theater of consciousness: The workspace of the mind.* New York, NY: Oxford University Press.

Banta, D. (2009). What is technology assessment? *International Journal of Technology Assessment in Health Care, 25*(S1), 7–9.

Barrat, J. (2013). *Our final invention: Artificial intelligence and the end of the human era.* New York, NY: Thomas Dunne Books.

Bates, S. (2010). Progress towards personalized medicine. *Drug Discovery Today, 15*(3–4), 115–120.

Batty, D. (2012, May 5). Japan shuts down last working nuclear reactor. *The Guardian*. Retrieved from http://www.theguardian.com/world/2012/may/05/japan-shuts-down-last-nuclear-reactor

BBC. (2005, January 23). US plans "robot troops" for Iraq. *BBC News*. Retrieved from http://news.bbc.co.uk/2/hi/americas/4199935.stm

Beauchamp, T. L., & Childress, J. F. (2001). *Principles of biomedical ethics* (5th ed.). Oxford, UK: Oxford University Press.

Beecher, H. K. (1966). Ethics and clinical research. *The New England Journal of Medicine, 274*(24), 1354–1360.

Bell, D. (2009). Communication technology: For better or worse? In J. L. Salvaggio (Ed.), *The information society: Economic, social, & structural issues* (pp. 89–104). New York, NY: Routledge.

Bennett, G. (2015). The malicious and the uncertain biosecurity, self-justification, and the arts of living. In L. Darash & P. Rabinow (Eds.), *Modes of uncertainty: Anthropological cases*. Chicago, IL: University of Chicago Press.

Bentham, J. (1907). *An introduction to the principles of morals and legislation*. Oxford, UK: Clarendon Press. (Original work published 1780)

Berg, P., Baltimore, D., Brenner S., Roblin, R. O., & Singer, M. F. (1975). Summary statement of the Asilomar conference on recombinant DNA molecules. *Proceedings of the National Academy of Sciences of the United States of America, 72*(6), 1981–1984.

Bernat, J. L. (1998). A defense of the whole-brain concept of death. *The Hastings Center Report, 28*(2), 14–23.

Berube, D.M. (2006). *Nano-hype: The truth behind the nanotechnology buzz*. Amherst, NY: Prometheus Books.

Bethe, H. A. (1991). *The road from Los Alamos*. New York, NY: Simon and Schuster.

Bianconi, E., Piovesan, A., Facchin, F., Beraudi, A., Casadei, R., Frabetti, F., ... Canaider, S. (2013). An estimation of the number of cells in the human body. *Annals of Human Biology, 40*(6), 463–471.

Billings, L. (2007, July 16). Rise of roboethics. *Seed Magazine*. Retrieved from http://seedmagazine.com/content/article/rise_of_roboethics

Blackford, R., & Broderick, D. (Eds.). (2014). *Intelligence unbound: The future of uploaded and machine minds*. West Sussex, UK: Wiley Blackwell.

Blackmore, S. (2004). *Consciousness: An introduction*. Oxford, UK: Oxford University Press.

Blount, P. J. (2012). The preoperational legal review of cyber capabilities: Ensuring the legality of cyber weapons. *Northern Kentucky Law Review, 39*(2), 211–220.

Boden, M. A. (1995). Could a robot be creative—how would we know? In K. M. Ford, C. N. Glymour, & P. J. Hayes (Eds.), *Thinking about android epistemology* (pp. 51–72). Menlo Park, CA: AAAI.

Bolinsky, D. (2007, March). Visualizing the wonder of a living cell [Video]. *TED*. Retrieved from https://www.ted.com/talks/david_bolinsky_animates_a_cell

Boorstin, D. (1978). *The republic of technology: Reflections on our future community*. New York, NY: Harper & Row.

Borge, J. L. (1998). *Collected Fictions*. (A. Hurley, Trans.). New York, NY: Penguin Books.

Bosso, C. J. (1987). *Pesticides and politics: The life cycle of a public issue*. Pittsburgh, PA: University of Pittsburgh Press.

Bosso, C. J. (2013). The enduring embrace: The regulatory *Ancien Régime* and governance of nanomaterials in the U.S. *Nanotechnology Law & Business, 9*(4), 381–392.

Bostrom, N. (2003). When machines outsmart humans. *Futures, 35*(7), 759–764.

Bostrom, N. (2014). *Superintelligence: Paths, dangers, strategies*. Oxford, UK: Oxford University Press.

Bostrom, N., & Cirkovic, M. (Eds.). (2008). *Global catastrophic risks*. Oxford, UK: Oxford University Press.

Bowler, P. J. (1989). *Evolution: The history of an idea*. Berkeley, CA: University of California Press.

Bowman, D. M., & Hodge, G. A. (2009). Counting on codes: An examination of transnational codes as a regulatory governance mechanism for nanotechnologies. *Regulation & Governance, 3*(2), 145–164.

Boyd, D. (2014). *It's complicated: The social lives of networked teens*. New Haven, CT: Yale University Press.

Braude, P., Pickering, S., Flinter, F., & Ogilvie, C. M. (2002). Preimplantation genetic diagnosis. *Nature Reviews Genetics, 3*(12), 941–955.

Brashear, R. (2013). Fixed: The science/fiction of human enhancement [Film]. Available from http://www.fixedthemovie.com/reviews

Breazeal, C. L. (2002). *Designing sociable robots*. Cambridge, MA: MIT Press.

Breggin, L. K., & Pendergrass, J. (2010). Regulation of nanoscale materials under media-specific environmental laws. In G. A. Hodge, D. M. Bowman, & A. D. Maynard (Eds.), *International handbook on regulating nanotechnologies* (pp. 342–371). Cheltenham, UK: Edward Elgar Publishing.

Brey, P. (1999). The ethics of representation and action in virtual reality. *Ethics and Information Technology, 1*(1), 5–14.

Brockman, J. (Ed.). (2002). *The next fifty years: Science in the first half of the twenty-first century*. New York, NY: Vintage Books.

Brockman, J. (Ed.). (2014). *What should we be worried about? Real scenarios that keep scientists up at night*. New York, NY: Harper Perennial.

Brooks, R. A. (2002). *Flesh and machines: How robots will change us*. New York, NY: Pantheon Books.

Brown, D. (2000, October 27). The "recipe for disaster" that killed 80 and left a £5bn bill. *The Telegraph*. Retrieved from http://www.telegraph.co.uk/news/uknews/1371964/the-recipe-for-disaster-that-killed–80-and-left-a–5bn-bill.html

Brown, E. M. (2013). The Deepwater Horizon disaster: Challenges in ethical decision making. In S. May (Ed.), *Case studies in organization communication: Ethical perspectives and practice* (2nd ed.) (pp. 233–246). Los Angeles, CA: Sage Publications.

Brown, J. S., & Duguid, P. (2001). A response to Bill Joy and the doom-and-gloom technofuturists. In A. H. Teich, S. D. Nelson, C. McEnaney, & S. J. Lita (Eds.), *Science and Technology Policy Yearbook 2001* (pp. 77–83). American Association for the Advancement of Science.

Brown, M. (2015). Genetics, science fiction, and the ethics of athletic enhancement. In M. McNamee & W. J. Morgan (Eds.), *The routledge handbook of the philosophy of sport*. New York, NY: Routledge.

Brumfiel, G. (2013). Fukushima: Fallout of fear. *Nature, 493*(7432), 290–293.

Brunn, S. D. (1998). The Internet as "the new world" of and for geography: Speed, structures, volumes, humility and civility. *GeoJournal, 45*(1), 5–15.

Brynjolfsson, E., & McAfee, A. (2012). *Race against the machines: How the digital revolution is accelerating innovation, driving productivity, and irreversibly transforming employment and the economy*. Lexington, MA: Digital Frontier Press.

Bryson, B. (2003). *A short history of nearly everything*. New York, NY: Broadway Books.

Buchanan, A. (2009). Human nature and enhancement. *Bioethics, 23*(3), 141–150.

Buchanan, A. (2011). *Better than human: The promise and perils of enhancing ourselves*. New York, NY: Oxford University Press.

Buck v. Bell 274 U.S. 200 (1927). Retrieved from http://supreme.justia.com/cases/federal/us/274/200/case.html

Burrill, D. R., Boyle, P. M., & Silver, P. A. (2011). A new approach to an old problem: Synthetic biology tools for human disease and metabolism. *Cold Spring Harbor Symposia on Quantitative Biology, 76,* 145–154.

Bush, G. W. (2003, December 3). Statement on signing the 21st century Nanotechnology Research and Development Act. *The American Presidency Project.* Retrieved from http://www.presidency.ucsb.edu/ws/?pid=799

Busza, W., Jaffe, R. L., Sandweiss, J., & Wilczek, F. (2000). Review of speculative "disaster scenarios" at RHIC. *Reviews of Modern Physics, 72*(4), 1125–1140.

Bynum, T. W. (2001). Computer ethics: Its birth and its future. *Ethics and Information Technology, 3*(2), 109–112.

Byrne, M. T., & Gun'ko, Y. K. (2010). Recent advances in research on carbon nanotube-polymer composites. *Advanced Materials, 22*(15), 1672–1688.

Caldwell, J. A., Caldwell, J. L., Smythe, N. K., & Hall, K. K. (2000). A double-blind, placebo-controlled investigation of the efficacy of Modafinil for sustaining the alertness and performance of aviators: A helicopter simulator study. *Psychopharmocology, 150*(3), 272–282.

Callahan, D. (2008). Health care costs and medical technology. In M. Crowley (Ed.), *From birth to death and bench to clinic: The Hastings Center bioethics briefing book for journalists, policymakers, and campaigns* (pp. 79–82). New York, NY: The Hastings Center.

Callahan, D. (2009). *Taming the beloved beast: How medical technology costs are destroying our health care system.* Princeton, NJ: Princeton University Press.

Callahan, D. (2012a). *In search of the good: A life in bioethics.* Cambridge, MA: MIT Press.

Callahan, D. (2012b). *The roots of bioethics: Health, progress, technology, death.* Oxford, UK: Oxford University Press.

Callahan, D., & Nuland, S. B. (2011, May 19). The quagmire: How American medicine is destroying itself. *New Republic.* Retrieved from http://www.newrepublic.com/article/economy/magazine/88631/american-medicine-health-care-costs

Calo, R. (2011). Open robotics. *Maryland Law Review, 70*(3), 571–613. Retrieved from http://papers.ssrn.com/sol3/papers.cfm?abstract_id=1706293

Calo, R. (2014, September). The case for a federal robotics commission. *Brookings.* Retrieved from http://www.brookings.edu/research/reports2/2014/09/case-for-federal-robotics-commission

Campbell, M., Hoane Jr., A. J., & Hsu, F. (2002). Deep blue. *Artificial Intelligence, 134*(1), 57–83.

Campos, L. (2012). The biobrick™ road. *BioSocieties, 7*(2), 115–139.

Capek, K. (1973). *R.U.R.: Rossum's universal robots.* New York, NY: Simon and Schuster. (Original work published 1920)

Caplan, A. (2007). *Smart mice, not-so-smart people: An interesting and amusing guide to bioethics.* Lanham, MD: Rowman & Littlefield Publishers.

Carlson, R. (2009). The changing economics of DNA synthesis [Commentary]. *Nature Biotechnology, 27*(12), 1091–1094.

Carlson, R. (2010). *Biology is technology: The promise, peril, and new business of engineering life.* Cambridge, MA: Harvard University Press.

Carpenter, C. (2013, June 19). How do Americans feel about fully autonomous weapons? *The Duck of Minerva.* Retrieved from http://www.whiteoliphaunt.com/duckofminerva/2013/06/how-do-americans-feel-about-fully-autonomous-weapons.html

Carr, N. (2010). *The shallows: What the internet is doing to our brains.* New York, NY: W.W. Norton.

Carroll, L. (1992). *Alice's adventures in wonderland & through the looking-glass.* Hertfordshire, UK: Wordsworth Classics.

Carson, R. L. (2002). *Silent spring*. New York, NY: Houghton Mifflin Company. (Original work published 1962)

Cattaneo, R., Miest, T., Shashkova, E. V., & Barry, M. A. (2008). Reprogrammed viruses as cancer therapeutics: Targeted, armed and shielded. *Nature Reviews Microbiology, 6*(7), 529–540.

CBS News. (2008, June 1). The pentagon's ray gun [Video]. *60 Minutes*. Retrieved from http://www.cbsnews.com/videos/the-pentagons-ray-gun–50038414

Center for Climate and Energy Solutions (C2ES). (2014, May). *Carbon pricing proposal of the 113th Congress*. Retrieved from http://www.c2es.org/publications/carbon-pricing-proposals–113th-congress

Centers for Disease Control and Prevention. (2012). *Prevention and treatment of avian influenza A viruses in people*. Retrieved from http://www.cdc.gov/flu/avianflu/prevention.htm

Centers for Disease Control and Prevention. (2013, December 30). *U.S. public health service syphilis study at Tuskegee*. Retrieved from http://www.cdc.gov/tuskegee

Chalmers, D. J. (1996). *The conscious mind: In search of a fundamental theory*. Oxford, UK: Oxford University Press.

Cho, A. (2012, July 13). Higgs Boson makes its debut after decades-long search. *Science, 337*(6091), 141–143.

Church, G. M., Elowitz, M. B., Smolke, C. D., Voigt, C. A., & Weiss, R. (2014). Realizing the potential of synthetic biology. *Nature Reviews Molecular Cell Biology, 15*, 289–294. doi:10.1038/nrm3767

Church, G. M., & Regis, E. (2012). *Regenesis: How synthetic biology will reinvent nature and ourselves*. New York, NY: Basic Books.

Clark, A. (1998). *Being there: Putting brain, body, and world together again*. Cambridge, MA: MIT Press.

Clark, A. (2003). *Natural-born cyborgs: Minds, technologies, and the future of human intelligence*. New York, NY: Oxford University Press.

Clarke, R. (1993). Asimov's laws of robotics: Implications for information technology (1). *IEEE Computer, 26*(12), 53–61.

Clarke, R. (1994). Asimov's laws of robotics: Implications for information technology (2). *IEEE Computer, 27*(1), 57–66.

Clean Air Taskforce (CATF). (2010). *The toll from coal: An updated assessment of death and disease from America's dirtiest energy source*. Boston, MA: CATF.

Cobb, J. (1998). *CyberGrace: The search for God in the digital world*. New York, NY: Crown Publishing Group.

Coeckelbergh, M. (2013). *Human being @ risk: Enhancement, technology, and the evaluation of vulnerability transformations*. Dordrecht, NL: Springer.

Coeytaux, F., Darnovsky, M., & Fogel, S.B. (2011). Assisted reproduction and choice in the biotech age: Recommendations for a way forward. *Contraception, 83*(1), 1–4.

Cohen, C. B. (2002). Protestant perspectives on the uses of the new reproductive technologies. *Fordham Urban Law Journal, 30*(1), 135–145.

Cohen, D. L. (2014). Fostering mainstream adoption of industrial 3D printing: Understanding the benefits and promoting organizational readiness. *3D Printing and Additive Manufacturing, 1*(2), 62–69.

Collingridge, D. (1980). *The social control of technology*. London, UK: Frances Pinter Publishers.

Collins, F. S. (1999). Medical and societal consequences of the human genome project. *New England Journal of Medicine, 341*(1), 28–37.

Conca, J. (2014, May 04). Cancer and death by radiation? Not from Fukushima. *Forbes*. Retrieved from http://www.forbes.com/sites/jamesconca/2014/05/04/cancer-and-death-by-radiation-not-from-fukushima

Conway, G., Dahlsten, D. L., Haskell, P., Herman, S., Kok, L. T., Newsom, L. D., …Van Den Bosch, R. (1969). DDT on balance. *Environment: Science and Policy for Sustainable Development, 11*(7), 2–5.

Coombs, A. (2008, November 13). Stem cells drafted for war on wounds. *Nature Reports Stem Cells.* Retrieved from http://www.nature.com/stemcells/2008/0811/081113/full/stemcells.2008.148.html

Craik, F. I., & Bialystok, E. (2006). Cognition through the lifespan: Mechanisms of change. *Trends in Cognitive Sciences, 10*(3), 131–138.

Cui, L., Morris, A., & Ghedin, E. (2013). The human mycobiome in health and disease. *Genome Medicine, 5*(7), 63.

Cunningham, T. V. (2013). What justifies the United States ban on federal funding for nonreproductive cloning? *Medicine, Health Care, and Philosophy, 16*(4), 825–841.

Curfman, G. D., & Redberg, R. F. (2011). Medical devices—Balancing regulation and innovation. *New England Journal of Medicine, 365*(11), 975–977.

Cyranoski, D. (2006). Verdict: Hwang's human stem cells were all fakes. *Nature, 439*(7073), 122–123.

Dabrock, P. (2009). Playing God? Synthetic biology as a theological and ethical challenge. *Systems and Synthetic Biology, 3*(1–4), 47–54.

Damasio, A. R. (1995). *Descartes' error: Emotion, reason, and the human brain.* London, UK: Picador.

Daniels, N. (2000). Normal functioning and the treatment-enhancement distinction. *Cambridge Quarterly of Healthcare Ethics, 9*(3), 309–322.

Darwin, C. (1860). *Origin of species* (Harvard Classics, 11th ed.). New York, NY: Bartleby Press.

Dautenhahn, K., Bond, A. M., Canamero, L., & Edmonds, B. (Eds.). (2002). *Socially intelligent agents: Creating relationships with computers and robots.* New York, NY: Springer.

Davis, E. (1998). *TechGnosis: Myth, magic, and mysticism in the age of information.* New York, NY: Harmony Books.

Davison, J. (2010). GM plants: Science, politics and EC regulations. *Plant Science, 178*(2), 94–98.

Dawkins, R. (1989). *The selfish gene.* Oxford, UK: Oxford University Press.

Dawkins, R. (2006). *The God delusion.* London, UK: Bantam Press.

Dawkins, R. (2009). *The greatest show on Earth: The evidence for evolution.* New York, NY: Free Press.

Dechesne, F., Warnier, M., & Van den Hoven, J. (2013). Ethical requirements for reconfigurable sensor technology: A challenge for value sensitive design. *Ethics and Information Technology, 15*(3), 173–181.

de Garis, H. (2005). *The artilect war: Cosmists vs. terrans: A bitter controversy concerning whether humanity should build Godlike massively intelligent machines.* Palm Springs, CA: ETC Publications.

de Grey, A. (2005, July). A roadmap to end aging [Video]. *TED.* Retrieved from http://www.ted.com/talks/aubrey_de_grey_says_we_can_avoid_aging

de Grey, A., & Rae, M. (2008). *Ending aging: The rejuvenation breakthroughs that could reverse human aging in our lifetime.* New York, NY: St. Martin's Griffin.

de Leon, J., Arranz, M. J., & Ruano, G. (2008). Pharmacogenetic testing in psychiatry: A review of features and clinical realities. *Clinics in Laboratory Medicine, 28*(4), 599–617.

del Moral, R., & Wood, D. M. (1988). Dynamics of herbaceous vegetation recovery on Mount St. Helens, Washington, USA, after a volcanic eruption. *Vegetatio, 74*(1), 11–27.

Dennett, D. C. (1996). When Hal kills, who's to blame? Computer ethics. In D. G. Stork (Ed.), *Hal's legacy: 2001's computer as dream and reality* (pp. 351–366). Cambridge, MA: MIT Press.

Dennett, D. C. (2003). *Freedom evolves.* New York, NY: Penguin Books.

Descartes, R. (1978). *The philosophical works of Descartes.* (E. S. Haldane & G. R. Ross, Trans.). Cambridge, UK: Cambridge University Press.

de Waal, F. (2006). *Primates and philosophers: How morality evolved.* S. Macedo & J. Ober (Eds.). Princeton, NJ: Princeton University Press.

Diamandis, P. H., & Kotler, S. (2012). *Abundance: The future is better than you think.* New York, NY: Free Press.

Diamond v. Chakrabarty, 447 U.S. 303 (1980). Retrieved from http://supreme.justia.com/cases/federal/us/447/303/case.html

Dillow, C. (2013, September 3). 5 reasons 3-D printing isn't quite ready for prime time. *Fortune.* Retrieved from http://fortune.com/2013/09/03/5-reasons-3-d-printing-isnt-quite-ready-for-prime-time

Dipert, R. R. (2010). The ethics of cyberwarfare. *Journal of Military Ethics, 9*(4), 384–410.

Dobzhansky, T. (1973). Nothing in biology makes sense except in the light of evolution. *The American Biology Teacher, 35*(3), 125–129.

Doidge, N. (2007). *The brain that changes itself: Stories of personal triumph from the frontiers of brain science.* New York, NY: Penguin Books.

Doom, J. (2013, December 6). U.S. eases turbine bird-death rule as cats kill millions. *Bloomberg.* Retrieved from http://www.bloomberg.com/news/2013-12-06/u-s-eases-turbine-bird-death-rule-as-cats-kill-millions.html

Doorn, N., Schuurbiers, D., van de Poel, I., & Gorman, M. E. (Eds.). (2014). *Early engagement and new technologies: Opening up the laboratory.* Dordrecht, NL: Springer.

Douglas, T., & Savulescu, J. (2010). Synthetic biology and the ethics of knowledge. *Journal of Medical Ethics, 36*(11), 687–693.

Drexler, K. E. (1986). *Engines of creation: The coming era of nanotechnology.* New York, NY: Anchor Books.

Drexler, K. E. (2013). *Radical abundance: How a revolution in nanotechnology will change civilization.* New York, NY: PublicAffairs.

Dreyfus, H. L. (1979). *What computers can't do: The limits of artificial intelligence.* New York, NY: Harper Colophon Books.

Duckrow, R. B., & Clark, C. C. (Unpublished). *An ethical analysis of role priorities and role conflicts in new research designed to treat intractable epilepsy.*

Duncan, D. E. (2010, June 8). On a mission to sequence the genomes of 100,000 people. *New York Times.*

Dvorsky, G. (2008). All together now: Developmental and ethical considerations for biologically uplifting nonhuman animals. *Journal of Evolution and Technology, 18*(1), 129–142.

Dyer, H. (2013). Ethical dimensions of renewable energy. In H. Dyer & M. J. Trombetta (Eds.), *International handbook of energy security* (pp. 443–461). Cheltenham, UK: Edward Elgar Publishing.

Ebersberger, I., Metzler, K., Schwarz, C., & Pääbo, S. (2002). Genomewide comparison of DNA sequences between humans and chimpanzees. *The American Journal of Human Genetics, 70*(6), 1490–1497.

Economist. (2007, March 15). Logical endings: Computers may soon be better than kin at predicting the wishes of the dying. *The Economist.* Retrieved from http://www.economist.com/node/8847826

Eden, A. H., Moor, J. H., Søraker, S. H., & Steinhart, E. (2012). *Singularity hypothesis: A scientific and philosophical assessment.* Heidelbergh, DE: Springer.

Effros, R. B. (2007). Telomerase induction in T cells: A cure for aging and disease? *Experimental Gerontology, 42*(5), 416–420.

Endy, D. (2005). Foundations for engineering biology. *Nature, 438,* 449–453.

Egan, T. M., Boychuk, J. E., Rosato, K., & Cooper, J. D. (1992). Whence the lungs? A study to assess suitability of donor lungs for transplantation. *Transplantation, 53*(2), 420–422.

Ehrenberg, R. (2012). The facts behind the frack: Scientists weigh in on the hydraulic fracturing debate. *Science News, 182*(5), 20–25.

Eilperin, J. (2010, October 29). Geoengineering sparks international ban, first-ever congressional report. *Washington Post.* Retrieved from http://www.washingtonpost.com/wp-dyn/content/article/2010/10/29/AR2010102906365.html

Ellis, J., Giudice, G., Mangano, M. L., Tkachev, I., & Wiedemann, U. (2008). Review of the safety of LHC collisions. *Journal of Physics G: Nuclear and Particle Physics, 35*(11), 1–18.

Emanuel, E. J. (2014). *Reinventing American health care: How the Affordable Care Act will improve our terribly complex, blatantly unjust, outrageously expensive, grossly inefficient, error prone system.* New York, NY: PublicAffairs.

Engelberger, J. F. (1989). *Robotics in service.* Cambridge, MA: MIT Press.

Enserink, M. (2008, April 25). Tough lessons from golden rice. *Science, 320*(5875), 468–471.

Ess, C., & Sudweeks, F. (2005). Culture and computer mediated communication: Toward new understandings. *Journal of Computer Mediated Communication, 11*(1), 179–191.

Evans, T. D. (2013). At war with the robots: Autonomous weapon systems and the Martens Clause. *Hofstra Law Review, 41*, 697–777.

Fantz, A. (2011, November 15). Lack of soap means illness, death for millions of children. *CNN.* Retrieved from http://www.cnn.com/2011/11/15/health/cnnheroes-soap-hygiene/

Farah, M. J., Illes, J., Cook-Deegan, R., Gardner, H., Kandel, E., King, P., … Wolpe, P. R. (2004). Neurocognitive enhancement: What can we do and what should we do? *Nature Reviews Neuroscience, 5*(5), 421–425.

Farmer, J. (2007, Spring/Summer). Gone native: California's love-hate relationship with eucalyptus trees. *Huntington Frontiers.* Retrieved from http://jaredfarmer.net/wp-content/uploads/2013/05/Gone-Native.pdf

Feigenson, N., & Spiesel, C. (2009). *Law on display: The digital transformation of legal persuasion and judgment.* New York, NY: New York University Press.

Feil, R., & Fraga, M. F. (2012). Epigenetics and the environment: Emerging patterns and implications. *Nature Reviews Genetics, 13*(2), 97–109.

Feinberg, A. P. (2007). Phenotypic plasticity and the epigenetics of human disease. *Nature, 447*(7143), 443–440.

Feldman, R., Weller, A., Zagoory-Sharon, O., & Levine, A. (2007). Evidence for a neuroendocrinological foundation of human affiliation: Plasma oxytocin levels across pregnancy and the postpartum period predict mother-infant bonding. *Psychological Science, 18*(11), 965–970.

Feynman, R. P. (1960). There's plenty of room at the bottom: An invitation to enter a new field of physics. *Engineering and Science, 23*(5), 22–36.

Feynman, R. P. (1985). *Surely you're joking, Mr. Feynman! Adventures of a curious character.* New York, NY: W.W. Norton & Company.

Finkel, M. (2007, July). Malaria. *National Geographic.* Retrieved from http://ngm.nationalgeographic.com/2007/07/malaria/finkel-text/1

Fisher, J. (1987). The history of landholding in Ireland. *Transactions of the Royal Historical Society, 5*, 228–236.

Flamm, E. L. (1991). How FDA approved Chymosin: A case history. *Nature Biotechnology, 9*(4), 349–351.

Flandreau, M., & Flores, J. H. (2009). Bonds and brands: Foundations of sovereign debt markets, 1820–1830. *Journal of Economic History, 69*(3), 646–684.

Fleming, N. (2009, September 30). Campaign asks for international treaty to limit war robots. *New Scientist.* Retrieved from http://www.newscientist.com/article/dn17887-campaign-asks-for-international-treaty-to-limit-war-robots.html#.VCY00imSxuA

Floridi, L. (Ed.). (2010). *The Cambridge handbook of information and computer ethics.* Cambridge, UK: Cambridge University Press.

Floridi, L. (2013). *The ethics of information.* Oxford, UK: Oxford University Press.

Floridi, L. (2014). *The 4th revolution: How the infosphere is reshaping human reality*. Oxford, UK: Oxford University Press.

Foot, P. (1967). The problem of abortion and the doctrine of double effect. *Oxford Review, 5*, 5–15.

Ford, D., Easton, D. F., Bishop, D. T., Narod, S. A., & Goldgar, D. E. (1994). Risks of cancer in BRCA1-mutation carriers. *The Lancet, 343*(8899), 692–695.

Ford, K. M., Glymour, C., & Hayes, P. J. (Eds.). (2006). *Thinking about android epistemology*. Cambridge, MA: MIT Press.

Ford, M. (2009). *The lights in the tunnel: Automation, accelerating technology and the economy of the future*. USA: Acculant Publishing.

Fox, J. L. (2010). Transgenic salmon inches toward finish line. *Nature Biotechnology, 28*(11), 1141–1142.

Franklin, S. (1995). *Artificial minds*. Cambridge, MA: MIT Press.

Frantz, S. (2012). Embryonic stem cell pioneer Geron exits field, cuts losses. *Nature Biotechnology, 30*(1), 12–13.

Freitas, R. A. (1999). *Nanomedicine* (Vol. 1): *Basic capabilities*. Boca Raton, FL: CRC Press.

Freitas, R. A. (2003). *Nanomedicine* (Vol. IIA): *Biocompatibility*. Boca Raton, FL: CRC Press.

Freitas, R. A. (2006). Pharmacytes: An ideal vehicle for targeted drug delivery. *Journal of Nanoscience and Nanotechnology, 6*(9–10), 2769–2775.

Friedman, B. (1996). Value sensitive design. *Interactions, 3*(6), 16–23.

Friedman, B., & Kahn, P. H. (1992). Human agency and responsible computing: Implications for computer system design. *Journal of Systems and Software, 17*(1), 7–14.

Friedman, B., Kahn, P. H., & Borning, A. (2008). Value sensitive design and information systems. In K. E. Himma & H. T. Tavani (Eds.), *The handbook of information and computer ethics* (pp. 69–102). Hoboken, NJ: John Wiley & Sons.

Friedman, B., & Nissenbaum, H. (1996). Bias in computer systems. *ACM Transactions on Information Systems, 14*(3), 330–347.

Frey, C. B., & Osborne, M. A. (2013). The future of employment: How susceptible are jobs to computerisation? Retrieved from http://www.oxfordmartin.ox.ac.uk/downloads/academic/The_Future_of_Employment.pdf

Frost, R. (1979). The road not taken. In E. C. Lathem (Ed.), *The poetry of Robert Frost: The collected poems* (p. 105). New York, NY: Henry Holt and Company.

Fukuyama, F. (2002, March/April). Gene regime. *Foreign Policy, 129*, 56–63.

Fukuyama, F. (2003). *Our posthuman future: Consequences of the biotechnology revolution*. New York, NY: Picador.

Fukuyama, F. (2004). Transhumanism. *Foreign Policy*, 42–43.

Future of Life Institute. (2015). Research priorities for robust and beneficial artificial intelligence: An open letter. Retrieved from http://futureoflife.org/misc/open_letter

Galton, F. (1883). *Inquiries into human faculty and its development*. London, UK: Macmillan.

Gardiner, S. M. (2011). Some early ethics of geoengineering the climate: A commentary on the values of the royal society report. *Environmental Values, 20*(2), 163–188.

Garreau, J. (2005). *Radical evolution: The promise and peril of enhancing our minds, our bodies— And what it means to be human*. New York, NY: Doubleday.

Garreau, J. (2007, May 6). Bots on the ground: In the field of battle (or even above it), robots are a soldier's best friend. *Washington Post*, p. D01.

Gazzaniga, M. S. (2005). *The ethical brain: The science of our moral dilemmas*. New York, NY: Dana Press.

Gibson, W. (1984). *Neuromancer*. New York, NY: The Berkley Publishing Group.

Gibson, D. G., Glass, J. I., Lartigue, C., Noskov, V. N., Chuang, R., Algire, M. A., … Venter, J. C. (2010, July 2). Creation of a bacterial cell controlled by a chemically synthesized genome. *Science, 329*(5987), 52–56.

Gilligan, C. (1982). *In a different voice: Psychological theory and women's development.* Cambridge, MA: Harvard University Press.

Giordano, J. (2013a). *Neuroethics: At the intersection of neuroscience, neurotechnology, nature, and nurture.* Cambridge, UK: Cambridge University Press.

Giordano, J. (2013b). *Neuroscience and neurotechnology in national security and defense: Practical capabilities, neuroethical considerations.* Boca Raton, FL: CRC Press.

Gladwell, M. (1996, January 22). Blowup. *The New Yorker.* Retrieved from http://www.newyorker.com/archive/1996/01/22/1996_01_22_032_TNY_CARDS_000374484

Glass, J. I., Assad-Garcia, N., Alperovich, N., Yooseph, S., Lewis, M. R., Maruf, M., ... Venter, J. C. (2006). Essential genes of a minimal bacterium. *Proceedings of the National Academy of Sciences of the United States of America, 103*(2), 425–430.

Gleick, J. (1999). *Faster: The acceleration of just about everything.* New York, NY: Pantheon.

Goertzel, B. (2002, May). Thoughts on AI morality. *Dynamic Psychology.* Retrieved from http://www.goertzel.org/dynapsyc/2002/AImorality.htm

Goertzel, B., & Pennachin, C. (Eds.). (2007). *Artificial general intelligence.* Berlin, DE: Springer.

Goleman, D. (1995). *Emotional intelligence: Why it can matter more than IQ.* New York, NY: Bantam Books.

González, J. G. (2012). *Xenotransplantation.* New York, NY: Humana Press.

Good, I. J. (1982). *Ethical machines.* Paper presented at the Tenth Machine Intelligence Workshop, Cleveland, OH.

Goodrich, J. (2013). Driving Miss Daisy: An automated chauffeur system. *Houston Law Review, 51*(1), 265–296.

Gormley, D. M. (2008). *Missile contagion: Cruise missile proliferation and the threat to international security.* Westport, CT: Praeger Security International.

Gormley, D. M. (2013). *Dealing with the threat of cruise missiles.* Hoboken, NJ: Taylor and Francis.

Gormley, D. M. (2014). Sixty minutes to strike: Assessing the risks, benefits, and arms. *Sicherheit & Frieden, 32,* 46–53.

Goss, R. J. (1992). The evolution of regeneration: Adaptive or inherent? *Journal of Theoretical Biology, 159*(2), 241–260.

Gotterbarn, D. (1992). The use and abuse of computer ethics. *Journal of Systems and Software, 17*(1), 75–80.

Gotterbarn, D., Miller, K. W., Impagliazzo, J., & Abu Bakar, A. Z. (Eds.). (2014). *Computing ethics: A multicultural approach.* Chapman & Hall/CRC.

Gottweis, H., & Triendl, R. (2006). South Korean policy failure and the Hwang debacle. *Nature Biotechnology, 24*(2), 141–143.

Gould, S. J. (1991). Fall in the house of Ussher. *Natural History, 100*(11), 12–19.

Graham, N. (2010, July 6). Jim Kramer, Erin Burnett freak out as Dow plunges 900 points, then rebounds [Video]. *The Huffington Post.* Retrieved from http://www.huffingtonpost.com/2010/05/06/jim-cramer-erin-burnett-f_n_566543.html

Granger, M., Nordhaus, R. R., & Gottlieb, P. (2013). Needed: Research guidelines for solar radiation management. *Issues in Science and Technology, 29*(3), 37–44.

Grant, R. A., Halpern, C. H., Baltuch, G. H., O'Reardon, J. P., & Caplan, A. (2014). Ethical considerations in deep brain stimulation for psychiatric illness. *Journal of Clinical Neuroscience, 21*(1), 1–5.

Gray, R. (2008, August 30). Legal bid to stop CERN atom smasher from "destroying the world." *The Telegraph.* Retrieved from http://www.telegraph.co.uk/news/worldnews/europe/2650665/Legal-bid-to-stop-CERN-atom-smasher-from-destroying-the-world.html.

Greely, H. (2007). The uneasy ethical and legal underpinnings of large-scale genomic biobanks. *Annual Review of Genomics and Human Genetics, 8,* 343–364.

Greely, H., Sahakian, B., Harris, J., Kessler, R. C., Gazzaniga, M., Campbell, P., & Farah, M. J. (2008). Towards responsible use of cognitive-enhancing drugs by the healthy. *Nature, 456*(7223), 702–705.

Greene, B. (1999). *The elegant universe: Superstrings, hidden dimensions, and the quest for the ultimate theory.* New York, NY: W.W. Norton.

Gubrud, M. (2014). Stopping killer robots. *Bulletin of the Atomic Scientists, 70*(1), 32–42.

Gunkel, D. J. (2012). *The machine question: Critical perspectives on AI, robots, and ethics.* Cambridge, MA: MIT Press.

Gupta, K. K., Rehman, A., & Sarviya, R. M. (2010). Bio-fuels for the gas turbine: A review. *Renewable and Sustainable Energy Reviews, 14*(9), 2946–2955.

Guston, D. H. (1999). Evaluating the first US consensus conference: The impact of the citizens' panel on telecommunications and the future of democracy. *Science, Technology & Human Values, 24*(4), 451–482.

Guston, D. H. (2010). The anticipatory governance of emerging technologies. *Journal of the Korean Vacuum Society, 19*(6), 432–441.

Guston, D. H. (2012). The pumpkin or the tiger? Michael Polanyi, Frederick Soddy, and anticipating emerging technologies. *Minerva, 50*(3), 363–379.

Guston, D. H. (2014). Understanding "anticipatory governance." *Social Studies of Science, 44*(2), 218–242.

Guston, D. H., & Sarewitz, D. (2002). Real-time technology assessment. *Technology in Society, 24*(1), 93–109.

Guston, D. H., & Sarewitz, D. (2006). *Shaping science and technology policy: The next generation of research.* Madison, WI: University of Wisconsin Press.

Gutmann, A. (2011). The ethics of synthetic biology: Guiding principles for emerging technologies. *Hastings Center Report, 41*(4), 17–22.

Gutmann, A., Wagner, J. W., Yolanda, A., Grady, C., Anita, L. A., Stephen, L. H., … Nelson, L. M. (2010). New directions: The ethics of synthetic biology and emerging technologies. *The Presidential Commission for the Study of Bioethical Issues,* 1–92. Retrieved from http://bioethics.gov/sites/default/files/PCSBI-Synthetic-Biology-Report–12.16.10_0.pdf

Haas, P. (2010, July). Haiti's disaster of engineering [Video]. *TED.* Retrieved from https://www.ted.com/talks/peter_haas_haiti_s_disaster_of_engineering?language=en

Habermas, J. (1975). *Legitimation crisis.* (T. McCarthy, Trans.). Boston, MA: Beacon Press.

Hagel, J., Brown, J. S., Samoylova, T., & Lui, M. (2013). *From exponential technologies to exponential innovation: Report 2 of the 2013 shift index.* Deloitte University Press. Retrieved from http://www2.deloitte.com/content/dam/Deloitte/es/Documents/sector-publico/Deloitte_ES_Sector-Publico_From-exponential-technologies-to-exponential-innovation.pdf

Haidt, J. (2001). The emotional dog and its rational tail: A social intuitionist approach to moral judgment. *Psychology Review, 108*(4), 814–834.

Haidt, J. (2007, May 18). The new synthesis in moral psychology. *Science, 316*(5827), 998–1002.

Hall, J. S. (2007). *Beyond AI: Creating the conscience of the machine.* Amherst, NY: Prometheus Books.

Hall, J. S. (2009). The changing shape of things to come. In J. R. Shook & L. S. Swan (Eds.), *Transformers and philosophy: More than meets the mind* (pp. 3–12). Peru, IL: Carus Publishing Company.

Hall, S. S. (2010). Revolution postponed. *Scientific American, 303*(4), 60–67.

Hamilton, C. (2013). *Earthmasters: The dawn of the age of climate engineering.* New Haven, CT: Yale University Press.

Hamlett, P., Cobb, M. D., & Guston, D. H. (2013). National citizens' technology forum: Nanotechnologies and human enhancement. In S. A. Hays, J. S. Robert, C. A. Miller, &

I. Bennett (Eds.), *Nanotechnology, the brain, and the future* (pp. 265–284). Dordrecht, NL: Springer.

Handyside, A. H., Lesko, J. G., Tarín, J. J., Winston, R. M., & Hughes, M. R. (1992). Birth of a normal girl after in vitro fertilization and preimplantation diagnostic testing for cystic fibrosis. *New England Journal of Medicine, 327*(13), 905–909.

Hanson, F. A. (1992). *Testing testing: Social consequences of the examined life*. Berkeley, CA: University of California Press.

Hanson, N., Harris, J., Joseph, L. A., Ramakrishnan, K., & Thompson, T. (2011, December 29). *EPA needs to manage nanomaterial risks more effectively*. Report Number 12-P-0162 for the Office of the Inspector General, U.S. EPA. Retrieved from http://www.epa.gov/oig/reports/2012/20121229–12-P-0162.pdf

Hanson, R. (2001). Economic growth given machine intelligence. *Journal of Artificial Intelligence Research*. 1–13.

Hardy, C., & Maguire, S. (2010). Discourse, field-configuring events, and change in organizations and institutional fields: Narratives of DDT and the Stockholm convention. *Academy of Management Journal, 53*(6), 1365–1392.

Hare, M., & Deadman, P. (2004). Further towards a taxonomy of agent-based simulation models in environmental management. *Mathematics and Computers in Simulation, 64*(1), 25–40.

Hargraves, R., & Moir, R. (2010). Liquid fluoride thorium reactors. *American Scientist, 98*(4), 304–313.

Harmon, A. (2013, August 24). Golden rice: Lifesaver? *The New York Times*. Retrieved from http://www.nytimes.com/2013/08/25/sunday-review/golden-rice-lifesaver.html

Harmon, A., & McDonald, B. (2010, July 9). Interview with a robot [Video]. *The New York Times*. Retrieved from http://www.youtube.com/watch?v=uvcQCJpZJH8

Harnad, S. (2003). Can a machine be conscious? How? *Journal of Consciousness Studies, 10*(4–5), 67–75.

Harrison, M., Peggs, S., & Roser, T. (2002). The RHIC accelerator. *Annual Review of Nuclear and Particle Science, 52*(1), 425–469.

Hatton, T. J. (2013). How have Europeans grown so tall? *Oxford Economic Papers*. doi:10.1093/oep/gpt030

Hauser, M. D. (2006). *Moral minds: How nature designed our universal sense of right and wrong*. New York, NY: Ecco.

Hawking, S. (1988). *A brief history of time*. New York, NY: Bantam Books.

Heinlein, R. A. (1960). *Methuselah's children*. New York, NY: New American Library.

Hellsten, I. (2005). From sequencing to annotating: Extending the metaphor of the book of life from genetics to genomics. *New Genetics and Society, 24*(3), 283–297.

Henry, M., Fishman, J. R., & Youngner, S. J. (2007). Propranolol and the prevention of post-traumatic stress disorder: Is it wrong to erase the "sting" of bad memories? *The American Journal of Bioethics, 7*(9), 12–20.

Herbert, F. (2008). *Dune messiah*. New York, NY: Penguin Group.

Herbert, F. (2010). *Children of Dune*. UK: Hachette.

Herlihy, D. V. (2004). *Bicycle: The history*. New Haven, CT: Yale University Press.

Herman, R. A., & Price, W. D. (2013). Unintended compositional changes in genetically modified (GM) crops: 20 years of research. *Journal of Agricultural and Food Chemistry, 61*(48), 11695–11701.

Herper, M. (2010, April 22). From satellites to pharmaceuticals. *Forbes*. Retrieved from http://www.forbes.com/forbes/2010/0510/second-acts-pharmaceuticals-orphan-drugs-pah-deep-breaths.html

Hesketh, T. (2011). Selecting sex: The effect of preferring sons. *Early Human Development, 87*(11), 759–761.

Hessel, A., Goodman, M., & Kotler, S. (2012, October 24). Hacking the President's DNA. *The Atlantic*. Retrieved from http://www.theatlantic.com/magazine/archive/2012/11/hacking-the-presidents-dna/309147

Hewitt, R. E. (2011). Biobanking: The foundation of personalized medicine. *Current Opinion in Oncology, 23*(1), 112–119.

Heyns, C. (2013, April 9). *Report of the special rapporteur on extrajudicial, summary or arbitrary executions*. Presented at the 23rd session of the U.N. General Assembly, Human Rights Council, New York, NY. Retrieved from http://www.ohchr.org/Documents/HRBodies/HRCouncil/RegularSession/Session23/A-HRC–23–47_en.pdf

Heyward, C. (2013). Situating and abandoning geoengineering: A typology of five responses to dangerous climate change. *PS: Political Science and Politics, 46*(1), 230–27.

Hibbard, B. (2003, May 9). *Critique of the SIAI guidelines on friendly AI*. Retrieved from http://www.ssec.wisc.edu/~billh/g/SIAI_critique.html

Himma, K. E. (2007). The concept of information overload: A preliminary step in understanding the nature of a harmful information-related condition. *Ethics and Information Technology, 9*(4), 259–272.

Himma, K. E., & Tavani, H. T. (Eds). (2008). T*he handbook of information and computer ethics*. Hoboken, NJ: Wiley.

Hodge, G. A., Bowman, D. M., & Maynard, A. D. (Eds.). (2010). *International handbook on regulating nanotechnologies*. Cheltenham, UK: Edward Elgar Publishing.

Hodges, A. (1992). *Alan Turing: The enigma*. New York, NY: Simon and Schuster.

Hoffman, R. R., & Woods, D. D. (2011). Beyond Simon's slice: Five fundamental trade-offs that bound the performance of macrocognitive work systems. *IEEE Intelligent Systems, 26*(6), 67–71.

Hoffman, W., Terschueren, C., & Richardson, D. B. (2007). Childhood leukemia in the vicinity of the Geesthacht nuclear establishments near Hamburg, Germany. *Environmental Health Perspectives, 115*(6), 947–952.

Holland, S., Lebacqz, K., & Zoloth, L. (Eds.). (2001). *The human embryonic stem cell debate: Science, ethics, and public policy*. Cambridge, MA: MIT Press.

Holland, O. (Ed.). (2003). *Machine consciousness* (Vol. 10). Thorverton, UK: Imprint Academic.

Hollnagel, E., Pariés, J., Woods, D. D., & Wreathall, J. (Eds.). (2011). *Resilience engineering in practice: A guidebook*. Surrey, UK: Ashgate Publishing.

Hollnagel, E., & Woods, D. D. (2005). *Joint cognitive systems: Foundations of cognitive systems engineering*. Boca Raton, FL: Taylor & Francis Group.

Hollnagel, E., Woods, D. D., & Leveson, N. (Eds.). (2006). *Resilience engineering: Concepts and precepts*. Hampshire, UK: Ashgate Publishing.

Hopkins, P. D. (1998). How popular media represent cloning as an ethical problem. *Hastings Center Report, 28*(2), 6–13.

Horvitz, E., & Selman, B. (2009). Interim report from the panel chairs: AAAI Presidential Panel on Long-Term AI Futures. Retrieved from https://www.aaai.org/Organization/Panel/panel-note.pdf

Howell, K. (2009, July 14). Exxon sinks $600m into algae-based biofuels in major strategy shift. *The New York Times*. Retrieved from http://www.nytimes.com/gwire/2009/07/14/14greenwire-exxon-sinks–600m-into-algae-based-biofuels-in–33562.html

Hughes, J. J. (2004). *Citizen cyborg: Why democratic societies must respond to the redesigned human of the future*. Cambridge, MA: Westview Press.

Hughes, J. J. (2010). Technoprogressive biopolitics and human enhancement. In J. D. Moreno & S. Berger (Eds.), *Progress in bioethics: Science, policy, and politics* (pp. 163–188). Cambridge, MA: MIT Press.

Hughes, J. J. (2012). The politics of transhumanism and the techno-millenial imagination, 1626–2030. *Zygon: Journal of Religion and Science, 47*(4), 757–776.

Hughes, J. J. (2014). How conscience apps and caring computers will illuminate and strengthen human morality. In R. Blackford & D. Broderick (Eds.), *Intelligence unbound: The future of uploaded and machine minds* (pp. 26–34). West Sussex, UK: John Wiley & Sons.

Hull, C. W. (1986). Apparatus for production of three-dimensional objects by stereolithography. *U.S. Patent No. 4,575,330.* Washington, DC: Patent and Trademark Office.

Human Events. (2005, May 31). Ten most harmful books of the 19th and 20th centuries. *Human Events: Powerful conservative voices.* Retrieved from http://humanevents.com/2005/05/31/ten-most-harmful-books-of-the–19th-and–20th-centuries

Hume, D. (2000). *A treatise on human nature.* Oxford, UK: Oxford University Press. (Original work published 1739–1740)

Huxford, J. (2000). Framing the future: Science fiction frames and the press coverage of cloning. *Continuum: Journal of Media & Cultural Studies, 14*(2), 187–199.

Ilieva, I., Boland, J., & Farah, M. J. (2013). Objective and subjective cognitive enhancing effects of mixed amphetamine salts in healthy people. *Neuropharmacology, 64,* 496–505. doi:10.1016/j.neuropharm.2012.07.021

Illes, J., Moser, M. A., McCormick, J. B., Racine, E., Blakeslee, S., Caplan, A., … Weiss, S. (2010). Neurotalk: Improving the communication of neuroscience research. *Nature Reviews Neuroscience, 11*(1), 61–69.

Ingram, V. M. (1957). Gene mutations in human haemoglobin: The chemical difference between normal and sickle cell haemoglobin. *Nature, 180*(4581), 326–328.

International Committee of the Red Cross (ICRC). (1949, August 12). *Fourth Geneva convention: Relative to the protection of civilian persons in time of war.* Geneva, CH: ICRC.

International Committee of the Red Cross (ICRC). (2011). *International humanitarian law and the challenges of contemporary armed conflicts.* Geneva, CH: ICRC. Retrieved from http://www.icrc.org/eng/assets/files/red-cross-crescent-movement/31st-international-conference/31-int-conference-ihl-challenges-report–11–5–1–2-en.pdf

International Energy Agency (IEA). (2012, December 18). Medium-term coal market report 2012 factsheet. *IEA.* Retrieved from http://www.iea.org/newsroomandevents/news/2012/december/name,34467,en.html

International Human Rights Clinic. (2012). *Losing humanity: The case against killer robots.* Human Rights Watch.

International Organization for Standardization (ISO). (2006). ISO robot safety standards 10218–1:2006. ISO.

Itskov, D. (2011, October 15). *Project "Immortality 2045"—Russian experience* [Transcript]. Report for the Singularity Summit 2011. Retrieved from http://2045.com/articles/29105.html

Jasanoff, S. (2003). Technologies of humility: Citizen participation in governing science. *Minerva, 41*(3), 223–244.

Jasanoff, S. (2011). Constitutional moments in governing science and technology. *Science and Engineering Ethics, 17*(4), 621–638.

Jensen, K., & Murray, F. (2005, October 14). Intellectual property landscape of the human genome. *Science, 310*(5746), 239–240.

Johnson, D. G. (2006). Computer systems: Moral entities but not moral agents. *Ethics and Information Technology, 8*(4), 195–204.

Johnson, D. G., & Noorman, M. (2013). Recommendations for future development of artifical agents. Retrieved from http://www.law.upenn.edu/live/files/3772-johnson-d-and-noorman-m-recommendations-for-future

Johnson, D. G., & Noorman, M. (2014). Artefactual agency and artefactual moral agency. In P. Kroes & P. Verbeek (Eds.), *The moral status of technical artefacts* (pp. 143–158). Dordrecht, NL: Springer.

Johnson, E. E., & Baram, M. (2014, February 10). New U.S. science commission should look at experiment's risk of destroying the earth. *International Business Times*. Retrieved from http://www.ibtimes.com/new-us-science-commission-should-look-experiments-risk-destroying-earth-1554380

Johnson, G. (2009). Plugging into the sun: Sunlight bathes us in far more energy than we could ever need—If we could just catch enough. *National Geographic, 216*(3), 28.

Johnson, M. E., & Lucey, J. A. (2006). Major technological advances and trends in cheese. *Journal of Dairy Science, 89*(4), 1174–1178.

Johnson, N. (2010). *Simply complexity: A clear guide to complexity theory*. Oxford, UK: Oneworld.

Johnson, S. (2001). *Emergence: The connected lives of ants, brains, cities, and software*. New York, NY: Scribner.

Johnston, J., & Parens, E. (Eds.). (2014). *Interpreting neuroimages: An introduction to the technology and its limits*. Hastings Center Special Report.

Jolie, A. (2013, May 14). My medical choice [Op-ed]. *The New York Times*. Retrieved from http://www.nytimes.com/2013/05/14/opinion/my-medical-choice.html?_r=0

Jonas, H. (1979). Toward a philosophy of technology. *Hastings Center Report, 9*(1), 34–43.

Jones, N. (2013, October 17). Geoengineering: One cool solution. *Nature, 502*(7471), 302.

Jones, S. (1999). *Almost like a whale: The origin of species updated*. London, UK: Doubleday.

Joy, B. (2000, April). Why the future doesn't need us. *WIRED*. Retrieved from http://archive.wired.com/wired/archive/8.04/joy.html

Juengst, E. T., Binstock, R. H., Mehlman, M., Post, S. G., & Whitehouse, P. (2003). Biogerontology, "anti-aging medicine," and the challenges of human enhancement. *Hastings Center Report, 33*(4), 21–30.

Kabachinski, J. (2014, April 10). 3D printing: A disruptive technology? *24x7 Mag*. Retrieved from http://www.24x7mag.com/2014/04/3d-printing-a-disruptive-technology

Kaebnick, G., & Murray, T. (Eds.). (2013). *Synthetic biology and morality: Artificial life and the bounds of nature*. Cambridge, MA: MIT Press.

Kahan, D., Brahman, D., Slovic, P., Gastil, J., & Cohen, G. (2009). Cultural cognition of the risks and benefits of nanotechnology. *Nature Nanotechnology, 4*(2), 87–90.

Kahn, P. (2011). *Technological nature: Adaptation and the future of human life*. Cambridge, MA: MIT Press.

Kahneman, D. (2011). *Thinking, fast and slow*. New York, NY: Farrar, Straus and Giroux.

Kahneman, D., Slovic, P., & Tversky, A. (Eds.). (1982). *Judgment under uncertainty: Heuristics and biases*. Cambridge, UK: Cambridge University Press.

Kaku, M. (2008a, September 13). Don't buy into the supercollider hype. *The Wall Street Journal*. Retrieved from http://online.wsj.com/news/articles/SB122126297197130483

Kaku, M. (2008b). *Physics of the impossible: A scientific exploration into the world of phasers, force fields, teleportation, and time travel*. New York, NY: Doubleday.

Kaku, M. (2011). *Physics of the future: How science will shape human destiny and our daily lives by the year 2100*. New York, NY: Doubleday.

Kamel, R. M. (2013). Assisted reproductive technology after the birth of Louise Brown. *Journal of Reproduction & Infertility, 14*(3), 96–109.

Kandel, E. R. (2001, November 2). The molecular biology of memory storage: A dialogue between genes and synapses. *Science, 294*(5544), 1030–1038.

Kandel, E. R., & Mack, S. (2003). A parallel between radical reductionism in science and in art. *Annals of the New York Academy of Sciences, 1001*(1), 272–294.

Kant, E. (1996). *Groundwork of the metaphysics of morals*. Cambridge, UK: Cambridge University Press. (Original work published 1785)

Kapoor, N., Liang, W., Marbán, E., & Cho, H. C. (2013). Direct conversion of quiescent cardiomyocytes to pacemaker cells by expression of Tbx18. *Nature Biotechnology, 31*(1), 54–62.

Kass, L. R. (1997). The wisdom of repugnance. *The New Republic, 216*(22), 17–26.

Kates, R. W., & Clark, W. C. (1996). Environmental surprise: Expecting the unexpected? *Environment: Science and Policy for Sustainable Development, 38*(2), 6–34.

Katz, L. (Ed.). (2000). *Evolutionary origins of morality: Cross-disciplinary perspectives.* Thorverton, UK: Imprint Academic.

Keith, D. W. (2000). Geoengineering the climate: History and prospect. *Annual Review of Energy and the Environment, 25*(1), 245–284.

Keith, D. W., Parson, E., & Morgan, M. G. (2010). Research on global sun block needed now. *Nature, 463*(7280), 426–427.

Kelly, D. (2011). *Yuck! The nature and moral significance of disgust.* Cambridge, MA: MIT Press.

Kelly, K. (2010). *What technology wants.* New York, NY: Penguin Group.

Kevles, D. J. (1985). *In the name of eugenics: Genetics and the uses of human heredity.* New York, NY: Alfred A. Knopf.

Keynes, J. M. (1963). *Essays in persuasion.* New York, NY: W. W. Norton.

Khushf, G. (2004). Systems theory and the ethics of human enhancement: A framework for NBIC convergence. *Annals of the New York Academy of Sciences, 1013*, 124–149.

King, D. S. (1999). Preimplantation genetic diagnosis and the "new" eugenics. *Journal of Medical Ethics, 25*(2), 176–182.

King, N. M., & Robeson, R. (2007). Athlete or guinea pig? Sports and enhancement research. *Studies in Ethics, Law, and Technology, 1*(1). doi:10.2202/1941–6008.1006

Kirilenko, A., Kyle, A. S., Samadi, M., & Tuzun, T. (n.d.). *The flash crash: The impact of high frequency trading on an electronic market.* Retrieved from http://www.fma.org/Istanbul/Papers/HFT_and_Flash_Crash.pdf

Kluge, E. W. (2011). Ethical and legal challenges for health telematics in a global world: Telehealth and the technological imperative. *International Journal of Medical Informatics, 80*(2), e1-e5.

Kluger, J. (1999, February 1). The suicide seeds. *Time.* Retrieved from http://content.time.com/time/magazine/article/0,9171,18814,00.html

Knaus, W. A., Zimmerman, J. E., Wagner, D. P., Draper, E. A., & Lawrence, D. E. (1981). APACHE- Acute physiology and chronic health evaluation: A physiologically based classification system. *Critical Care Medicine, 9*(8), 591–597.

Koch, B., Bleicher, M., & Stöcker, H. (2009). Exclusion of black hole disaster scenarios at the LHC. *Physics Letters B, 672*(1), 71–76.

Koepp, M. J., Gunn, R. N., Lawrence, A. D., Cunningham, V. J., Dagher, A., Jones, T., … Grasby, P. M. (1998). Evidence for striatal dopamine release during a video game. *Nature, 393*(6682), 266–268.

Kohlberg, L. (1981). *Essays on moral development: The philosophy of moral development* (Vol. 1). San Francisco, CA: Harper & Row.

Kohlberg, L. (1984). *Essays on moral development: The psychology of moral development* (Vol. 2). San Francisco, CA: Harper & Row.

Kolata, G. (2007, November 22). Man who helped start stem cell war may end it. *The New York Times.* Retrieved from http://www.nytimes.com/2007/11/22/science/22stem.html

Konopinski, E. J., Marvin, C., & Teller, E. (1946). *Ignition of the atmosphere with nuclear bombs (Report LA602).* Los Alamos, NM: National Laboratory.

Konrad, K., Coenen, C., Dijkstra, A., Milburn, C., & van Lente, H. (Eds.). (2013). *Shaping emerging technologies: Governance, innovation, discourse.* Amsterdam, NL: IOS Press.

Kramer, D. B., Xu, S., & Kesselheim, A. S. (2012). Regulation of medical devices in the United States and European Union. *The New England Journal of Medicine, 366*(9), 848–855.

Krimsky, S., & Gruber, J. (Eds.). (2014). *The GMO deception: What you need to know about the food, corporations, and government agencies putting our families and our environment at risk.* New York, NY: Skyhorse Publishing.

Kris, M. G. (2013, April 08). How Memorial-Sloan Kettering is training Watson to personalize cancer care. *The Atlantic*. Retrieved from http://www.theatlantic.com/sponsored/ibm-watson/archive/2013/04/how-memorial-sloan-kettering-is-training-watson-to-personalize-cancer-care/274556

Kuflik, A. (1999). Computers in control: Rational transfer of authority or irresponsible abdication of autonomy? *Ethics and Information Technology, 1*(3), 173–184.

Kurzweil, R. (1999). *The age of spiritual machines: When computers exceed human intelligence*. New York, NY: Penguin Group.

Kurzweil, R. (2005). *The singularity is near: When humans transcend biology*. New York, NY: Penguin Group.

Kurzweil, R. (2012). *How to create a mind: The secret of human thought revealed*. New York, NY: Penguin Group.

Kuzma, J., & Fatehi, L. (2012). Policy innovation in synthetic biology governance. *Institute on Science for Global Policy*. Retrieved from http://www.scienceforglobalpolicy.org/LinkClick.aspx?fileticket=-j2UfjRLd4E%3D&tabid=158

Kuzma, J., Paradise, J., Ramachandran, G., Kim, J., Kokotovich, A., & Wolf, S. M. (2008). An integrated approach to oversight assessment for emerging technologies. *Risk Analysis, 28*(5), 1197–1220.

Lakoff, G. (1995). Metaphor, morality, and politics, or, why conservatives have left liberals in the dust. *Social Research, 62*(2), 177–213.

Lanier, J. (2011). *You are not a gadget: A manifesto*. New York, NY: Vintage Books.

Lanier, J. (2014). *Who owns the future?* New York, NY: Simon & Schuster.

Latour, B., & Venn, C. (2002). Morality and technology: The ends of the means. *Theory, Culture, & Society, 19*(5–6), 247–260.

Lebacqz, K. (1996). Diversity, disability, and designer genes: On what it means to be human. *Studia Theologica, 50*(1), 3–14.

Lederer, S. E. (1995). *Subjected to science: Human experimentation in America before the second world war*. Baltimore, MD: Johns Hopkins University Press.

Lederer, S. E. (2008). *Flesh and blood: Organ transplantation and blood transfusion in twentieth-century America*. Oxford, UK: Oxford University Press.

LeDoux, J. E., Debiec, J., & Moss, H. (Eds.). (2003). *The self: From soul to brain*. New York, NY: New York Academy of Sciences.

Lee, B. (2014). Where Gutenberg meets guns: The liberator, 3D-printed weapons, and the first amendment. *North Carolina Law Review, 92*, 1393–1425.

Lentzos, F., Bennett, G., Boeke, J., Endy, D., Rabinow, P. (2008). Visions and challenges in redesigning life. *Biosocieties, 3*(3), 311–323.

Levin, I. (1976). *The boys from Brazil*. New York, NY: Pegasus Books.

Levine, C., Dubler, N. N., & Levine, R. J. (1991). Building a new consensus: Ethical principles and policies for clinical research on HIV/AIDS. *IRB: Ethics and Human Research, 13*(2), 1–17.

Levine, R. (1988). *Ethics and regulation of clinical research* (2nd ed.). New Haven, CT: Yale University Press.

Levy, D. (2007). *Love and sex with robots: The evolution of human-robot relationships*. New York, NY: HarperCollins.

Lewontin, R. (2000). *The triple helix: Gene, organism, and environment*. Cambridge, MA: Harvard University Press.

Li, F. C., Choi, B. C., Sly, T., & Pak, A. W. (2008). Finding the real case-fatality rate of H5N1 avian influenza. *Journal of Epidemiology & Community Health, 62*(6), 555–559.

Liao, S. M., Sandberg, A., & Roache, R. (2012). Human engineering and climate change. *Ethics, Policy & Environment, 15*(2), 206–221.

Lin, P. (2013, April 15). Pain rays and robot swarms: The radical new war games the DOD plays. *The Atlantic*. Retrieved from http://www.theatlantic.com/technology/archive/2013/04/pain-rays-and-robot-swarms-the-radical-new-war-games-the-dod-plays/274965

Lin, P. (2013, July 30). The ethics of saving lives with autonomous cars is far murkier than you think. *Wired*. Retrieved from http://www.wired.com/2013/07/the-surprising-ethics-of-robot-cars/

Lin, P., Abney, K., & Bekey, G. A. (Eds.). (2012). *Robot ethics: The ethical and social implications of robotics*. Cambridge, MA: MIT Press.

Lin, P., Mehlman, M., & Abney, K. (2013). *Enhanced warfighters: Risk, ethics, and policy*. The Greenwall Foundation.

Lin, P., Melhman, M., Abney, K., French, S., Vallor, S., Galliott, J., ... Schuknect, S. (2014). Super soldiers (part 2): The ethical, legal and operational implications. In S. J. Thompson (Ed.), *Global Issues and ethical considerations in human enhancement technologies* (pp. 139–160). Hershey, PA: IGI Global.

Lin, P., Mehlman, M., Abney, K., & Galliott, J. (2014). Super soldiers (part 1): What is military enhancement? In S. J. Thompson (Ed.), *Global issues and ethical considerations in human enhancement technologies* (pp. 119–138). Hershey, PA: IGI Global.

Lombardo, P. A. (1985). Three generations, no imbeciles: New light on Buck v. Bell. *New York University Law Review, 60*, 30–63.

Lorenz, E. (1972). Predictability: Does the flap of a butterfly's wings in Brazil set off a tornado in Texas? *American Association for the Advancement of Science*, Washington, DC.

Low, K. G., & Gendaszek, A. E. (2002). Illicit use of psychostimulants among college students: A preliminary study. *Psychology, Health & Medicine, 7*(3), 283–287.

Lu, E, T., Reitsema, H., Troeltzsch, T., & Hubbard, S. (2013). The B612 foundation sentinel space telescope. *New Space, 1*(1), 42–45.

Lucas, G. (2013). Jus in silico: Moral restrictions on the use of cyberwarfare. In F. Allhoff, N. Evans, & A. Henschke (Eds.), *Routledge handbook of ethics and war* (pp. 367–381). New York, NY: Routledge.

Lucas, G. (2014). Legal and ethical precepts governing emerging military technologies: Research and use. *Amsterdam Law Forum, 6*(1), 23–34.

Lyall, S. (2010, July 12). In BP's record, a history of boldness and costly blunders. *The New York Times*. Retrieved from http://www.nytimes.com/2010/07/13/business/energy-environment/13bprisk.html

Macoubrie, J. (2004). Public perceptions about nanotechnology: Risks, benefits and trust. *Journal of Nanoparticle Research, 6*(4), 395–405.

Maddox, B. (2002). *Rosalind Franklin: The dark lady of DNA*. New York, NY: HarperCollins.

Madrigal, A. C. (2012, August 1). Huh, another rogue algorithm may have thrown off trading in 148 stocks. *The Atlantic*. Retrieved from http://www.theatlantic.com/technology/archive/2012/08/huh-another-rogue-algorithm-may-have-thrown-off-trading-in—148-stocks/260622/

Maes, P. (1989). How to do the right thing. *Connection Science, 1*(3), 291–323.

Mahnken, T. G. (2008). *Technology and the American way of war since 1945*. New York, NY: Columbia University Press.

Malinowski, B. (1944). *A scientific theory of culture and other essays*. Raleigh, NC: University of North Carolina Press.

Mandel, G. N. (2009). Regulating emerging technologies. *Law, Innovation and Technology, 1*(1), 75–92.

Marchant, G. (2013). Impact of the precautionary principle on feeding current and future generations. *Council for Agricultural Science and Technology, 52*, 1–20.

Marchant, G., Abbott, K. W., & Allenby, B. (Eds.). (2013). *Innovative governance models for emerging technologies*. Cheltenham, UK: Edward Elgar Publishing.

Marchant, G., Allenby, B., & Herkert, J. R. (Eds.). (2011). *The growing gap between emerging technologies and legal-ethical oversight: The pacing problem.* Dordrecht, NL: Springer.

Marchant, G., Meyer, A., & Scanlon, M. (2010). Integrating social and ethical concerns into regulatory decision-making for emerging technologies. *Minnesota Journal of Law, Science & Technology, 11*(1), 345–363.

Marchant, G., & Wallach, W. (2013). Governing the governance of emerging technologies. In G. Marchant, K. W. Abbott., & B. Allenby (Eds.), *Innovative governance models for emerging technologies* (pp. 136–152). Cheltenham, UK: Edward Elgar Publishing.

Mariotto, A. B., Yabroff, K. R., Shao, Y., Feuer, E. J., & Brown, M. L. (2011). Projections of the cost of cancer care in the United States: 2010–2020. *Journal of the National Cancer Institute, 103*(2), 117–128.

Markoff, J. (2011, February 16). Computer wins on "Jeopardy!": Trivial, it's not. *The New York Times.* Retrieved from http://www.cs.hmc.edu/~cs5grad/cs5/watson.pdf

Markoff, J. (2014, November 11). Fearing bombs that can pick whom to kill. *The New York Times.* Retrieved from http://nyti.ms/1pNYvFi

Marks, J. (in press). Heaven can't wait: A critical look at public and private sector responses to the risk of impact by asteroids and comets. In J. Galliott (Ed.), *Commercial space exploration: Ethics, policy and governance.* Farnham, UK: Ashgate.

Marks, P. (2006, September 21). Robot infantry get ready for the battlefield. *New Scientist.* Retrieved from http://www.newscientist.com/article/mg19125705.600-robot-infantry-get-ready-for-the-battlefield.html

Martin, D., & Caldwell, S. (2011, July 22). 150 human animal hybrids grown in UK labs: Embryos have been produced secretively for the past three years. *The Daily Mail.* Retrieved from http://www.dailymail.co.uk/sciencetech/article–2017818/Embryos-involving-genes-animals-mixed-humans-produced-secretively-past-years.html

Massingham, E. (2012). Conflict without casualties… A note of caution: Non-lethal weapons and international humanitarian law. *International Review of the Red Cross, 94*(886), 673–685.

Masson, M. F. (2013, May 22). 3D printed splint saves the life of a baby [Article and video]. *Michigan Engineering Website.* Retrieved from http://www.engin.umich.edu/college/about/news/stories/2013/may/3d-printed-splint-saves-life

Matsumura, A. (2013, April 3). Fukushima Daiichi site: Cesium–137 is 85 times greater than at Chernobyl accident. *Finding the Missing Link: New Conflicts. New security.New perspectives.* Retrieved from http://akiomatsumura.com/2012/04/682.html

Matsuyama, K. (2013, September 15). Shutdown of Japan's last nuclear reactor raises power concerns. *Bloomberg.* Retrieved from http://www.bloomberg.com/news/2013-09–16/shutdown-of-japan-s-last-nuclear-reactor-raises-power-concerns.html

Mauron, A. (2001, February 2). Is the genome the secular equivalent of the soul? *Science, 291*(5505), 831–832.

Maynard, A., Bowman, D., & Hodge, G. (2011). The problem of regulating sophisticated materials. *Nature materials, 10*(8), 554–557.

Mayr, E. & Provine, W. B. (Eds.). (1998). *The evolutionary synthesis: Perspectives on the unification of biology.* Cambridge, MA: Harvard University Press.

McCarthy, J. (1995). *Making robots conscious of their mental states.* Retrieved from http://www-formal.stanford.edu/jmc/consciousness-submit/consciousness-submit.html

McGinn, C. (1999). *The mysterious flame: Conscious minds in a material world.* New York, NY: Basic Books.

McGonigle, P., & Ruggeri, B. (2014). Animal models of human disease: Challenges in enabling translation. *Biochemical Pharmacology, 87*(1), 162–171.

McKibben, B. (2003). *Enough: Staying human in an engineered age.* New York, NY: Henry Holt.

Mehlman, M. J. (2009). *The price of perfection: Individualism and society in the era of biomedical enhancement.* Baltimore, MD: Johns Hopkins University Press.

Mehlman, M. J., Lin, P., & Abney, K. (2013). Enhanced warfighters: A policy framework. In M. L. Gross & D. Carrick (Eds.), *Military medical ethics for the 21st century* (pp. 113–126). Surrey, UK: Ashgate Publishing.

Merrill, R. A. (1997). Food safety regulation: Reforming the Delaney Clause. *Annual Review of Public Health, 18*(1), 313–340.

Miah, A. (2006). Rethinking enhancement in sport. *Annals of the New York Academy of Sciences, 1093*(1), 301–320.

Mill, J. S. (1998). *Utilitarianism.* Oxford, UK: Oxford University Press. (Original work published 1864)

Miller, H. I. (2003). Vox populi and public policy: Why should we care? *Nature Biotechnology, 21*(12), 1431–1432.

Miller, J., & Page, S. (2007). *Complex adaptive systems: An introduction to computational models of social life.* Princeton, NJ: Princeton University Press.

Mimeault, M., Hauke, R., & Batra, S. K. (2007). Stem cells: A revolution in therapeutics—Recent advances in stem cell biology and their therapeutic applications in regenerative medicine and cancer therapies. *Clinical Pharmacology & Therapeutics, 82*(3), 252–264.

Miyasaka, M., Sasaki, S., Tanaka, M., & Kikunaga, J. (2012). Use of brain-machine interfaces as prosthetic devices: An ethical analysis. *Journal of Philosophy and Ethics in Health Care and Medicine, 6,* 29–38.

Møldrop, C., & Morgall, J. M. (2001). Risks of future drugs: A Danish expert Delphi. *Technological Forecasting and Social Change, 67*(2–3), 273–289.

Montgomery, C. T., & Smith, M. B. (2010). Hydraulic fracturing: History of an enduring technology. *Journal of Petroleum Technology, 62*(12), 26–41.

Moor, J. H. (1979). Are there decisions computers should never make? *Nature and System, 1*(4), 217–229.

Moor, J. H. (2006). The nature, importance, and difficulty of machine ethics. *IEEE Intelligent Systems, 21*(4), 18–21.

Moravec, H. (2000). *Robot: Mere machine to transcendent mind.* Oxford, UK: Oxford University Press.

More, M. (2000, February 21). Embrace, don't relinquish, the future. *Kurzweil: Accelerating Intelligence.* Retrieved from http://www.kurzweilai.net/embrace-dont-relinquish-the-future

More, M., & Vita-More, N. (Eds.). (2013). *The transhumanist reader: Classical and contemporary essays on the science, technology, and philosophy of the human future.* West Sussex, UK: Wiley-Blackwell.

Moreno, J. (1995). *Deciding together: Bioethics and moral consensus.* New York, NY: Oxford University Press.

Moreno, J. (2000). *Undue risk: Secret state experiments on humans.* New York, NY: Henry Holt.

Moreno, J. (2006). *Mind wars: Brain research and national defense.* New York, NY: Dana Press.

Moreno, J. (2011). *The body politic: The battle over science in America.* New York, NY: Bellevue Literary Press.

Moreno, J., & Berger, S. (Eds.). (2010). *Progress in bioethics: Science, policy, and politics.* Cambridge, MA: MIT Press.

Morisette, P. M. (1991). The Montreal protocol: Lessons for formulating policies for global warming. *Policy Studies Journal, 19*(2), 152–161.

Morozov, E. (2011). *The net delusion: The dark side of internet freedom.* New York, NY: PublicAffairs.

Morozov, E. (2013). *To save everything, click here: The folly of technological solutionism.* New York, NY: PublicAffairs.

Moses, L. B. (2007). Why have a theory of law and technological change? *Minnesota Journal of Law, Science & Technology, 8*(2), 589–606.

Moskowitz, C. (2013, October 28). United Nations to adopt asteroid defense plan. *Scientific American*. Retrieved from http://www.scientificamerican.com/article/un-asteroid-defense-plan

Muehlhauser, L. (2013). *Facing the intelligence explosion*. Berkeley, CA: Machine Intelligence Research Institute.

Murray, T. H. (2010). Making sense of fairness in sports. *Hastings Center Report, 40*(2), 13–15.

Naam, R. (2005). *More than human: Embracing the promise of biological enhancement*. New York, NY: Broadway Books.

Naam, R. (2012). *Nexus: Mankind gets an upgrade*. Nottingham, UK: Angry Robot.

Naam, R. (2013). *The infinite resource: The power of ideas on a finite planet*. Hanover, NH: University Press of New England.

National Aeronautics and Space Administration (NASA). (n.d.). *Global warming*. Retrieved from http://earthobservatory.nasa.gov/Features/GlobalWarming/page2.php

National Aeronautics and Space Administration (NASA). (2003, August 22). *Study to determine the feasibility of extending the search for near-earth objects to smaller limiting diameters*. Report of the near-earth object science definition team. Retrieved from http://neo.jpl.nasa.gov/neo/neoreport030825.pdf

National Institutes of Health. (2012, June 13). *NIH human microbiome project defines normal bacterial makeup of the body*. Retrieved from http://www.nih.gov/news/health/jun2012/nhgri-13.htm

National Nanotechnology Coordination Office (NNCO). (2013). *The national nanotechnology initiative: Supplement to the President's 2014 budget*. NNCO.

Nehaniv, C. L., & Dautenhahn, K. (Eds.). (2007). *Imitation and social learning in robots, humans and animals: Behavioural, social and communicative dimensions*. Cambridge, UK: Cambridge University Press.

Nelson, B., Bly, K., & Magaña, S. (2014). *Freezing people is (not) easy: My adventures in cryonics*. Guilford, CT: Lyons Press.

Nicolia, A., Manzo, A., Veronesi, F., & Rosellini, D. (2013). An overview of the last 10 years of genetically engineered crop safety research. *Critical Reviews in Biotechnology, 34*(1), 77–88.

Nietzsche, F. (1996). *On the genealogy of morals*. (D. Smith, Trans.). New York, NY: Oxford University Press. (Original work published 1887)

Niiler, E. (1999). Terminator technology temporarily terminated. *Nature Biotechnology, 17*(11), 1054.

Nisbet, M. C. (2004). Public opinion about stem cell research and human cloning. *Public Opinion Quarterly, 68*(1), 131–154.

Nissenbaum, H. (1996). Accountability in a computerized society. *Science and Engineering Ethics, 2*(1), 25–42.

Nissenbaum, H. (2001). How computer systems embody values. *Computer, 34*(3), 118–119.

Noetic Corporation. (2013, June). *Technology as dialectic: Understanding game changing technology*. Office of the Secretary of Defense: Rapid Fielding.

Nöggerath, J., Geller, R. J., & Gusiakov, V. K. (2011). Fukushima: The myth of safety, the reality of geoscience. *Bulletin of the Atomic Scientists, 67*(37), 39.

Nordmann, A. (in press). Discussion paper: Responsible innovation, the art and craft of anticipation. *Journal of Responsible Innovation*.

Nordmann A., & Schwarz A. (2010). Lure of the "yes": The seductive power of technoscience. In M. Kaiser, M. Kurath, S. Maasen, & C. Rehmann-Sutter (Eds.), *Governing future technologies: Nanotechnology and the rise of an assessment regime* (pp. 255-278). Dordrecht, NL: Springer.

Nourbakhsh, I. R. (2013). *Robot futures*. Cambridge, MA: MIT Press.

Nuffield Council on Bioethics. (2002). *Genetics and human behaviour: The ethical context*. London, UK: Nuffield Council on Bioethics.

Nussbaum, R. H. (2009). Childhood leukemia and cancers near German nuclear reactors: Significance, context, and ramifications of recent studies. *International Journal of Occupational and Environmental Health, 15*(3), 318–323.

O'Connell, M. E. (2014). Banning autonomous killing: The legal and ethical requirement that humans make near-time lethal decisions. In M. Evangelista & H. Shue (Eds.), *The American way of bombing: Changing ethical and legal norms, from flying fortresses to drones* (pp. 224-236). Ithaca, NY: Cornell University Press.

Office of Legislative Policy and Analysis. (2003). *21st century nanotechnology research and development act.* Retrieved from http://olpa.od.nih.gov/legislation/108/publiclaws/nanotechnology.asp

Office of Science and Technology Policy (OSTP). (2013, April 02). *President Obama launches the "BRAIN" initiative.* Retrieved from http://www.whitehouse.gov/blog/2013/04/02/president-obama-launches-brain-initiative

Office of the Surgeon Multinational Force-Iraq and Office of the Surgeon General United States Army Medical Command. (2006, November 17). *Mental health advisory team (MHAT) IV: Operation Iraqi Freedom 05-07* (Final Report). Retrieved from http://www.motherjones.com/documents/551721-mental-health-advisory-teammhat-iv

O'Hara, A. M., & Shanahan, F. (2006). The gut flora as a forgotten organ. *EMBO Reports, 7*(7), 688–693.

Okrent, D. (1987, April 17). The safety goals of the US Nuclear Regulatory Commission. *Science, 236*(4799), 296–300.

Oliver, R. (2000). *The coming biotech age: The business of bio-materials.* New York, NY: McGraw-Hill.

Orlando, G., Wood, K. J., Stratta, R. J., Yoo, J. J., Atala, A., & Shoker, S. (2011). Regenerative medicine and organ transplantation: Past, present, and future. *Transplantation, 91*(12), 1310–1317.

Owen, R., Bessant, J., & Heintz, M. (Eds.). (2013). *Responsible innovation: Managing the responsible emergence of science and innovation in society.* West Sussex, UK: Wiley.

Owen, R., Macnaghten, P., & Stilgoe, J. (2012). Responsible research and innovation: From science in society to science for society, with society. *Science and Public Policy, 39*(6), 751–760.

Pacholczyk, A. (2011). Moral enhancement: What is it and do we want it? *Law, Innovation and Technology, 3*(2), 251–277.

Parens, E. (Ed.). (1998). Enhancing human traits: Ethical and social implications. Washington, DC: Georgetown University Press.

Parens, E. (Ed.). (2006). *Surgically shaping children: Technology, ethics, and the pursuit of normality.* Baltimore, MD: Johns Hopkins University Press.

Parens, E. (2013, 3rd Quarter). The need for moral enhancement. *The Philosopher's Magazine, 62*, 114–117.

Parens, E. (2014). *Shaping our selves: On technology, flourishing, and a habit of thinking.* Oxford, UK: Oxford University Press.

Pearson, H. (2004, November 22). UN ditches cloning ban. *Nature.* doi:10.1038/news041122–2

Peck, M. (2003, May). Global hawk crashes: Who's to blame? *National Defense, 87*(594). Retrieved from http://www.questia.com/magazine/1P3–333006161/global-hawk-crashes-who-s-to-blame

Pelly, J. & Saner, M. (2009). *International approaches to the regulatory governance of nanotechnology.* Regulatory Governance Initiative, Carleton University.

Pennisi, E. (2012, September 7). ENCODE project writes eulogy for junk DNA. *Science, 337*(6099), 1159–1161.

Perlroth, N., Larson, J., & Shane, S. (2013, September 5). N.S.A. able to foil basic safeguards of privacy on the web. *The New York Times.* Retrieved from http://www.nytimes.com/2013/09/06/us/nsa-foils-much-internet-encryption.html?pagewanted=all

Perrow, C. (1999). *Normal accidents: Living with high-risk technologies.* Princeton, NJ: Princeton University Press.

Perry, J. E., Churchill, L. R., & Kirshner, H. S. (2005). The Terri Schiavo case: Legal, ethical, and medical perspectives. *Annals of Internal Medicine, 143*(10), 744–748.

Persson, I., & Savulescu, J. (2012). *Unfit for the future: The need for moral enhancement.* Oxford, UK: Oxford University Press.

Pfoutz, A. (2014, January 31). Saving heirloom corn from GMO contamination. *The Organic & Non-GMO Report.* Retrieved from http://www.non-gmoreport.com/articles/february2014/saving-heirloom-corn-from-GMO-contamination.php

Phillips, M. (2010, May 11). Nasdaq: Here's our timeline of the flash crash. *The Wall Street Journal.* Retrieved from http://blogs.wsj.com/marketbeat/2010/05/11/nasdaq-heres-our-timeline-of-the-flash-crash

Phillips, M. (2012, August 12). Knight shows how to lose $440 Million in 30 minutes. *Business Week.* Retrieved from http://www.businessweek.com/articles/2012-08-02/knight-shows-how-to-lose–440-million-in–30-minutes

Piaget, J. (1972). *Judgment and reasoning in the child.* Totowa, NJ: Littlefield, Adams & Company.

Picard, R. (1997). *Affective computing.* Cambridge, MA: MIT Press.

Pierrehumbert, R. T. (2006). Climate change: A catastrophe in slow motion. *Chicago Journal of International Law, 6*(2), 1–24.

Pierrehumbert, R. T. (2010). *Principles of planetary climate.* Cambridge, UK: Cambridge University Press.

Pimple, K. D. (Ed.). (2013). *Emerging pervasive information and communication technologies (PICT): Ethical challenges, opportunities and safeguards.* Dordrecht, NL: Springer.

Pinker, S. (1997). *How the mind works.* New York, NY: W. W. Norton.

Pinker, S. (2004). Why nature & nurture won't go away. *Daedalus, 133*(4), 5–17.

Plotz, D. (2006). *The genius factory: The curious history of the nobel prize sperm bank.* New York, NY: Random House.

Poincaré, H. (2003). On the three-body problem and the equations of dynamics. In S. G. Brush (Ed.), *The kinetic theory of gases: An anthology of classic papers with historical commentary* (pp. 368–376). London, UK: Imperial College Press.

Pollack, A. (2010, September 4). His corporate strategy: The scientific method. *The New York Times.* Retrieved from http://www.nytimes.com/2010/09/05/business/05venter.html?_r=0

Posner, R. A. (2004). *Catastrophe: Risk and response.* Oxford, UK: Oxford University Press.

Poston, D. L., & Glover, K. S. (2005). Too many males: Marriage market implications of gender imbalances in China. *Genus, 61*(2), 119–140.

Powers, T. (2006). Prospects for a Kantian machine. *IEEE Intelligent Systems, 21*(4), 46–51.

President's Council on Bioethics. (2003). *Beyond therapy: Biotechnology and the pursuit of happiness.* Washington, DC: Dana Press.

Prince, J. D. (2014). 3D printing: An industrial revolution. *Journal of Electronic Resources in Medical Libraries, 11*(1), 39–45.

Public Service Commission. (2003, November). Michigan public service commission report on August 14th Blackout. Retrieved from http://www.michigan.gov/documents/mpsc_blackout_77423_7.pdf

Purnick, P. E., & Weiss, R. (2009). The second wave of synthetic biology: From modules to systems. *Nature Reviews Molecular Cell Biology, 10*(6), 410–422.

Qin, J., Li, R., Raes, J., Arumugam, M., Burgdorf, K. S., Manichanh, C., ... Weissenbach, J. (2010). A human gut microbial gene catalogue established by metagenomic sequencing. *Nature, 464*(7285), 59–65.

Quammen, D. (2007). *The reluctant Mr. Darwin: An intimate portrait of Charles Darwin and the making of his theory of evolution.* New York, NY: W. W. Norton.

Quinn, J. (2008, September 15). Lehman Brothers files for bankruptcy as credit crisis bites. *The Telegraph.* Retrieved from http://www.telegraph.co.uk/finance/newsbysector/banksand-finance/4676621/Lehman-Brothers-files-for-bankruptcy-as-credit-crisis-bites.html

Rabinow, P. (1997). *Making PCR: A story of biotechnology.* Chicago, IL: Chicago University Press.

Rabinow P., & Bennett, G. (2009). Synthetic biology: Ethical ramifications 2009. *Systems and Synthetic Biology, 3,* 99–108.

Rabinow, P., & Stavrianakis, A. (2014). *Designs on the contemporary: Anthropological tests.* Chicago, IL: University of Chicago Press.

Ramachandran, G., Wolf, S. M., Paradise, J., Kuzma, J., Hall, R., Kokkoli, E., & Fatehi, C. (2011). Recommendations for oversight of nanobiotechnology: Dynamic oversight for complex and convergent technology. *Journal of Nanoparticle Research, 13*(4), 1345–1371.

Rawls, J. (1999). *A theory of justice.* Cambridge, MA: Harvard University Press.

Ray, T. S. (2000). Evolution of complexity: Tissue differentiation in network tierra. *ATR Journal, 40*(8), 12–13.

Reason, J. (2000). Human error: Models and management. *BMJ: British Medical Journal, 320*(7237), 768–770.

Rees, M. J. (2003). *Our final hour: A scientist's warning: How terror, error, and environmental disaster threaten humankind's future in this century—On Earth and beyond.* New York, NY: Basic Books.

Regan, T. (1987). *The case for animal rights.* Dordrecht, NL: Springer.

Reid, A. J. (2010). It's alive! *Nature Reviews Microbiology, 8*(7), 468.

Reilly, P. R. (1991). *The surgical solution: A history of involuntary sterilization in the United States.* Baltimore, MD: Johns Hopkins University Press.

Reiss, T. (2001). Drug discovery of the future: The implications of the human genome project. *Trends in Biotechnology, 19*(12), 496–499.

Relman, D. A. (2013). The increasingly compelling moral responsibilities of life scientists. *Hastings Center Report, 43*(2), 34–35.

Renn, O., & Roco, M. C. (2006). Nanotechnology and the need for risk governance. *Journal of Nanoparticle Research, 8*(2), 153–191.

Resnik, D. B. (2007). The price of truth: How money affects the norms of science. New York, NY: Oxford University Press.

Reynolds, C., & Picard, R. (2004). Ethical evaluation of displays that adapt to affect. *CyberPsychology and Behavior, 7*(6), 662–666.

Rheingold, H. (2012). *Net smart: How to thrive online.* Cambridge, MA: MIT Press.

Rhodes, R. (1986). *The making of the atomic bomb.* New York, NY: Simon and Schuster.

Rich, N. (2004, February 27). The mammoth cometh. *New York Times Magazine.* Retrieved from http://www.nytimes.com/2014/03/02/magazine/the-mammoth-cometh.html

Richards, J. W., Kurzweil, R., & Gilder, G. (Eds.). (2002). *Are we spiritual machines: Ray Kurzweil vs. the critics of strong A.I.* Seattle, WA: Discovery Institute.

Richter, S. (2013, September 19). Stock ownership: Who benefits? *Salon.* Retrieved from http://www.salon.com/2013/09/19/stock_ownership_who_benefits_partner/

Ridley, M. (1999). *Genome: The autobiography of a species in 23 chapters.* New York, NY: HarperCollins.

Rifkin, J. (1998). *The biotech century: Harnessing the gene and remaking the world.* New York, NY: Jeremy P. Tarcher/Putnam.

Rip, A. (1995). Introduction of new technology: Making use of recent insights from sociology and economics of technology. *Technology Analysis & Strategic Management, 7*(4), 417–432.

Robert, J. S., & Baylis, F. (2003). Crossing species boundaries. *American Journal of Bioethics, 3,* 1–13.

Roco, M. (2008). Possibilities for global governance of converging technologies. *Journal of Nanoparticle Research, 10*(1), 11–29.

Roco, M., & Bainbridge, W. (Eds.). (2002). *Converging technologies for improving human performance: Nanotechnology, biotechnology, information technology, and cognitive science.* Dordrecht, NL: Springer.

Ropeik, D. (2010). *How risky is it, really: Why our fears don't always match the facts.* New York, NY: McGraw-Hill Companies.

Rose, D. S. (2014). *Angel investing: The gust guide to making money and having fun investing in startups.* Hoboken, NJ: John Wiley & Sons.

Rosenfeld, J. A., & Mason, C. E. (2013). Pervasive sequence patents cover the entire human genome. *Genome Medicine, 5*(27), 1–7.

Rosenfeld, P. E., & Feng, L. G. (2011). *Risks of hazardous wastes.* Oxford, UK: Elsevier.

Ross, W. D. (1930). *The right and the good.* Oxford, UK: Clarendon Press.

Rössler, O. (2012, March 24). CERN cannot continue the LHC experiment [Blog]. *Lifeboat Foundation: Safeguarding humanity.* Retrieved from http://lifeboat.com/blog/2012/03/cern-cannot-continue-the-lhc-experiment

Rothblatt, M. (2014). *Virtually human: The promise—And the peril—Of digital immortality.* New York, NY: St. Martin's Press.

Rothenberg, D. (1993). Hand's end: Technology and the limits of nature. Berkeley, CA: University of California Press.

Rothenberg, D. (2013, May 6). What the drone debate is really about. *Slate.* Retrieved from *http://www.slate.com/articles/technology/future_tense/2013/05/drones_in_the_united_states_what_the_debate_is_really_about.html*

Rothstein, J. (2006). Soldiers bond with battlefield robots: Lessons learned in Iraq may show up in future homeland "avatars." *MSNBC/Reuters.* Retrieved from http://www.msnbc.msn.com/id/12939612

Roubini, N. (2014, December 8). *Rise of the machines: Downfall of the economy?* Retrieved from http://www.roubinisedge.com/nouriel-unplugged/rise-of-the-machines-downfall-of-the-economy

Rozin, P., Haidt, J., & McCauley, C. (2000). Disgust. In M. Lewis & J. M. Haviland-Jones (Eds.), *Handbook of emotions* (2nd ed.) (pp. 637–653). New York, NY: Guilford Press.

Ruano, G. (2004). Quo vadis personalized medicine? *Personalized medicine, 1*(1), 1–7.

Rudd, J. (2008). Regulating the impacts of engineered nanoparticles under TSCA: Shifting authority from industry to government. *Columbia Journal of Environmental Law, 33*(2), 215–282.

Ruder, W. C., Lu, T., & Collins, J. J. (2011, September 2). Synthetic biology moving into the clinic. *Science, 333*(6047), 1248–1252.

Russell, E. P. (1999). The strange career of DDT: Experts, federal capacity, and environmentalism in world war II. *Technology and Culture, 40*(4), 770–796.

Russell, S., & Norvig, P. (1995). *Artificial intelligence: A modern approach.* Upper Saddle River, NJ: Prentice Hall.

Rysewyk, S. P., & Pontier, M. (Eds.). (2014). *Machine medical ethics.* Cham, CH: Springer.

Sanchez, R. (2014, April 3). Experts: Strict building codes saved lives in powerful Chile earthquake. *CNN.* Retrieved from http://edition.cnn.com/2014/04/02/world/americas/chile-earthquake/

Sandberg, A., & Bostrom, N. (2008). *Whole brain emulation: A roadmap* (technical report #2008–3). Future of Humanity Institute, Oxford University. Retrieved from http://www.fhi.ox.ac.uk/brain-emulation-roadmap-report.pdf

Sandel, M. (2004). The case against perfection. *The Atlantic Monthly, 293*(3), 51–62.

Sandin, P. (1999). Dimensions of the precautionary principle. *Human and Ecological Risk Assessment: An International Journal, 5*(5), 889–907.

Sandler, R. (Ed.). (2013). *Ethics and emerging technologies*. Hampshire, UK: Palgrave MacMillan.

Saner, M. (2010). Ethics as problem and ethics as solution. *International Journal of Bioetechnologies, 2*(1), 239–256.

Sapolsky, R. M. (2004). *Why zebras don't get ulcers* (3rd ed.). New York, NY: Holt Paperbacks.

Sapolsky, R. M. (2008). Biology and human behavior: The neurological origins of individuality (2nd ed.) [24 lecture series]. The Great Courses.

SARA Title III – Emergency Planning and Community Right-to-Know Act, 42 USC 9601: Public Law 99–499. In R. C. Barth, P. D. George & R. H. Hill (Eds.), *Environmental health and safety for hazardous waste sites* (pp. 24, Appendix C). Fairfax, VA: AIHA Press.

Sarewitz, D. (2004). How science makes environmental controversies worse. *Environmental Science & Policy, 7*(5), 385–403.

Sarewitz, D. (2011). Anticipatory governance of emerging technologies. In G. E. Marchant, B. R. Allenby, & J. R. Herkert (Eds.), *The growing gap between emerging technologies and legal-ethical oversight: The pacing problem* (pp. 95-106). New York, NY: Springer.

Savulescu, J. (2003). Human-animal transgenesis and chimeras might be an expression of our humanity. *The American Journal of Bioethics, 3*(3), 22–25.

Savulescu, J., Foddy, B., & Clayton, M. (2004). Why we should allow performance enhancing drugs in sport. *British Journal of Sports Medicine, 38*(6), 666–670.

Schell, J. (2000). *The fate of the Earth and the abolition*. Stanford, CA: Stanford University Press.

Sclove, R. E. (1995). *Democracy and technology*. New York, NY: The Guilford Press.

Sclove, R. E. (2000). Town meetings on technology: Consensus conferences as democratic participation. In D. L. Kleinman (Ed.), *Science, technology, and democracy* (pp. 33–48). Albany, NY: State University of New York Press.

Searle, J. R. (1980). Minds, brains, and programs. *Behavioral and Brain Sciences, 3*(3), 417–424.

Shacthman, N. (2007, August 16). Armed robots pushed to police. *Wired Magazine*. Retrieved from http://blog.wired.com/defense/2007/08/armed-robots-so.html

Shachtman, N. (2007, October 18). Robot cannon kills 9, wounds 14. *Wired Magazine*. Retrieved from http://blog.wired.com/defense/2007/10/robot-cannon-ki.html

Sharkey, N. (2010). Saying "no!" to lethal autonomous targeting. *Journal of Military Ethics, 9*(4), 369–383.

Sharkey, N., & Sharkey, A. (2010). The crying shame of robot nannies: An ethical appraisal. *Interaction Studies, 11*(2), 161–190.

Shetty, R. P., Endy, D., & Knight, T. F. (2008). Engineering biobrick vectors from biobrick parts. *Journal of Biological Engineering, 2*(1), 1–12.

Shnayerson, M. (2004, January 1). The code warrior. *Vanity Fair*. Retrieved from http://www.vanityfair.com/culture/features/2004/01/virus-hunters–200401

Siegfried, T. (2000). *The bit and the pendulum: From quantum computing to M theory—The new physics of information*. New York, NY: John Wiley & Sons.

Simkovic, M. (2009). Secret liens and the financial crisis of 2008. *American Bankruptcy Law Journal, 83*, 253–296.

Simon, H. A. (1982). *Models of bounded rationality*. Cambridge, MA: MIT Press.

Singer, P. W. (2009). *Wired for war: The robotics revolution and conflict in the 21st Century*. New York, NY: Penguin Press.

Singer, P. W. (2013, November 3). Get ready for a whole new way of war. *Pittsburgh Post Gazette*. Retrieved from http://www.post-gazette.com/opinion/Op-Ed/2013/11/03/A-WHOLE-NEW-WAY-OF-WAR/stories/201311030028

Singh, S., & Thayer, S. (2001). *ARMS (Autonomous Robots for Military Systems): A survey of collaborative robotics core technologies and their military applications*. Carnegie Mellon University, The Robotics Institute.

Singularity Institute. (2001). *SIAI guidelines on friendly AI*. Retrieved from http://www.singinst.org/ourresearch/publications/guidelines.html

Skoro-Sajer, N., Lang, I., & Naeije, R. (2008). Treprostinil for pulmonary hypertension. *Vascular Health and Risk Management, 4*(3), 507–513.

Skyrms, B. (2000). Game theory, rationality and evolution of the social contract. In L. Katz (Ed.), *Evolutionary origins of morality: Cross disciplinary perspectives* (pp. 269–285). Thorverton, UK: Imprint Academic.

Slovic, P. (1987, April 17). Perception of risk. *Science, 236*(4799), 280–285.

Smart J. (2009). Evo devo universe?: A framework for speculations on cosmic culture. In S. J. Dick & M. L. Lupisella (Eds.), *Cosmos and culture: Cultural evolution in a cosmic context* (pp. 201–296). Washington, DC: U.S. Government Printing Office.

Smith, J. D., & Washburn, D. A. (2005). Uncertainty monitoring and metacognition by animals. *Current Directions in Psychological Science, 14*(1), 19–24.

Smith, S. (2013, May 23). 3-D printer helps save dying baby. *CNN Health*. Retrieved from http://www.cnn.com/2013/05/22/health/baby-surgery

Smith-Smart, L., & Boulden, J. (2013, May 3). UK lifts ban on fracking to exploit shale gas reserves. *CNN*. Retrieved from http://edition.cnn.com/2012/12/13/business/uk-fracking

Snapper, J. W. (1985). Responsibility for computer-based errors. *Metaphilosophy, 16*(4), 289–295.

Snippert, H. J., & Clevers H. (2011). Tracking adult stem cells. *EMBO Reports, 12*(2), 113–122.

Sofair, A. N., & Kaldjian, L. C. (2000). Eugenic sterilization and a qualified Nazi analogy: The United States and Germany, 1930–1945. *Annals of Internal Medicine, 132*(4), 312–319.

Sofge, E. (2014, May 12). The mathematics of murder: Should a robot sacrifice your life to save two? *Popular Science*. Retrieved from http://www.popsci.com/blog-network/zero-moment/mathematics-murder-should-robot-sacrifice-your-life-save-two

Sonnenburg, J. L., & Fischbach, M. A. (2011). Community health care: Therapeutic opportunities in the human microbiome. *Science Translational Medicine, 3*(78), 78ps12.

Sparrow, R. (2007). Killer robots. *Journal of Applied Philosophy, 24*(1), 62–77.

Sparrow, R. (2011). A not-so-new eugenics: Harris and Savulescu on human enhancement. *Hastings Center Report, 41*(1), 32–42.

Sparrow, R., & Sparrow, L. (2006). In the hands of machines? The future of aged care. *Minds and Machines, 16*(2), 141–161.

Spencer, H. (1896). *The principles of biology* (Vol. 1). D. Appleton.

Stephens, J., Wilson, E., & Peterson, T. R. (2015). *Smart grid (r)evolution: Electric power struggles*. New York, NY: Cambridge University Press.

Stephenson, N. (2000). *Snow crash*. New York, NY: Bantam Spectra.

Stilgoe, J., Owen, R., & Macnaghten, P. (2013). Developing a framework for responsible innovation. *Research Policy, 42*(9), 1568–1580.

Stross, C. (2006). *Accelerando*. New York, NY: Ace.

Sullins, J. (2011). Introduction: Open questions in roboethics. *Philsophy & Technology, 24*(3), 233–238.

Sullins, J. (2012). Robots, love, and sex: The ethics of building a love machine. *IEEE Transactions on Affective Computing, 3*(4), 398–409.

Summers, L. H. (2014, July 7). Lawrence H. Summers on the economic challenge of the future: Jobs. *The Wall Street Journal*. Retrieved from http://www.wsj.com/articles/lawrence-h-summers-on-the-economic-challenge-of-the-future-jobs-1404762501

Sutton, V. (2011a). *Nanotechnology law and policy: Cases and materials*. Durham, NC: Carolina Academic Press.

Sutton, V. (2011b). Wind energy law and ethics: A meeting of Kant, Leopold and cultural relativism. *Seattle Journal of Environmental Law, 1*, 69–80.

Szasz, F. M. (1984). *The day the sun rose twice: The story of the Trinity site nuclear explosion, July 16, 1945*. Albuquerque, NM: University of New Mexico Press.

Taddeo, M. (2012). Information warfare: A philosophical perspective. *Philosophy and Technology, 25*(1), 105–120.

Tait, J. (2001). More faust than Frankenstein: The European debate about the precautionary principle and risk regulation for genetically modified crops. *Journal of Risk Research, 4*(2), 175–189.

Talbot, M. (2009, April 27). Brain gain: The underground world of "neuroenhancing" drugs. *The New Yorker.* Retrieved from http://www.newyorker.com/reporting/2009/04/27/090427fa_fact_talbot?currentPage=all

Taleb, N. (2007). *The black swan: The impact of the highly improbable.* New York, NY: Random House.

Tamburrini, C. M. (2007). What's wrong with genetic inequality? The impact of genetic technology on elite sports and society. *Sports, Ethics and Philosophy, 1*(2), 229–238.

Tateno, C., Yoshizane, Y., Saito, N., Kataoka, M., Utoh, R., Yamasaki, C., … Yoshizato, K. (2004). Near completely humanized liver in mice shows human-type metabolic responses to drugs. *The American Journal of Pathology, 165*(3), 901–912.

Taubenberger, J. K., & Morens, D. M. (2006). 1918 influenza: The mother of all pandemics. *Emerging Infectious Diseases, 12*(1), 15–22.

Tavani, H. T., & Grodzinsky, F. S. (2002). Cyberstalking, personal privacy, and moral responsibility. *Ethics and Information Technology, 4*(2), 123–132.

Taylor, K., Gordon, N., Langley, G., & Higgins, W. (2008). Estimates for worldwide laboratory animal use in 2005. *ATLA-Alternatives to Laboratory Animals, 36*(3), 327–342.

Ten Hoeve, J. E., & Jacobson, M. Z. (2012). Worldwide health effects of the Fukushima Daiichi nuclear accident. *Energy and Environmental Science, 5*(9), 8743–8757. doi:10.1039/c2ee22019a

Tennyson, A. (1849). *In Memoriam A. H. H.*

Terbeck, S., Kahane, G., McTavish, S., Savulescu, J., Cowen, P. J., & Hewstone, M. (2012). Propranolol reduces implicit negative racial bias. *Psychopharmacology, 222*(3), 419–424.

Tetlock, P. E. (1998). Close-call counterfactuals and belief-system defenses: I was not almost wrong but I was almost right. *Journal of Personality and Social Psychology, 75*(3), 639–652.

Thaler, R. H., & Sunstein, C. R. (2003). Libertarian paternalism. *The American Economic Review, 93*(2), 175–179.

Thayyil, N. (2014). *Biotechnology regulation and GMOs: Law, technology and public contestations in Europe.* Cheltenham, UK: Edward Elgar Publishing.

Thompson, P. B. (2008). The agricultural ethics of biofuels: A first look. *Journal of Agricultural and Environmental Ethics, 21*(2), 183–198.

Thompson, W. I. (1991). *The American replacement of nature.* New York, NY: Doubleday.

Thomson, J. A., Itskovitz-Eldor, J., Shapiro, S. S., Waknitz, M. A., Swiergiel, J. J., Marshall, V. S., & Jones, J. M. (1998, November 6). Embryonic stem cell lines derived from human blastocysts. *Science, 282*(5391), 1145–1147.

Timmermans, J., Zhao, Y., & van den Hoven, J. (2011). Ethics and nanopharmacy. *Nanoethics, 5*(3), 269–283.

Tollefson, J. (2012). Ocean-fertilization project off Canada sparks furore. *Nature, 490*(7421), 458–459.

Torrance, S. (2011). Machine ethics and the idea of a more-than-human moral world. In M. Anderson & S. Anderson (Eds.), *Machine ethics* (pp. 115-137). New York, NY: Cambridge University Press.

Toxic Substances Control Act, Pub. L. No. 94–469, 90 Stat. 2003 (1976).

Troy, A. (2012). *The very hungry city: Urban energy efficiency and the economic fate of cities.* New Haven, CT: Yale University Press.

Tucker, J. B., & Zilinskas, R. A. (2006). The promise and perils of synthetic biology. *New Atlantis, 12*(1), 25–45.

Tucker, P. (2014). *The naked future: What happens in a world that anticipates your every move?* New York, NY: Penguin Group.

Turing, A. (1950). Computing machinery and intelligence. *Mind, 59*(236), 433–460.

Turkle, S. (1984). *The second self: Computers and the human spirit.* New York, NY: Simon & Schuster.

Turkle, S. (2011). *Alone together: Why we expect more from technology and less from each other.* New York, NY: Basic Books.

Turner, L. (2004). Science, politics and the President's council on bioethics [Commentary]. *Nature Biotechnology, 22*(5), 509–510.

Turner, P. (2014, May 14). The military wants to teach robots right from wrong. *The Atlantic.* Retrieved from http://www.theatlantic.com/technology/archive/2014/05/the-military-wants-to-teach-robots-right-from-wrong/370855

Tversky, A., & Kahneman, D. (1974, September 27). Judgment under uncertainty: Heuristics and biases. *Science, 185*(4157), 1124–1131.

Twin, A. (2010, May 6). Glitches send Dow on wild ride. *CNN Money.* Retrieved from http://money.cnn.com/2010/05/06/markets/markets_newyork

Tyrrell, T. (1994). An evaluation of Maes's bottom-up mechanism for behavior selection. *Adaptive Behavior, 2*(4), 307–348.

UNDP and UNICEF with the support of UN-OCHA and WHO. (2002, January 22). *The human consequences of the Chernobyl nuclear accident: A strategy for recovery.* Minsk, Belarus: UNDP and UNICEF. Retrieved from http://www.unicef.org/newsline/chernobylreport.pdf

United Nations. (n.d.). *Background on the UNFCCC: The international response to climate change.* Retrieved from http://unfccc.int/essential_background/items/6031.php

United Nations. (n.d.). *The convention on certain conventional weapons.* Geneva, CH: UN. Retrieved from http://www.unog.ch/80256EE600585943/(httpPages)/4F0DEF093B4860B-4C1257180004B1B30?OpenDocument

United Nations. (1992). *United Nations framework convention on climate change.* Rio de Janeiro, BR: UN. Retrieved from http://unfccc.int/resource/docs/convkp/conveng.pdf

United Nations Environment Programme. (2010). *Global biodiversity outlook 3.* Montreal, CA: Secretariat of the Convention on Biological Diversity. Retrieved from http://www.cbd.int/doc/publications/gbo/gbo3-final-en.pdf

United Nations Secretary-General. (2013). *United Nations mission to investigate allegations of the use of chemical weapons in the Syrian Arab Republic: Report on the alleged use of chemical weapons in the Ghouta area of Damascus on 21 August 2013.* Retrieved from http://www.un.org/disarmament/content/slideshow/Secretary_General_Report_of_CW_Investigation.pdf

United Therapeutics Corporation. (n.d.). *Lung transplantation.* Retrieved from http://www.unither.com/engineered-lungs-for-transplantation.aspx

Urmson, C. (2014, April 28). The latest chapter for the self-driving car: Mastering city street driving [Blog]. Retrieved from http://googleblog.blogspot.com.au/2014/04/the-latest-chapter-for-self-driving-car.html

U.S. Department of Defense (DoD). (2012, November 21). *Autonomy in weapon systems* (DoD Directive 3000.09). Washington, DC: U.S. Government Printing Office. Retrieved from http://www.dtic.mil/whs/directives/corres/pdf/300009p.pdf

U.S. Department of Health and Human Services (HHS). (n.d.). *Federal policy for the protection of human subjects ("common rule").* Retrieved from http://www.hhs.gov/ohrp/humansubjects/commonrule/

U.S. Department of Justice (DOJ). (2008, September 29). *Biopharmaceutical company, Cephalon, to pay $425 million & enter plea to resolve allegations of off-label marketing.* Retrieved from http://www.justice.gov/opa/pr/2008/September/08-civ-860.html

U.S. Environmental Protection Agency (EPA). (2011, December 29). *EPA needs to manage nanomaterial risks more effectively.* Report Number 12-P-0162 for the Office of the Inspector General.

Washington, DC: N. Hanson, J. Harris, L. A. Joseph, K. Ramakrishnan, & T. Thompson. Retrieved from http://www.epa.gov/oig/reports/2012/20121229–12-P-0162.pdf

U.S. Environmental Protection Agency (EPA). (2012). *Toxic Substances Control Act (TSCA)*. Retrieved from http://www.epa.gov/oecaagct/lsca.html#

U.S. Food and Drug Administration (FDA). (1997, February 12). *Colloidal silver not approved.* Retrieved from http://www.fda.gov/AnimalVeterinary/NewsEvents/CVMUpdates/ucm127976.htm

U.S. Food and Drug Administration (FDA). (2014a). *FDA's response to public comment on the animal cloning risk assessment, risk management plan, and guidance for industry.* Retrieved from http://www.fda.gov/AnimalVeterinary/SafetyHealth/AnimalCloning/ucm055491.htm#top

U.S. Food and Drug Administration (FDA). (2014b). *Significant dates in U.S. food and drug law history.* Retrieved from http://www.fda.gov/AboutFDA/WhatWeDo/History/Milestones/ucm128305.htm

U.S. Food and Drug Administration Center for Veterinary Medicine. (2010, September 20). Briefing packet: AquAdvantage salmon. *Veterinary Medicine Advisory Committee.* Retrieved from http://www.fda.gov/downloads/AdvisoryCommittees/CommitteesMeetingMaterials/VeterinaryMedicineAdvisoryCommittee/UCM224762.pdf

U.S. House of Representatives. (2005–2006). *Making appropriations for science, the departments of state, justice, and commerce, and related agencies for the fiscal year ending September 30, 2006, and for other purposes: House report 109–272.* Washington, DC: U.S. Government Printing Office. Retrieved from http://thomas.loc.gov/cgibin/cpquery/10?cp109:temp/~cp109Id-9ky&sid=cp109Id9ky&item=10&sel=TOCLIST&l_f=1&l_file=list/cp109co.lst&report=hr272.109&hd_count=50&20&&&l_t=30&&&

U.S. Nuclear Regulatory Commission (USNRC). (2014). *Location of proposed new nuclear power reactors.* Retrieved from http://www.nrc.gov/reactors/new-reactors/col/new-reactor-map.html

Vallor, S. (2011). Carebots and caregivers: Sustaining the ethical ideal of care in the twenty-first century. *Philosophy & Technology, 24*(3), 251–268.

Van den Berg, B. (2010). *The situated self: Identity in a world of ambient intelligence.* Nijmegen, NL: Wolf Legal Publishers.

Van den Hoven, J., Lokhorst, G., & van de Poel, I. (2012). Engineering and the problem of moral overload. *Science and Engineering Ethics, 18*(1), 143–155.

Van den Hoven, J., & Manders-Huits, N. (2009). Value-sensitive design. In J. K. Olsen, S. A. Pedersen, & V. F. Hendricks (Eds.), *A companion to the philosophy of technology* (pp. 477–480). Oxford, UK: Wiley-Blackwell.

Van den Hoven, J., & Vermaas, P. E. (2007). Nano-technology and privacy: On continuous surveillance outside the panopticon. *Journal of Medicine and Philosophy, 32*(3), 283–297.

Van der Loos, H. F. (2007). *Ethics by design: A conceptual approach to personal and service robot systems.* Paper presented at the IEEE ICRA '07 Workshop on Roboethics, Rome, Italy.

Van Dijck, J. (1999). Cloning humans, cloning literature: Genetics and the imagination deficit. *New Genetics and Society, 18*(1), 9–22.

Van Eenennaam, A. L., & Muir, W. M. (2011). Transgenic salmon: A final leap to the grocery shelf? *Nature Biotechnology, 29*(8), 706–710.

Van Wynsberghe, A. (2013). Designing robots for care: Care centered value-sensitive design. *Science and Engineering Ethics, 19*(2), 407–433.

Velmans, M., & Schneider, S. (Eds.). (2007). *The blackwell companion to consciousness.* Malden, MA: Blackwell Publishing.

Venter, J. (2007). *A life decoded: My genome: My life.* New York, NY: Viking.

Veruggio, G. (2005, April 18). *The birth of roboethics.* Paper presented at the International Conference on Robotics and Automation, Barcelona, Spain: Institute of Electrical and Electronics Engineers.

Veruggio, G. (2006). *The EURON roboethics roadmap.* Paper presented at the International Conference on Humanoid Robots, Geona, Italy: EURON.

Viens, A. M., & Selgelid, M. J. (Eds.). (2012). *Emergency ethics* (Vol. 1). Ashgate.

Villagra, D., Goethe, J., Schwartz, H. I., Szarek, B., Kocherla, M., Gorowski, K., … Ruaño, G. (2011). Novel drug metabolism indices for pharmacogenetic functional status based on combinatory genotyping of CYP2C9, CYP2C19 and CYP2D6 genes. *Biomarkers in Medicine, 5*(4), 427–438.

Vinge, V. (1983, January). First word. *OMNI,* 10.

Vinge, V. (1993, March 30–31). *The coming technological singularity: How to survive in the post-human era.* Presented at *VISION–21* Symposium, NASA Lewis Research Center.

Visscher, P. M. (2008). Sizing up human height variation. *Nature Genetics, 40*(5), 489–490.

Volodin, E. M., Kostrykin, S. V., & Ryaboshapko, A. G. (2011). Simulation of climate change induced by injection of sulfur compounds into the stratosphere. *Izvestiya, Atmospheric and Oceanic Physics, 47*(4), 430–438.

Wade, N. (2010, May 20). Researchers say they created a "synthetic cell." *The New York Times.* Retrieved from http://www.nytimes.com/2010/05/21/science/21cell.html

Wakeford, R. (2011). And now, Fukushima [Editorial]. *Journal of Radiological Protection, 31*(2), 167–176.

Walker, J. S. (2006). *Three Mile Island: A nuclear crisis in historical perspective.* Berkeley, CA: University of California Press.

Wallace, M. (1989). Brave new workplace: Technology and work in the new economy. *Work and Occupations, 16*(4), 363–392.

Wallach, W. (2010). Cognitive models of moral decision making. *TopiCS: Topics in Cognitive Science, 2*(3), 420–429.

Wallach, W. (2011). From robots to techno sapiens: Ethics, law, and public policy in the development of robotics and neurotechnologies. *Law, Innovation and Technology, 3*(2), 185–207.

Wallach, W. (2012). Establishing limits on autonomous weapons capable of initiating lethal force. (Unpublished proposal)

Wallach, W. (2013, January 29). Terminating the terminator: What to do about autonomous weapons. *Science Progress.* Retrieved from http://scienceprogress.org/2013/01/terminating-the-terminator-what-to-do-about-autonomous-weapons

Wallach, W., & Allen, C. (2009). *Moral machines: Teaching robots right from wrong.* Oxford, UK: Oxford University Press.

Wallach W., & Allen, C. (2013). Framing robot arms control. *Ethics and Information Technology, 15*(2), 125–135.

Wallach, W., Allen, C., & Franklin, S. (2011). Consciousness and ethics: Artificially conscious moral agents. *International Journal of Machine Consciousness, 30*(1), 177–192.

Wallach, W., Franklin, S., & Allen, C. (2010). A conceptual and computational model of moral decision making in human and artificial agents. *TopiCS: Topics in Cognitive Science, 2*(3), 454–485.

Warrick, J. (2013, August 30). More than 1,400 killed in Syrian chemical weapons attack, U.S. says. *The Washington Post.* Retrieved from http://www.washingtonpost.com/world/national-security/nearly–1500-killed-in-syrian-chemical-weapons-attack-us-says/2013/08/30/b2864662–1196–11e3–85b6-d27422650fd5_story.html

Warwick, K. (2004). *I, cyborg.* Chicago, IL: University of Illinois Press.

Weckert, J. (1997). Intelligent machines, dehumanisation and professional responsibility. In J. van den Hoven (Ed.), *Computer ethics: Philosophical enquiry* (pp. 179–192). Rotterdam, NL: Erasmus University Press.

Weckert, J. (2005). Trusting agents. *Proceedings of the sixth International Conference of Computer Ethics: Philosophical Enquiry.* Enschede, Holland.

Weinman, J. (2001). Autonomous agents: Motivations, ethics, and responsibility. (Unpublished manuscript)

Wells, H. G. (2009). *The island of Doctor Moreau*. Ontario, CA: Broadview Press.

Wertheim, M. (1999). *The pearly gates of cyberspace: A history of space from Dante to the internet*. New York, NY: W.W. Norton.

West, M. J., & Gundersen, H. J. (1990). Unbiased stereological estimation of the number of neurons in the human hippocampus. *Journal of Comparative Neurology, 296*(1), 1–22.

Whitbeck, C. (1995). Teaching ethics to scientists and engineers: Moral agents and moral problems. *Science and Engineering Ethics, 1*(3), 299–308.

Whitbeck, C. (1996). Ethics as design: Doing justice to moral problems. *Hastings Center Report, 26*(3), 9–16.

Wieloch, T., & Nikolich, K. (2006). Mechanisms of neural plasticity following brain injury. *Current Opinion in Neurobiology, 16*(3), 258–264.

Williams, N. (2003). Death of Dolly marks cloning milestone. *Current Biology, 13*(6), R209–R210.

Williams, N. (2004). UN stalls on human cloning. *Current Biology, 14*(22), R937–R938.

Wilson, J. A., Onorati, K., Mishkind, M., Reger, M. A., & Gahm, G. A. (2008). Soldier attitudes about technology-based approaches to mental health care. *CyberPsychology & Behavior, 11*(6), 767–769.

Winchester, M. R., Sturgeon, R. E., & Costa-Fernández, J. M. (2010). Chemical characterization of engineered nanoparticles [Editorial]. *Analytical and Bioanalytical Chemistry, 396*(3), 951–952.

Winner, L. (1980). Do artifacts have politics? *Daedalus, 109*(1), 121–136.

Winner, L. (2003). Societal implications of nanotechnology: Testimony to the committee on science of the U.S. House of Representatives, April 9. Retrieved from http://ethics.iit.edu/NanoEthicsBank/node/1187

Wittneben, B. B. (2012). The impact of the Fukushima nuclear accident on European energy policy. *Environmental Science & Policy, 15*(1), 1–3.

Wolbring, G. (2010). Nanoscale science and technology and social cohesion. *International Journal of Nanotechnology, 7*(2), 155–172.

Wolens, D. (2013). *The singularity: Will we Survive our technology* [Film]. Available from http://thesingularityfilm.com

Wolpe, P. R. (2007). Ethical and social challenges of brain-computer interfaces. *Virtual Mentor, 9*(2), 128–131.

Woodhouse, E. (2013). *The future of technological civilization* (Pre-release). Retrieved from http://www.academia.edu/4271925/The_Future_of_Technological_Civilization

Woods, D., Dekker, S., Cook, R., Johannesen, L., & Sarter, N. (2010). *Behind human error* (2nd ed.). Surrey, UK: Ashgate.

Woods, D., & Hollnagel, E. (2006). *Joint cognitive systems: Patterns in cognitive systems engineering*. Boca Raton, FL: CRC Press.

World Health Organization (WHO). (n.d.). *Children's environmental health: Lack of water and inadequate sanitation*. Retrieved from http://www.who.int/ceh/risks/cehwater/en

World Health Organization (WHO). (2006). *Health effects of the Chernobyl accident and special health care programmes: Report of the UN Chernobyl forum expert group "health."* Geneva, Switzerland: WHO. Retrieved from http://whqlibdoc.who.int/publications/2006/9241594179_eng.pdf

World Health Organization (WHO). (2013). *World malaria report 2013*. Geneva, Switzerland: WHO. Retrieved from http://www.who.int/malaria/publications/world_malaria_report_2013/report/en

Wynne, B. (1988). Unruly technology: Practical rules, impractical discourses and public understanding. *Social Studies of Science, 18*(1), 147–167.

Yasunari, T. J., Stohl, A., Hayano, R. S., Burkhart, J. F., Eckhardt, S., & Yasunari, T. (2011). Cesium–137 deposition and contamination of Japanese soils due to the Fukushima nuclear accident: *Proceedings of the National Academy of Sciences of the United States of America, 108*(49), 19530–19534.

Yong, E. (2013, March 8). Will we ever… Regenerate limbs? *BBC.* Retrieved from http://www.bbc.com/future/story/20130307-will-we-ever-regenerate-limbs

Yong, E. (2013, June 6). Bird flu mutation risk. *The Scientist.* Retrieved from http://www.the-scientist.com/?articles.view/articleNo/35879/title/Bird-Flu-Mutation-Risk

Yudkowsky, E. (2001, May 3). What is friendly AI? *Kurzweill: Accelerating Intelligence.* Retrieved from http://www.kurzweilai.net/what-is-friendly-AI

Yudkowsky, E. (2008a). Artificial intelligence as a positive and negative factor in global risk. In N. Bostrom & M. Cirkovic (Eds.), *Global catastrophic risks* (pp. 308–345). Oxford, UK: Oxford University Press.

Yudkowsky, E. (2008b). Cognitive bases potentially affecting judgment of global risks. In N. Bostrom & M. Cirkovic (Eds.), *Global catastrophic risks* (pp. 91–119). Oxford, UK: Oxford University Press.

Zak, P. J., Kurzban, R., & Matzner, W. T. (2004). The neurobiology of trust. *Annals of the New York Academy of Sciences, 1032*(1), 224–227.

Zak, P. J., Kurzban, R., & Matzner, W. T. (2005). Oxytocin is associated with human trustworthiness. *Hormones and Behavior, 48*(5), 522–527.

Zimmer, C. (2003, February 14). Tinker, tailor: Can Venter stitch together a genome from scratch? *Science, 299*(5609), 1006–1007.

Zimmer, C. (2009). *Microcosm: E. coli and the new science of life.* New York, NY: Vintage.

Zimmer, C. (2013). Bringing them back to life. *National Geographic, 223(4),* 28–41.

Zimmer, S. (2013). The right to print arms: The effect on civil liberties of government restrictions on computer-aided design files shared on the internet. *Information & Communications Technology Law, 22*(3), 251–263.

Zittrain, J. (2009). *The future of the internet and how to stop it.* New Haven, CT: Yale University Press.

Živanović, S., Pavic, A., & Reynolds, P. (2005). Vibration serviceability of footbridges under human-induced excitation: A literature review. *Journal of Sound and Vibration, 279*(1), 1–74.

Zolli, A., & Healy, A. (2012). *Resilience: Why things bounce back.* New York, NY: Simon & Schuster.

Zoloth, L. (2009). Second life: Some ethical issues in synthetic biology and the recapitulation of evolution. In M. Bedau & E. Parke (Eds.), *The ethics of protocells* (pp. 143–164). Cambridge, MA: MIT Press.

# INDEX